ASPIRE
SUCCEED
PROGRESS

Complete Pure Mathematics 1 for Cambridge International AS & A Level

Second Edition

Jean Linsky
James Nicholson
Brian Western

Oxford excellence for Cambridge AS & A Level

OXFORD
UNIVERSITY PRESS

OXFORD
UNIVERSITY PRESS
Great Clarendon Street, Oxford, OX2 6DP, United Kingdom

Oxford University Press is a department of the University of Oxford. It furthers the University's objective of excellence in research, scholarship, and education by publishing worldwide. Oxford is a registered trade mark of Oxford University Press in the UK and in certain other countries

First published in 2018

British Library Cataloguing in Publication Data
Data available

978-0-19-842510-6

11

Paper used in the production of this book is a natural, recyclable product made from wood grown in sustainable forests.
The manufacturing process conforms to the environmental regulations of the country of origin.

Printed in China by Golden Cup.

The questions, all example answers and comments that appear in this book were written by the authors.

Acknowledgements

The publishers would like to thank the following for permissions to use their photographs:

Cover Image: Maksim Kabakou/Shutterstock; p2: Sue Stokes/ Shutterstock; p24: Detlev Van Ravenswaay/Science Photo Library; p48: MO_SES Premium/Shutterstock; p72t: DEA/Veneranda Bilioteca Ambrosiana; p72b: Rex/SIPA Press; p73t: Colouria Media/Alamy; p73b: imageBROKER/Alamy; p74: Angelina Dimitrova/Shutterstock; p86: Stocktrek Images/Glow Images; p109: E+/Getty Images; p120: Aleksandar Mijatovic/Shutterstock; p136t: Robert Gendler/Science Photo Library; p136b: Remus Stefan Cucu/Shutterstock; p137: Nagel Photography/Shutterstock; p138: Westend61/Getty Images; p156: Gordon Garradd/Science Photo Library; p173: Hong Xia/Shutterstock; p206: Mark Bowler/Photo Researchers/Getty Images; p207t: NASA/ Science Photo Library; p207b: dream designs/Shutterstock

This Student Book refers to the Cambridge International AS & A Level Mathematics (9709) Syllabus published by Cambridge Assessment International Education.

This work has been developed independently from and is not endorsed by or otherwise connected with Cambridge Assessment International Education.

Contents

Introduction

About this book

This book has been written to cover the **Cambridge International AS & A Level Mathematics (9709)** syllabus, and is fully aligned to the syllabus.

In addition to the main curriculum content, you will find:

- 'Maths in real-life', showing how principles learned in this course are used in the real world.
- Chapter introductions, which outline how each topic in the Cambridge 9709 syllabus is used in real-life.

The book contains the following features:

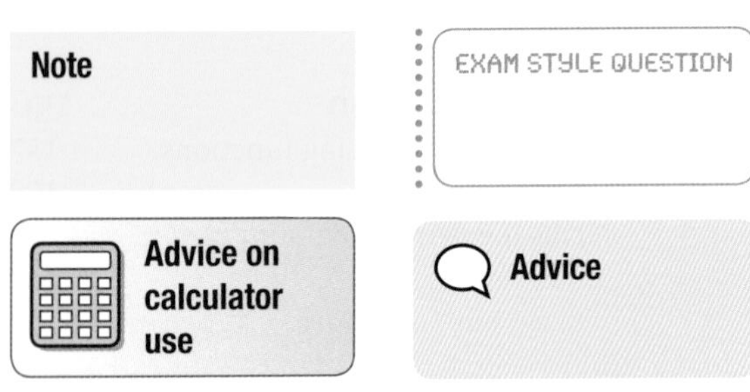

Throughout the book, you will encounter worked examples and a host of rigorous exercises. The examples show you the important techniques required to tackle questions. The exercises are carefully graded, starting from a basic level and going up to exam standard, allowing you plenty of opportunities to practise your skills. Together, the examples and exercises put maths in a real-world context, with a truly international focus.

At the start of each chapter, you will see a list of objectives that are covered in the chapter. These objectives are drawn directly from the Cambridge AS & A Level syllabus. Each chapter then begins with a *Before you start* section, which you should complete to ensure you have all the requisite skills for the chapter, and finishes with a *Summary exercise* and *Chapter summary*, ensuring that you fully understand each topic.

Each chapter contains key mathematical terms to improve understanding, highlighted in colour, with full definitions provided in the Glossary of terms at the end of the book.

The answers given at the back of the book are concise. However, when answering exam-style questions, you should show as many steps in your working as possible. All exam-style questions, as well as *Paper A* and *Paper B* have been written by the authors.

About the authors

Brian Western has over 40 years of experience in teaching Mathematics up to A level and beyond, and is also a highly experienced examiner. He taught Mathematics and Further Mathematics, and was an assistant head teacher in a large state school. Brian has written and consulted on a number of Mathematics textbooks.

Jean Linsky has been a Mathematics teacher for over 30 years, as well as head of Mathematics in Watford, Hertfordshire, in the United Kingdom. She currently teaches A Level Mathematics in a large mixed state school. Jean has authored and consulted on numerous Mathematics textbooks, is a very experienced examiner and has delivered many training courses to Mathematics teachers.

James Nicholson is an experienced teacher of Mathematics at secondary level; he has taught for 12 years at Harrow School and spent 13 years as Head of Mathematics in a large Belfast grammar school. He is the author of two A-level statistics texts, and editor of the *Concise Oxford Dictionary of Mathematics*. He has also contributed to a number of other sets of curriculum and assessment materials, is an experienced examiner and has acted as a consultant for UK government agencies on accreditation of new specifications.

James ran school workshops for the Royal Statistical Society for many years, and has been a member of the Schools and Further Education Committee of the Institute of Mathematics and its Application since 2000, including six years as chair; and is currently a member of the Outer Circle group for the Advisory Committee on Mathematics Education. He has served as a Vice-President of the International Association for Statistics Education for four years, and is currently Chair of the Advisory Board to the International Statistical Literacy Project.

A note from the authors

The aim of this book is to help students prepare for the Pure 1 unit of the Cambridge International AS & A Level Mathematics syllabus, though it may also be found to be useful in providing support material for other AS and A Level courses. The book contains a large number of practice questions, many of which are exam-style.

In writing the book we have drawn on our experiences of teaching Mathematics and Further Mathematics to A Level over the past 40 years as well as on our experience as examiners.

We are grateful to Robert Linsky and Graham Carter for their support during the completion of this project and to the many colleagues and students we have worked with over the years.

Student Book: *Complete Pure Mathematics 1 for Cambridge International AS & A Level*

Syllabus: Cambridge International AS & A Level Mathematics: Pure (9709)

PURE MATHEMATICS 1	Student Book
Syllabus overview for 9709, first examined in 2020.	
Pure Mathematics 1 (Paper 1)	
1. Quadratics	
• Carry out the process of completing the square for a quadratic polynomial $ax^2 + bx + c$, and use this form, e.g. to locate the vertex of the graph of $y = ax^2 + bx + c$ or to sketch the graph	Pages 2–20
• Find the discriminant of a quadratic polynomial $ax^2 + bx + c$ and use the discriminant, e.g. to determine the number of real roots of the equation $ax^2 + bx + c = 0$	Pages 2–20
• Solve quadratic equations, and linear and quadratic inequalities, in one unknown	Pages 2–20
• Solve by substitution a pair of simultaneous equations of which one is linear and one is quadratic	Pages 2–20
• Recognise and solve equations in x which are quadratic in some function of x, e.g. $x^4 - 5x^2 + 4 = 0$	Pages 2–20
2. Functions	
• Understand the terms function, domain, range, one-one function, inverse function and composition of functions	Pages 24–42
• Identify the range of a given function in simple cases, and find the composition of two given functions	Pages 24–42
• Determine whether or not a given function is one-one, and find the inverse of a one-one function in simple cases	Pages 24–42
• Illustrate in graphical terms the relation between a one-one function and its inverse	Pages 24–42
• Understand and use the transformations of the graph of $y = \mathrm{f}(x)$ given by $y = \mathrm{f}(x) + a$, $y = \mathrm{f}(x + a)$, $y = a\mathrm{f}(x)$, $y = \mathrm{f}(ax)$ and simple combinations of these	Pages 24–42

3. Coordinate geometry	
• Find the equation of a straight line given sufficient information	Pages 48–67
• Interpret and use any of the forms $y = mx + c$, $y - y_1 = m(x - x_1)$, $ax + by + c = 0$ in solving problems	Pages 48–67
• Understand that the equation $(x - a)^2 + (y - b)^2 = r^2$ represents the circle with centre (a, b) and radius r	Pages 48–67
• Use algebraic methods to solve problems involving lines and circles	Pages 48–67
• Understand the relationship between a graph and its associated algebraic equation, and use the relationship between points of intersection of graphs and solutions of equations	Pages 48–67
4. Circular measure	
• Understand the definition of a radian, and use the relationship between radians and degrees	Pages 74–81
• Use the formulae $s = r\,\theta$ and $A = \frac{1}{2}r^2\,\theta$ in solving problems concerning the arc length and sector area of a circle	Pages 74–81
5. Trigonometry	
• Sketch and use graphs of the sine, cosine and tangent functions (for angles of any size, and using either degrees or radians)	Pages 86–104
• Use the exact values of the sine, cosine and tangent of 30°, 45°, 60°, and related angles, e.g. $\cos 150° = -\frac{\sqrt{3}}{2}$	Pages 86–104
• Use the notations $\sin^{-1}x$, $\cos^{-1}x$, $\tan^{-1}x$ to denote the principal values of the inverse trigonometric relations	Pages 86–104
• Use the identities $\frac{\sin\theta}{\cos\theta} \equiv \tan\theta$ and $\sin^2\theta + \cos^2\theta \equiv 1$	Pages 86–104
• Find all the solutions of simple trigonometrical equations lying in a specified interval (general forms of solution are not included)	Pages 86–104
6. Series	
• Use the expansion of $(a + b)^n$, where n is a positive integer (knowledge of the greatest term and properties of the coefficients are not required, but the notations $\binom{n}{r}$ and $n!$ should be known)	Pages 109–116
• Recognise arithmetic and geometric progressions	Pages 120–133
• Use the formulae for the nth term and for the sum of the first n terms to solve problems involving arithmetic or geometric progressions	Pages 120–133
• Use the condition for the convergence of a geometric progression, and the formula for the sum to infinity of a convergent geometric progression	Pages 120–133

7. Differentiation	
• Understand the gradient of a curve at a point as the limit of the gradients of a suitable sequence of chords, and use the notations $f'(x), f''(x), \frac{dy}{dx}, \frac{d^2}{dx^2}$ for first and second derivatives	Pages 138–153
• Use the derivative of x^n (for any rational n), together with constant multiples, sums, differences of functions, and of composite functions using the chain rule	Pages 138–153
• Apply differentiation to gradients, tangents and normals, increasing and decreasing functions and rates of change (including connected rates of change)	Pages 138–153, 156–167
• Locate stationary points, and use information about stationary points in sketching graphs (the ability to distinguish between maximum points and minimum points is required, but identification of points of inflexion is not included)	Pages 156–167

8. Integration	
• Understand integration as the reverse process of differentiation, and integrate $(ax + b)n$ (for any rational n except −1), together with constant multiples, sums and differences	Pages 173–200
• Solve problems involving the evaluation of a constant of integration, e.g. to find the equation of the curve through (1, −2) for which $\frac{dy}{dx} = 2x + 1$	Pages 173–200
• Evaluate definite integrals (including simple cases of 'improper' integrals, such as $\int_0^1 x^{-\frac{1}{2}}\,dx$ and $\int_0^\infty x^{-2}\,dx$)	Pages 173–200
• Use definite integration to find: – the area of a region bounded by a curve and lines parallel to the axes, or between two curves – a volume of revolution about one of the axes	Pages 173–200

1 Quadratics

The Quadracci Pavilion is part of the Milwaukee Art Museum; it opened in 2001 and contains a movable, wing-like structure. The building, designed by Santiago Calatrava, received an outstanding structure award in 2004. The 'wings' open for a wingspan of 66 metres during the day and fold over the arched structure at night or during bad weather. Designing and constructing this building required the use of quadratic curves and parabolas. These same techniques are used in modelling bridges and a huge number of other structures.

Objectives

- Carry out the process of completing the square for a quadratic polynomial $ax^2 + bx + c$, and use a completed square form.
- Find the discriminant of a quadratic polynomial $ax^2 + bx + c$ and use the discriminant.
- Solve quadratic equations and quadratic inequalities, in one unknown.
- Solve by substitution a pair of simultaneous equations of which one is linear and one is quadratic.
- Recognise and solve equations in x which are quadratic in some function of x.

Before you start

You should know how to:

1. Square a number that has a square root sign,

e.g. $\sqrt{3} \times \sqrt{3} = 3$

e.g. $(2\sqrt{5})^2 = 4 \times 5 = 20.$

2. Factorise quadratic expressions,

e.g. $2x^2 - 13x + 20 = (2x - 5)(x - 4)$

e.g. $16x^2 - 49 = (4x + 7)(4x - 7).$

3. Solve linear inequalities,

e.g. $5x - 2 < 8x + 4$

$-6 < 3x$

$x > -2.$

4. Solve simultaneous linear equations,

e.g. $2x + 3y = 5$ $\quad$ $5x + 2y = -4$

$10x + 15y = 25$ $\quad$ $10x + 4y = -8$

Subtracting: $11y = 33$ $\quad$ $y = 3$

Substituting: $2x + 3(3) = 5$

$2x = -4 \Rightarrow x = -2.$

Skills check:

1. Simplify

a) $\sqrt{7} \times \sqrt{7}$ $\quad$ **b)** $(\sqrt{13})^2$

c) $3\sqrt{11} \times 2\sqrt{11}$ $\quad$ **d)** $(10\sqrt{2})^2.$

2. Factorise

a) $6x^2 - x - 1$ $\quad$ **b)** $1 - 100x^2$

c) $21 + 11x - 2x^2$ $\quad$ **d)** $2x^2 - 18.$

3. Solve

a) $2x + 3 \geq 7x - 1$ $\quad$ **b)** $5(2x - 3) < 2(1 - x)$

c) $6 - 7x > 3(2x + 5).$

4. Solve these simultaneous equations

a) $3x - 4y = -5$ $\quad$ $7x + 20y = 3$

b) $2x - y = 3$ $\quad$ $6x + 5y = 49$

c) $4x + 3y = -11$ $\quad$ $9x - 7y = -11$

d) $6x - 2y = 1$ $\quad$ $8x - 5y = -8.$

1.1 Solve quadratic equations by factorising

You should know how to solve a quadratic equation by factorising.
Here are some examples.

Example 1

Solve **a)** $6x^2 + 11x - 35 = 0$

b) $20x^2 + 80 = 82x$.

Always look to see if we can first divide each term by a common factor.

a) $6x^2 + 11x - 35 = 0$

$(3x - 5)(2x + 7) = 0$

$x = \frac{5}{3}$ or $x = \frac{-7}{2}$ ← Solve $(3x - 5) = 0$ and $(2x + 7) = 0$.

b) $20x^2 + 80 = 82x$

$20x^2 - 82x + 80 = 0$ ← Rearrange as $ax^2 + bx + c = 0$.

$10x^2 - 41x + 40 = 0$ ← Divide through by any integer common factor.

$(5x - 8)(2x - 5) = 0$

$x = \frac{8}{5}$ or $x = \frac{5}{2}$

Example 2

A rectangle has length $(x + 3)$ metres and width $(2x + 1)$ metres.

The area of the rectangle is 12 m^2.

Find the length of the rectangle.

$2x + 1$

$x + 3$

Area $= (2x + 1)(x + 3) = 2x^2 + 7x + 3 = 12$

$2x^2 + 7x - 9 = 0$

$(2x + 9)(x - 1) = 0$

$x = \frac{-9}{2}$ or $x = 1$

But the solution cannot be $x = \frac{-9}{2}$ as this will lead to a negative length and width, so $x = 1$.

Length $= x + 3 = 1 + 3 = 4$ metres.

Exercise 1.1

1. Solve each of these quadratic equations.

a) $x^2 + 2x - 35 = 0$ **b)** $x^2 - 7x + 10 = 0$ **c)** $x^2 - x - 12 = 0$

d) $x^2 + 8x - 9 = 0$ **e)** $x^2 - 3x = 18$ **f)** $x^2 = x + 6$

g) $0 = x^2 + 8x + 12$ h) $5x + x^2 = 24$ i) $x^2 = 4x$
j) $x^2 - 16 = 0$ k) $2x^2 - 9x + 4 = 0$ l) $3x^2 + 19x + 6 = 0$
m) $15x^2 - x - 2 = 0$ n) $8x^2 - 18x + 9 = 0$ o) $5x^2 = 20x$
p) $5x^2 + 23x = 10$ q) $8 - 2x - x^2 = 0$ r) $2x^2 + 7x - 9 = 0$
s) $5x^2 - 8x = 4$ t) $100x^2 - 3 = 20x$ u) $6x^2 + 20 - 23x = 0$
v) $7x^2 + 14x = 0$ w) $18 - 27x - 5x^2 = 0$ x) $21 - 11x - 2x^2 = 0$
y) $5 - 5x^2 = 0$ z) $24x^2 + 40x + 6 = 0$

2. A rectangle has length $(x + 4)$ cm and width $(3x + 4)$ cm. The area of the rectangle is 11 cm^2. Find x.

3. A piece of card has a length of $(2x - 1)$ cm and a width of $(x + 2)$ cm. A square of side x cm is removed from the card. The area of the card that is left is 68 cm^2. Find the area of the card that has been removed.

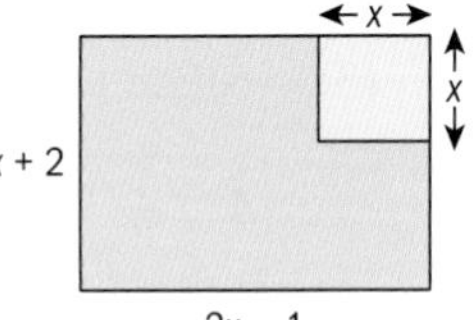

4. Two numbers differ by 4. Their product is 21. Write down a quadratic equation and solve it to find the two numbers.

5. Solve $\dfrac{2x^2 + 5x + 3}{x^2 + 3x + 2} = 4$.

1.2 Solving quadratic inequalities

To solve a quadratic inequality it is useful to **sketch** the curve.

> If the coefficient of x^2 is positive, the curve is ∪ shaped.
>
> If the coefficient of x^2 is negative, the curve is ∩ shaped.

Note: Unlike the linear inequality, where the solution range has only **one boundary**, e.g. $x > -2$, the solution for the unknown variable in a quadratic inequality is a range of values with **two boundaries**, e.g. $x < -1$ or $x > 3$.

Example 3

Solve the inequality $x^2 - 3x - 4 \geq 0$.

$x^2 - 3x - 4 \geq 0$

$(x + 1)(x - 4) \geq 0$

We want the values of x when the curve is **above** the x-axis ($\geq$).

Looking at the sketch we can see that this is true for all values of x shown by the red arrows.

Answer: $x \leq -1$ or $x \geq 4$

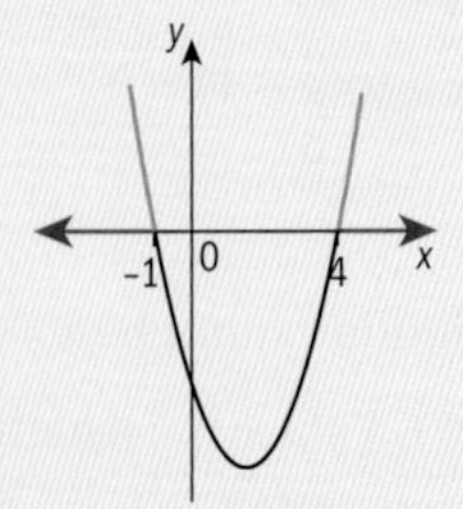

Note: There are **two** regions so we write the answer as **two** inequalities.

Example 4

Solve the inequality $6 + x - x^2 > 0$.

$6 + x - x^2 > 0$

$(3 - x)(2 + x) > 0$

We want the values of x when the curve is **above** the x-axis (>).

Looking at the sketch we can see that this is true for the values of x shown by the red arrows.

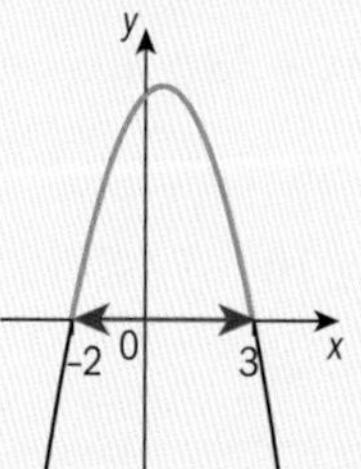

Answer: $-2 < x < 3$

Note: There is only **one** region so we write the answer as **one** inequality. We read $-2 < x < 3$ as 'x is between -2 and 3'.

Or

$6 + x - x^2 > 0$ can be written as $x^2 - x - 6 < 0$

$(x - 3)(x + 2) < 0$

Remember: We change the symbol when multiplying each side by -1.

Answer: $-2 < x < 3$

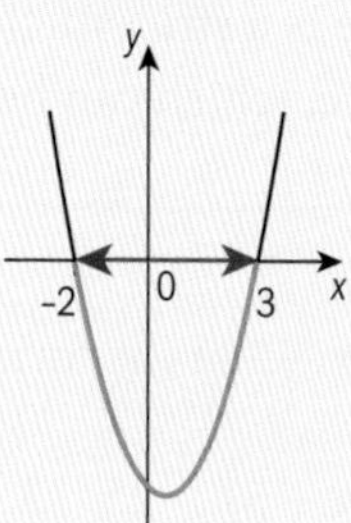

Example 5

Solve the inequality $(x + 5)(2x + 1) \leq 0$.

$(x + 5)(2x + 1) \leq 0$

As we have $+2x^2$, the curve is ∪ shaped.

We want the values of x when the curve is **on or below** the x-axis ($\leq$).

Looking at the sketch we can see that this is the region shown by the red arrows.

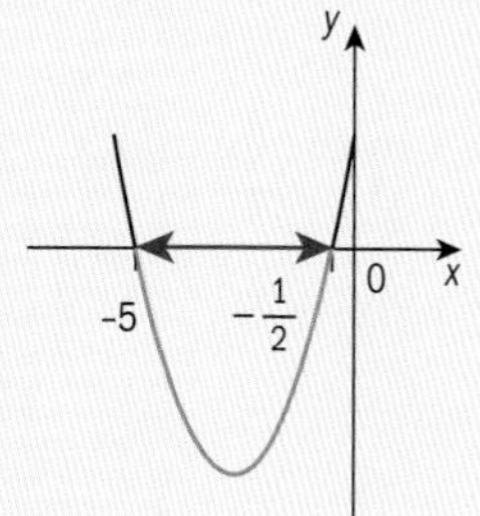

Answer: $-5 \leq x \leq -\frac{1}{2}$.

Exercise 1.2

Solve each of these inequalities.

Sketch the curve for each, showing the interval that satisfies each inequality.

1. **a)** $(x + 2)(x - 5) \leq 0$ **b)** $(x - 1)(x - 3) > 0$ **c)** $(3x + 2)(x + 4) > 0$

 d) $(x - 7)(x + 10) < 0$ **e)** $(6 - x)(3 - x) \leq 0$ **f)** $(5 + x)(1 - 2x) \geq 0$

2. a) $x^2 + 7x + 12 \leq 0$ b) $x^2 - x - 30 < 0$ c) $x^2 + 2x - 48 > 0$
d) $x^2 - 5x + 6 \geq 0$ e) $2x^2 - 11x + 12 > 0$ f) $3x^2 + 14x - 5 \leq 0$
g) $15 + 2x - x^2 \leq 0$ h) $16 - 6x - x^2 < 0$ i) $21 - x - 2x^2 > 0$
j) $10x^2 + 43x + 28 \geq 0$ k) $x^2 - 9 \geq 0$ l) $50 - 2x^2 < 0$

3. a) $x^2 - 2x > 35$ b) $x^2 + 6 \leq 5x$ c) $x^2 \leq x + 20$
d) $x(x + 3) \geq 10$ e) $x^2 < 9x + 10$ f) $x^2 - 4(x + 6) > 21$
g) $24 > 11x - x^2$ h) $7 + 2(4x^2 - 15x) \leq 0$ i) $\frac{x^2+12}{2} \geq 4x$
j) $(x + 5)^2 > 1$

1.3 The method of completing the square

We can express a quadratic **polynomial** of the form $ax^2 + bx + c$ in the form $a(x + p)^2 + q$, where p and q are constants, by the method of **completing the square**. There is a strict method for expressing a quadratic expression as a perfect square, which can be seen in the following examples.

Example 6

Express $x^2 + 10x - 3$ in the form $(x + p)^2 + q$, where p and q are constants.

$x^2 + 10x - 3 = (x + 5)^2 - 25 - 3$ ← Always halve the coefficient of x to give the value of p.

We subtract 25 because $(x + 5)^2 = x^2 + 10x + 25$ therefore, $(x + 5)^2 - 25 = x^2 + 10x$.

$= (x + 5)^2 - 28$

Example 7

Express $2x^2 - 12x + 1$ in the form $a(x + p)^2 + q$, where a, p and q are constants.

$2x^2 - 12x + 1 = 2[x^2 - 6x] + 1$ ← Divide the 1st two terms by the coefficient of x^2 and use square brackets.

$= 2[(x - 3)^2 - 9] + 1$ ← Complete the square for $x^2 - 6x$.

$= 2(x - 3)^2 - 18 + 1$ ← Multiply out the square brackets.

$= 2(x - 3)^2 - 17$

Note: We cannot divide $2x^2 - 12x + 1$ by 2 as it is not an equation. If we had $2x^2 - 12x + 1 = 0$, we could write $x^2 - 6x + \frac{1}{2} = 0$.

Example 8

Express $3 + 4x - x^2$ in the form $q - (x + p)^2$, where p and q are constants.

$$3 + 4x - x^2 = -[x^2 - 4x - 3]$$

We want the coefficient of x^2 to be +1.

$$= -[(x - 2)^2 - 4 - 3]$$

Complete the square for $x^2 - 4x$.

$$= -(x - 2)^2 + 7$$

Multiply the terms in the square brackets by −1.

$$= 7 - (x - 2)^2$$

A quadratic equation of the form $ax^2 + bx + c = 0$, where a, b and c are constants and $c \neq 0$, can also be solved by the method of completing the square.

Example 9

Solve the equation $x^2 - 6x + 2 = 0$, giving your answer in the form $x = p \pm \sqrt{q}$.

$$(x - 3)^2 - 9 + 2 = 0$$

$x^2 - 6x = (x - 3)^2 - 9$

$$(x - 3)^2 - 7 = 0$$

$$(x - 3)^2 = 7$$

$$x - 3 = \pm\sqrt{7}$$

Do not forget ± when taking the square root of each side.

$$x = 3 \pm \sqrt{7}$$

Example 10

Solve the equation $x^2 - 3x - 1 = 0$ by completing the square.

Give your answer in the form $p + \sqrt{\frac{q}{2}}$.

$$\left(x - \frac{3}{2}\right)^2 - \frac{9}{4} - 1 = 0$$

$$\left(x - \frac{3}{2}\right)^2 - \frac{13}{4} = 0$$

Leave $\frac{13}{4}$ as an improper fraction.

$$\left(x - \frac{3}{2}\right)^2 = \frac{13}{4}$$

$$x - \frac{3}{2} = \pm\sqrt{\frac{13}{4}}$$

$$x - \frac{3}{2} = \pm\frac{1}{2}\sqrt{13}$$

$\sqrt{\frac{13}{4}} = \frac{\sqrt{13}}{\sqrt{4}} = \frac{\sqrt{13}}{2} = \frac{1}{2}\sqrt{13}$.

$$x = \frac{3}{2} \pm \frac{1}{2}\sqrt{13}$$

$$x = \frac{3 \pm \sqrt{13}}{2}$$

Example 11

Solve the inequality $3x^2 + 24x + 2 < 0$ by completing the square. Leave your answer in **surd** form.

First consider $3x^2 + 24x + 2 = 0$

$$x^2 + 8x + \frac{2}{3} = 0$$

As we are considering an equation, we can ensure the coefficient of x^2 is 1 by dividing each term by 3.

$$(x + 4)^2 - 16 + \frac{2}{3} = 0$$

$$(x + 4)^2 - \frac{46}{3} = 0$$

$$(x + 4)^2 = \frac{46}{3}$$

$$x + 4 = \pm\sqrt{\frac{46}{3}}$$

$$x = -4 \pm \sqrt{\frac{46}{3}}$$

We want the answer in surd form.

Now consider $3x^2 + 24x + 2 < 0$

Answer: $-4 - \sqrt{\frac{46}{3}} < x < -4 + \sqrt{\frac{46}{3}}$

Exercise 1.3

1. Write each of these expressions in the form $(x + p)^2 + q$ or $q - (x + p)^2$, where p and q are constants.

   a) $x^2 + 2x - 5$ **b)** $x^2 - 10x + 20$ **c)** $x^2 - 4x + 1$ **d)** $6 - 8x - x^2$

   e) $10 - 16x - x^2$ **f)** $x^2 + 20x - 45$ **g)** $x^2 + 12x + 16$ **h)** $3 - 6x - x^2$

2. Solve each of these equations by completing the square. Leave your answers in surd form.

   a) $x^2 + 6x + 1 = 0$ **b)** $x^2 - 4x - 8 = 0$ **c)** $4 - 2x - x^2 = 0$ **d)** $x^2 - 20x + 30 = 0$

   e) $3 + 8x - x^2 = 0$ **f)** $x^2 + 14x - 1 = 0$ **g)** $x^2 - 3x - 2 = 0$ **h)** $x^2 - x - 3 = 0$

3. Write each of these expressions in the form $a(x + b)^2 + c$ or $c - a(x + b)^2$.

   a) $6x^2 + 12x - 3$ **b)** $3x^2 - 6x - 15$ **c)** $3x^2 - 18x + 4$ **d)** $4x^2 + 24x - 9$

   e) $5 - 2x - 2x^2$ **f)** $5x^2 - 20x + 2$ **g)** $4 + 3x - 2x^2$ **h)** $3x^2 + 5x + 1$

4. Solve each of these equations by completing the square. Leave your answers in surd form.

   a) $3x^2 + 12x + 2 = 0$ **b)** $3x^2 + 6x - 5 = 0$ **c)** $5x^2 + 50x - 7 = 0$ **d)** $3 + 20x - 2x^2 = 0$

   e) $2x^2 - 8x + 1 = 0$ **f)** $4x^2 - 48x - 26 = 0$ **g)** $2x^2 - x - 12 = 0$ **h)** $9 + 2x - 3x^2 = 0$

5. Solve each of these inequalities by completing the square. Leave your answers in surd form.

   a) $x^2 + 4x + 2 < 0$ **b)** $x^2 - 6x - 3 \geq 0$ **c)** $x^2 - 2x - 1 > 0$ **d)** $x^2 + 10x + 7 \leq 0$

   e) $2x^2 - 12x - 7 \geq 0$ **f)** $3x^2 + 6x + 1 < 0$ **g)** $5x^2 + 10x - 2 \leq 0$ **h)** $2x^2 - 8x - 3 > 0$

1.4 Solving quadratic equations using the formula

A quadratic equation can be expressed in the form $ax^2 + bx + c = 0$, where a, b and c are constants and $a \neq 0$.

We can use the method of completing the square to find the general formula for solving a quadratic equation.

$$ax^2 + bx + c = 0$$

$$x^2 + \frac{b}{a}x + \frac{c}{a} = 0$$

$$\left(x + \frac{b}{2a}\right)^2 - \left(\frac{b}{2a}\right)^2 + \frac{c}{a} = 0$$

$$\left(x + \frac{b}{2a}\right)^2 - \left(\frac{b}{2a}\right)^2 = -\frac{c}{a}$$

$$\left(x + \frac{b}{2a}\right)^2 = \frac{b^2}{4a^2} - \frac{c}{a}$$

$$\left(x + \frac{b}{2a}\right)^2 = \frac{b^2 - 4ac}{4a^2}$$

$$x + \frac{b}{2a} = \pm\frac{\sqrt{b^2 - 4ac}}{2a}$$

$$x = \frac{-b}{2a} \pm \frac{\sqrt{b^2 - 4ac}}{2a}$$

$$x = \frac{-b \pm \sqrt{b^2 - 4ac}}{2a}$$

This leads us to a formula for solving a quadratic equation of the form $ax^2 + bx + c = 0$. This is often called the **quadratic formula**.

$$x = \frac{-b \pm \sqrt{b^2 - 4ac}}{2a}$$

Example 12

Solve the equation $3x^2 + x - 5 = 0$. Give your answer correct to 3 significant figures.

$a = 3$, $b = 1$, $c = -5$

$$x = \frac{-(1) \pm \sqrt{(1)^2 - 4(3)(-5)}}{2(3)}$$

$$= \frac{-1 + \sqrt{61}}{6} \text{ or } \frac{-1 - \sqrt{61}}{6}$$

$= 1.135042...$ or $-1.468375...$

$= 1.14$ or -1.47

It is useful to write brackets when substituting to ensure we do not make errors with the negative values.

Example 13

Solve the inequality $2 - 6x - x^2 \leq 0$. Leave your answer in surd form.

Consider $2 - 6x - x^2 = 0$

$$x = \frac{-(-6) \pm \sqrt{(-6)^2 - 4(-1)(2)}}{2(-1)}$$

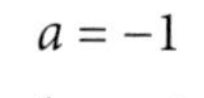

$a = -1$
$b = -6$
$c = 2$

$$= \frac{6 + \sqrt{44}}{-2} \text{ or } \frac{6 - \sqrt{44}}{-2}$$

$$= \frac{6 + 2\sqrt{11}}{-2} \text{ or } \frac{6 - 2\sqrt{11}}{-2}$$

$\sqrt{44} = \sqrt{4 \times 11} = 2\sqrt{11}$

$= -3 - \sqrt{11}$ or $-3 + \sqrt{11}$

Now $2 - 6x - x^2 \leq 0$

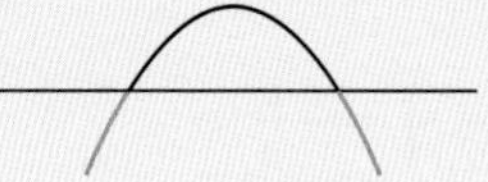

$x \leq -3 - \sqrt{11}$ or $x \geq -3 + \sqrt{11}$

Example 14

Solve the equation $2x^2 = 5x - 1$. Write your answer correct to 2 decimal places.

When asked to solve to 2 decimal places, it is likely that the quadratic formula will be appropriate.

$2x^2 - 5x + 1 = 0$

$a = 2, b = -5, c = 1$

We must first rearrange to $ax^2 + bx + c = 0$.

$$x = \frac{-(-5) \pm \sqrt{(-5)^2 - 4(2)(1)}}{2(2)}$$

$$= \frac{5 + \sqrt{17}}{4} \text{ or } \frac{5 - \sqrt{17}}{4}$$

$= 2.28$ or 0.22 (2 d.p.)

In examinations, unless otherwise indicated, answers to all questions on all topics should be rounded to 3 significant figures.

However, do **not** round your answers to 3 significant figures until the final answer. A significant number of candidates lose marks in examinations by rounding earlier in the question which then leads to an inaccurate answer.

You should therefore keep all numbers unrounded until the final answer.

Exercise 1.4

1. Solve each of these quadratic equations. Write your answers correct to 2 decimal places.

a) $2x^2 - 3x - 4 = 0$ **b)** $5x^2 - 11x + 4 = 0$ **c)** $3x^2 + 12x + 5 = 0$ **d)** $x^2 + 5x - 2 = 0$
e) $4x^2 + 2x - 5 = 0$ **f)** $x^2 + x - 4 = 0$ **g)** $3x^2 = 6x - 2$ **h)** $9x = 6x^2 + 2$
i) $3 - 8x = 2x^2$ **j)** $x^2 + 3x = 1$

2. Use the formula to solve each of these quadratic equations. Leave your answers in surd form.

a) $2x^2 - x - 5 = 0$ **b)** $3x^2 - 6x + 1 = 0$ **c)** $5x^2 - 3x - 7 = 0$ **d)** $6x^2 - x - 4 = 0$
e) $x^2 + 8x - 3 = 0$ **f)** $x^2 + 7x + 3 = 0$ **g)** $x^2 + 10x - 5 = 0$ **h)** $2x^2 + 5x - 4 = 0$
i) $4x^2 = 5 - 10x$ **j)** $4x - 1 - x^2 = 0$

3. Use the formula to solve the following inequalities. Leave your answers in surd form.

a) $3x^2 + 10x + 5 < 0$ **b)** $x^2 - 6x + 7 \geq 0$ **c)** $4x^2 + 3x - 2 > 0$ **d)** $2x^2 + x - 2 \leq 0$
e) $5x^2 - 8x - 2 \geq 0$ **f)** $6x^2 - 6x + 1 \geq 0$ **g)** $5 + 3x - x^2 \leq 0$ **h)** $2x^2 + 2x - 3 < 0$
i) $2x + 1 - 2x^2 > 0$ **j)** $x^2 < 1 - x$

1.5 Solve more complex quadratic equations

You can now adapt the techniques you have learnt to solve more complex quadratic equations.

Example 15

Solve $x^4 - 5x^2 + 4 = 0$.

Hint: Try to adapt the equation to quadratic form $(\ldots)^2 - 5(\ldots) + 4 = 0$ by appropriate substitution.

$x^4 - 5x^2 + 4 = 0$ Let $y = x^2$

Hence $y^2 - 5y + 4 = 0$ ← Substitute y for x^2.

$(y - 4)(y - 1) = 0$

$y = 4$ or $y = 1$

$x^2 = 4$ or $x^2 = 1$ ← Substitute x^2 for y.

$x = \pm 2$ or $x = \pm 1$

Or

Factorise straight away $(x^2 - 4)(x^2 - 1) = 0$

$x^2 = 4$ or $x^2 = 1$

$x = \pm 2$ or $x = \pm 1$

Example 16

Solve the equation $5x^4 - 20x^2 - 1 = 0$ by completing the square.

Give your answer correct to 3 significant figures.

$5x^4 - 20x^2 - 1 = 0$ Let $y = x^2$

Hence $5y^2 - 20y - 1 = 0$ ← Substitute y for x^2.

$$y^2 - 4y - \frac{1}{5} = 0$$ ← Divide each term by 5.

$$(y - 2)^2 - 4 - \frac{1}{5} = 0$$

$$(y - 2)^2 = \frac{21}{5}$$

$$y - 2 = \pm\sqrt{\frac{21}{5}}$$ ← Work out as a decimal to at least 5 s.f.

$$y = 2 \pm \sqrt{\frac{21}{5}} = 4.04939.... \text{ or } -0.04939...$$

$x^2 = 4.04939....$ or $-0.04939...$ ← Substitute x^2 for y.

$x = \pm\sqrt{4.04939...}$ or ← Cannot find the square root of a negative number.

$x = \pm 2.0123... = \pm 2.01$ (3 s.f.)

Example 17

Solve the equation $2x^6 - 3x^3 = 8$. Write your answer correct to 2 decimal places.

$2x^6 - 3x^3 - 8 = 0$

Let $y = x^3$ Hence $2y^2 - 3y - 8 = 0$

We must first rearrange to $ax^2 + bx + c = 0$.

$a = 2$

$b = -3$

$c = -8$

$$y = \frac{-(-3) \pm \sqrt{(-3)^2 - 4(2)(-8)}}{2(2)}$$

$$= \frac{3+\sqrt{73}}{4} \quad \text{or} \quad \frac{3-\sqrt{73}}{4}$$

$$= 2.886\ldots \quad \text{or} \quad -1.386\ldots$$

$$x^3 = 2.886\ldots \quad \text{or} \quad -1.386\ldots$$

Substitute x^3 for y.

$$x = \sqrt[3]{2.886\ldots} \quad \text{or} \quad \sqrt[3]{-1.386\ldots}$$

$$= 1.4237\ldots \quad \text{or} \quad -1.1149\ldots$$

Exercise 1.5

1. Solve each of these quadratic equations. Give your answers as exact answers.

a) $x^4 - 4x^2 - 21 = 0$ **b)** $6x^4 - x^2 = 2$

c) $x^6 + 7x^3 + 10 = 0$ **d)** $4 + 11x^2 - 3x^4 = 0$

e) $6x^8 + 6 = 13x^4$ **f)** $x^6 + 3x^3 = 40$

2. Solve each of these equations by completing the square.
Write your answers correct to 2 decimal places.

a) $x^6 - 8x^3 + 1 = 0$ **b)** $3 - 6x^2 - x^4 = 0$

c) $x^4 + 2x^2 = 10$ **d)** $x^6 + x^3 - 4 = 0$

e) $2x^8 - 20x^4 - 7 = 0$ **f)** $3x^4 + 1 = 12x^2$

3. Use the formula to solve each of these quadratic equations.
Write your answers correct to 3 significant figures.

a) $x^4 + 3x^2 - 5 = 0$ **b)** $2x^6 - 4x^3 + 1 = 0$

c) $3x^6 - 7x^3 = 2$ **d)** $5x^4 + 10x^2 + 2 = 0$

e) $4x^{10} + x^5 - 2 = 0$ **f)** $x^8 - x^4 - 1 = 0$

1.6 The discriminant of a quadratic equation

We sometimes call the solutions of a quadratic equation the **roots** of the equation. This also tells us where the quadratic graph crosses the x-axis. When we have an equation of the form $ax^2 + bx + c = 0$, we can tell the nature of the roots by looking at the **discriminant** of the quadratic equation.

The discriminant is the value of $b^2 - 4ac$ in $x = \dfrac{-b \pm \sqrt{b^2 - 4ac}}{2a}$.

If $b^2 - 4ac > 0$ there are two distinct **real roots**.
If $b^2 - 4ac = 0$ there are equal roots (one repeated root).
If $b^2 - 4ac < 0$ there are no real roots.

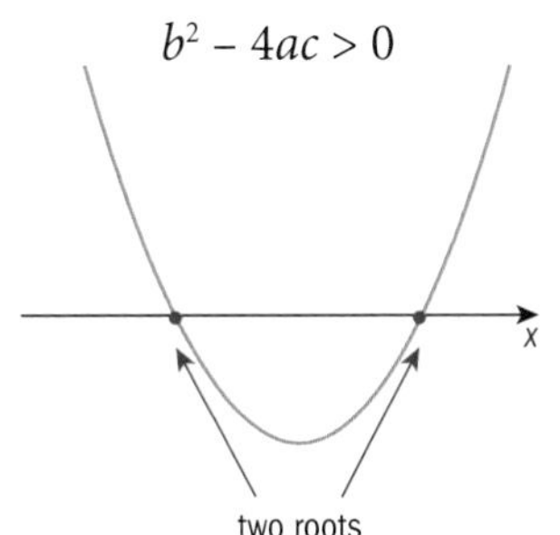

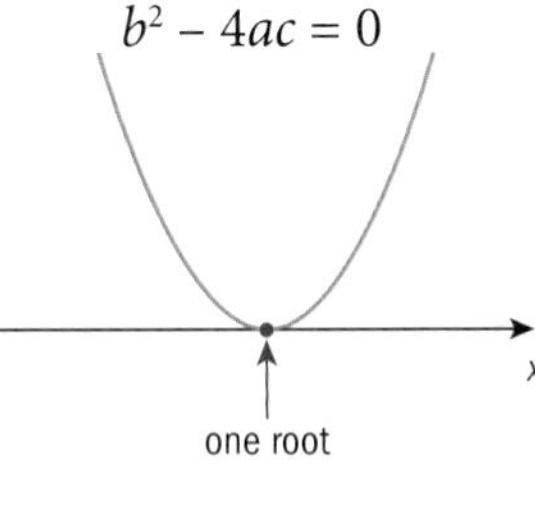

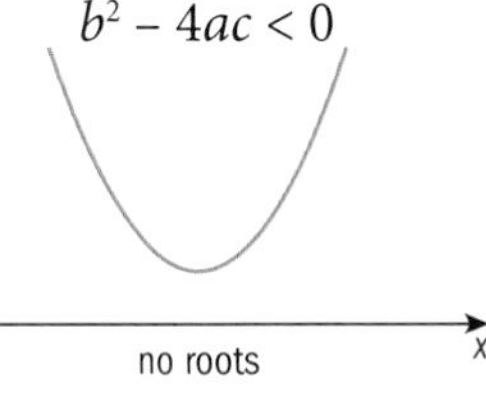

Example 18

Work out whether each of these quadratic equations has two distinct roots, equal roots or no real roots.

a) $x^2 - 3x + 5 = 0$

b) $3x^2 + x - 6 = 0$

c) $25x^2 + 20x + 4 = 0$

Note: We know from **1.3** that if $y = ax^2 + bx + c$ and $a > 0$, then the curve is ∪ shaped.
If $b^2 - 4ac < 0$, then there are no real roots and the curve does not cross the x-axis, so y is positive for all values of x.

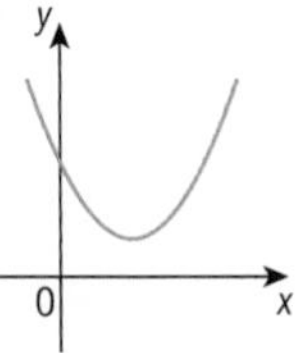

a) $b^2 - 4ac = (-3)^2 - 4(1)(5) = 9 - 20 = -11 < 0$ no real roots

b) $b^2 - 4ac = (1)^2 - 4(3)(-6) = 1 + 72 = 73 > 0$ two distinct roots

c) $b^2 - 4ac = (20)^2 - 4(25)(4) = 400 - 400 = 0$ equal roots

Example 19

The equation $x^2 + px + q = 0$, where p and q are constants, has roots -1 and 4.

a) Find the value of p and q.

b) Using these values of p and q, find the constant r for which the equation $x^2 + px + q + r = 0$ has equal roots.

▶ Continued on the next page

a) $x = -1$ and $x = 4$ ← The roots are the values of x.

$(x + 1)(x - 4) = 0$

$x^2 - 3x - 4 = 0$

$p = -3$ and $q = -4$ ← Comparing with $x^2 + px + q = 0$.

b) $x^2 - 3x - 4 + r = 0$ ← Substitute the values of p and q from part **(a)**.

If $x^2 - 3x - 4 + r = 0$ has equal roots then $b^2 - 4ac = 0$

where $a = 1, b = -3, c = -4 + r$

$(-3)^2 - 4(1)(-4 + r) = 0$

$9 + 16 - 4r = 0$

$r = \frac{25}{4}$

Exercise 1.6

1. Work out the discriminant for each equation and then **determine** whether each equation has two distinct roots, equal roots or no real roots.

   a) $2x^2 - 3x + 4 = 0$
   b) $5x^2 - 11x + 4 = 0$
   c) $x^2 + 12x + 36 = 0$
   d) $2x^2 + x - 2 = 0$
   e) $4x^2 + 5x + 2 = 0$
   f) $4x^2 - 12x + 9 = 0$
   g) $7x^2 - 2x + 1 = 0$
   h) $x^2 - 6x + 7 = 0$
   i) $2 - x - 4x^2 = 0$
   j) $6x^2 = 6x - 1$
   k) $16x^2 = 24x - 9$
   l) $(x + 2)^2 - 8x - 2 = 0$

2. **Show that** $x^2 + 8x + 16 \geq 0$ for all values of x.
3. Show that $1 + 100x^2 - 20x \geq 0$ for all values of x.
4. If $2x^2 - ax + 8 = 0$ has no real roots, find the range of possible values of a.
5. If $6 - 2x - kx^2 = 0$ has a repeated root, find the value of k.
6. The equation $x^2 + px + q = 0$, where p and q are constants, has roots -3 and 2. Find the value of p and q.
7. Use the discriminant to find the nature of the roots of the equation $3x + 4 = \frac{5}{x}$.
8. The quadratic equation $kx^2 + 5x + 2 = 0$ has two distinct real roots. Find the range of possible values of k.
9. Find the value of p for which the quadratic equation $px^2 - 4px + 2 - p = 0$ has equal roots.
10. Prove that the quadratic equation $(q - 5)x^2 + 5x - q = 0$ has real roots for any value of q.
11. The quadratic equation $x + k + \frac{9}{x} = 0$ has equal roots. Find the two possible values of k.
12. The equation $px^2 + qx + r = 0$, where p, q and r are constants, has roots $-\frac{1}{2}$ and $\frac{3}{4}$. Find the smallest possible integer values of p, q and r.

1.7 Solving simultaneous equations

We already know how to solve **simultaneous equations** where both equations are **linear**.

We are now going to solve simultaneous equations where one equation is linear and the other is quadratic.

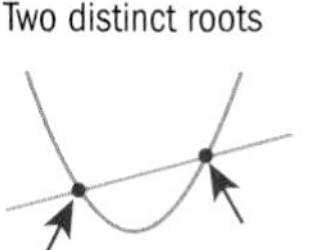

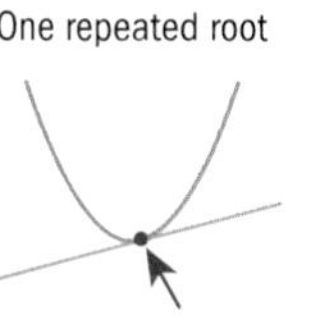

We will either get two distinct roots, one repeated root, or no real roots.

Example 20

Solve simultaneously $y = x^2 - 3x - 1$ and $y = 2x - 7$.

$x^2 - 3x - 1 = 2x - 7$ ← The y-values must be the same.

$x^2 - 5x + 6 = 0$

$(x - 3)(x - 2) = 0$

$x = 3$ or $x = 2$ ← Two distinct roots

When $x = 3$, $y = 2(3) - 7 = -1$

When $x = 2$, $y = 2(2) - 7 = -3$

Solutions are $x = 3$, $y = -1$ and $x = 2$, $y = -3$

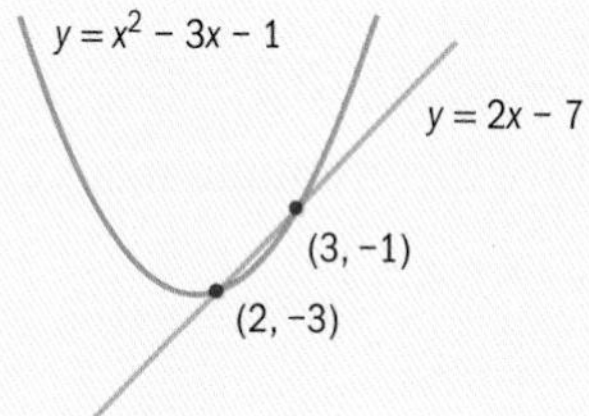

Example 21

Show that there are no real roots for the simultaneous equations $y = x^2 - 2x - 1$ and $y = x - 5$.

The curve $y = x^2 - 2x - 1$ meets the line $y = x - 5$

when $x^2 - 2x - 1 = x - 5$

$x^2 - 3x + 4 = 0$ ← It is not possible to factorise this.

$b^2 - 4ac = (-3)^2 - 4(1)(4) = -7 < 0$

There are no real roots to the equation $x^2 - 3x + 4 = 0$.

Thus, there are no real roots for the simultaneous equations.

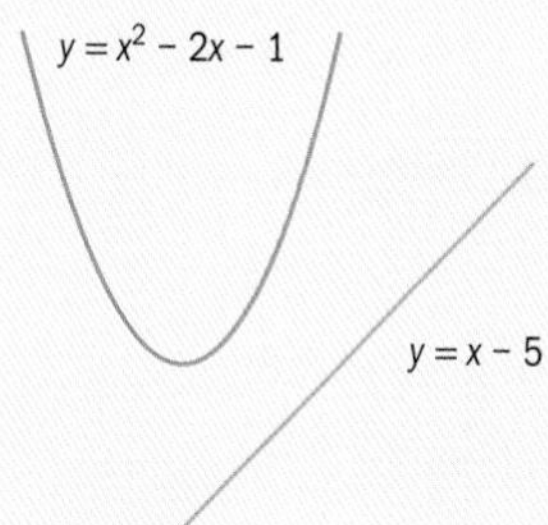

Example 22

Solve simultaneously $y^2 + xy + 4x = 7$ and $x - y = 3$.

We can use the same method to solve other pairs of simultaneous equations where we do not have one linear equation and one quadratic equation.

$x - y = 3 \qquad x = 3 + y$ ← We could also substitute $y = x - 3$.

$y^2 + (3 + y)y + 4(3 + y) = 7$ ← Substitute for x in the non-linear equation.

$y^2 + 3y + y^2 + 12 + 4y - 7 = 0$

$2y^2 + 7y + 5 = 0$

$(2y + 5)(y + 1) = 0$

$y = -\frac{5}{2}$ or $y = -1$

$x = 3 + -\frac{5}{2}$ or $x = 3 + -1$ ← Substitute for y in $x = 3 + y$.

$x = \frac{1}{2}$ or $x = 2$

Solutions: $x = \frac{1}{2}, y = -\frac{5}{2}$ or $x = 2, y = -1$

Exercise 1.7

Solve these simultaneous equations.

1. $y = x + 1$
$y = x^2 - 1$

2. $y = 4x + 7$
$y = 2x^2 + 1$

3. $y = x^2 - 6x + 5$
$y = x - 1$

4. $y = x^2 - x - 2$
$x + y = 7$

5. $y = 2x$
$y = x^2 - x + 2$

6. $y = x^2 - 2x - 5$
$y = x + 5$

7. $y = x$
$y = 6 - x^2$

8. $y = 4x + 5$
$y = x^2$

9. $y + 3 = 2x$
$y = x^2 - 2x + 1$

10. $y = 2x + 5$
$y = x^2 - 3x - 1$

11. $x^2 + y = 3$
$y + 3x = 5$

12. $y = x - 4$
$y = 2 - x^2$

13. $x - 2y = 8$
$xy = 24$

14. $y = 6x + 5$
$y = 12x^2 - 5x$

15. $y = \frac{1}{2}(1 - x^2)$
$y = x - 1$

16. $x^2 + y^2 = 8$
$2x = y + 2$

17. $3x + 4y = 15$
$2xy = 9$

18. $y = 1 + 2x$
$y^2 = 2x^2 + x$

19. $y = 2x + 1$
$x^2 - 2xy + y^2 = 1$

20. $3x = 1 + 2y$
$3x^2 - 2y^2 + 5 = 0$

21. Show that there are no real solutions to the simultaneous equations $y = 1 + 2x - x^2$ and $y = \frac{1}{2}x + 5$.

22. The diagram shows a rectangle.

The area of the rectangle is 3.5 m^2.

The perimeter of the rectangle is 7.8 m.

Find the dimensions of the rectangle.

y metres

x metres

23. The solid cuboid has a volume of 54cm^3 and a total surface area of 96cm^2 and $x > y$.

a) Use this information to write down two equations in x and y.

b) Solve the equations simultaneously to find the value of x and the value of y.

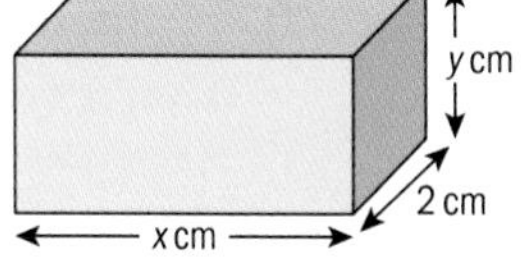

24. The diagram shows an L-shape.
All measurements are in centimetres.

a) Write down two equations in x and y.

b) Solve the equations simultaneously and hence find the perimeter of the L-shape.

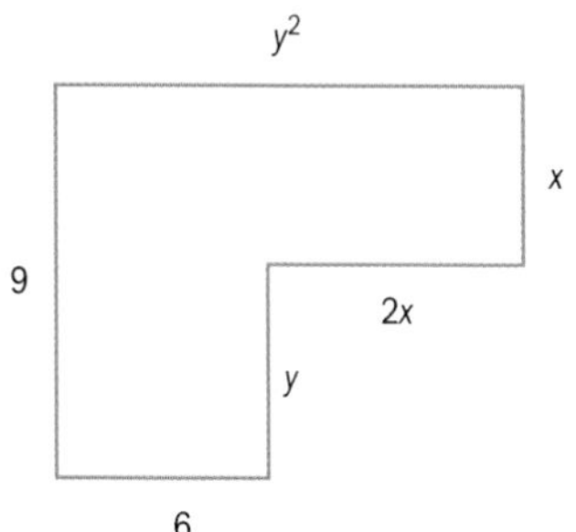

25. The diagram shows a trapezium.
All measurements are in centimetres.

The area of the trapezium is 60 cm^2.
The perimeter of the trapezium is 36 cm.

Find the value of x and the value of y.

Note: Area of a trapezium $= \frac{1}{2}$ (sum of parallel sides $\times$ distance between them).

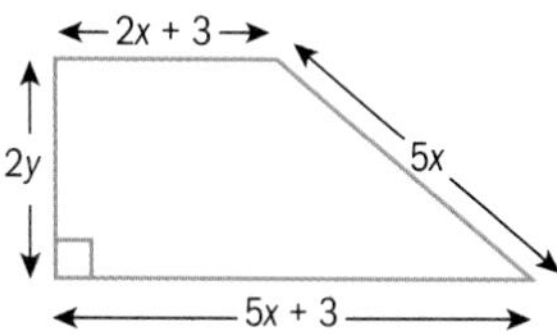

1.8 Graphs of quadratic functions

You can sketch a graph of the quadratic equation $y = ax^2 + bx + c$, where a, b and c are constants and where $a \neq 0$, by considering the following:

If $a > 0$, the curve is ◡ shaped and has a **minimum** value.

If $a < 0$, the curve is ◠ shaped and has a **maximum** value.

If $b^2 - 4ac > 0$,   there are **two distinct real roots.**

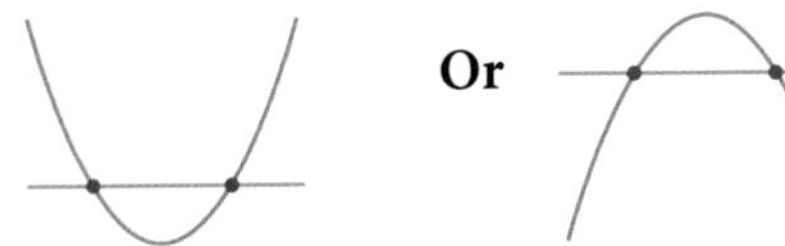

If $b^2 - 4ac = 0$,   there are **equal roots** (one repeated root).

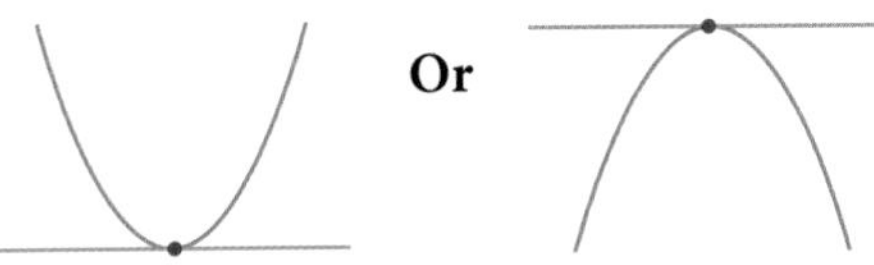

If $b^2 - 4ac < 0$,   there are **no real roots.**

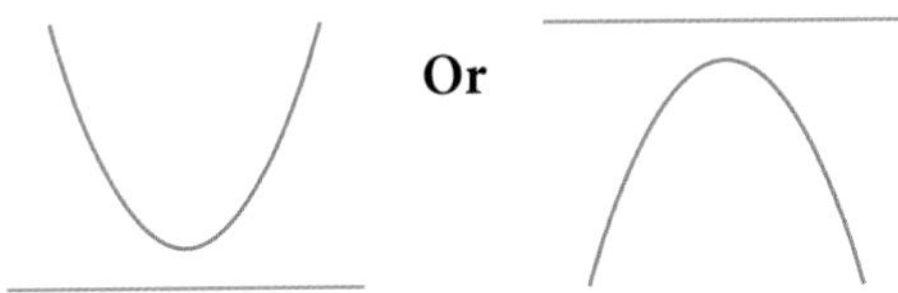

When $ax^2 + bx + c = 0$ is expressed in the form $(x + p)^2 + q$, then the **minimum value** is **when $x = -p$** and the **minimum value is q.**

When $ax^2 + bx + c = 0$ is expressed in the form $-(x + p)^2 + q$, then the **maximum value** is **when $x = -p$** and the **maximum value is q.**

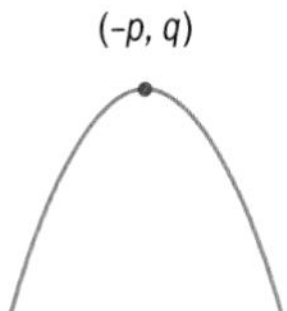

The coordinates of the vertex of the curve are $(-p, q)$.
We can find these values by the method of completing the square.

The equation of the line of symmetry is $\boxed{x = \frac{-b}{2a}}$

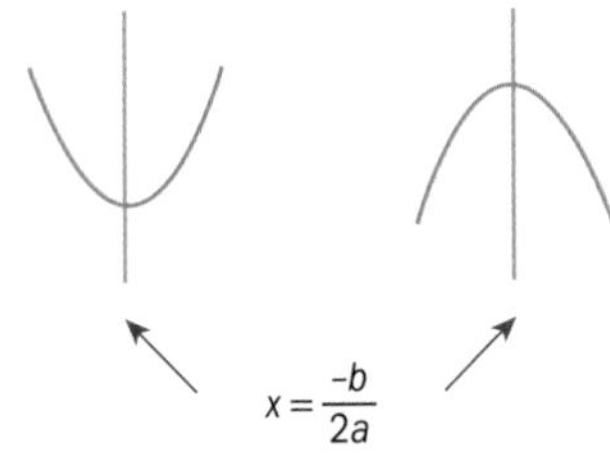

A vertex is a defining point in geometrical situations. It can be where the sides of a polygon intersect, where the edges of a polyhedron intersect or, in this case, where the curve and its line of symmetry intersect.

Example 23

Express $y = x^2 - 6x + 10$ in the form $y = (x + p)^2 + q$.

Sketch the curve, stating the coordinates of the vertex.

$y = (x - 3)^2 - 9 + 10 = (x - 3)^2 + 1$

Coordinates of the vertex are (3, 1).

The **minimum** value of y is 1.

$a > 0$ therefore the graph is ∪ shaped.

The line of symmetry: $x = \frac{-b}{2a} = \frac{-(-6)}{2(1)}$ $x = 3$

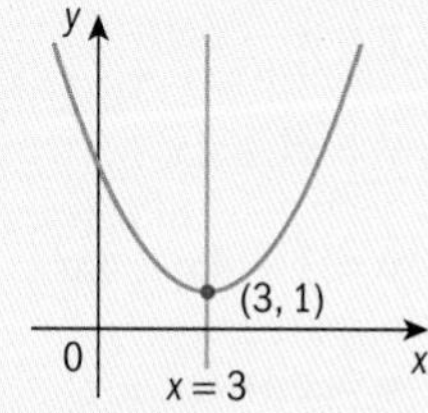

$b^2 - 4ac = (-6)^2 - 4(1)(10)$
$= -4 < 0$
therefore no real roots.

Example 24

Express $y = 3 - 8x - 2x^2$ in the form $y = a(x + b)^2 + c$.

Sketch the curve, stating the coordinates of the vertex.

$y = -2[x^2 + 4x] + 3$
$= -2[(x + 2)^2 - 4] + 3$
$= -2(x + 2)^2 + 8 + 3$
$= -2(x + 2)^2 + 11$

Coordinates of the vertex are $(-2, 11)$.

The **maximum** value of y is 11.

$a < 0$ therefore the graph is ∩ shaped.

The line of symmetry: $x = \frac{-b}{2a} = \frac{-(-8)}{2(-2)}$ $x = -2$

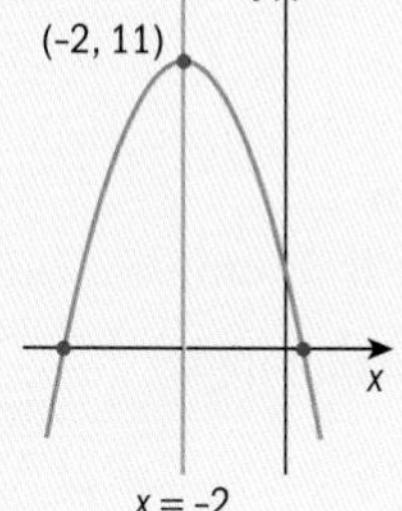

$b^2 - 4ac = (-8)^2 - 4(-2)(3)$
$= 88 > 0$
therefore two distinct roots.

Example 25

A sheep pen is in the shape of a rectangle.
One of the sides of the pen is a wall.
A farmer puts fencing on the other three sides of the pen.
The two sides that touch the wall are each x metres.
He uses 40 m of fencing.

Find the maximum area of the pen.

The third side of the pen $= 40 - 2x$

Area of the pen $= x(40 - 2x)$
$= 40x - 2x^2$
$= -2[x^2 - 20x]$
$= -2[(x - 10)^2 - 100]$
$= -2(x - 10)^2 + 200$

The **maximum** area is 200 m^2 (when $x = 10$).

Exercise 1.8

In questions **1** to **10,** express each of the quadratic equations in the form $y = a(x + b)^2 + c$. Sketch the curve, stating the coordinates of the vertex and whether there is a maximum or minimum value of y.

1. $y = x^2 + 2x$

2. $y = x^2 - 6x + 10$

3. $y = 3 - 2x - 4x^2$

4. $y = x^2 + 4x - 3$

5. $y = 5 + 2x - x^2$

6. $y = 3 + 4x - 4x^2$

7. $y = x^2 - 4x - 8$

8. $y = x^2 + 6x + 7$

9. $y = 5x^2 - 4x - 2$

10. $y = 2x^2 - 5x - 3$

11. A right-angled triangle has a width of x cm.
The length of the hypotenuse is 10 cm.
The perimeter of the triangle is 24 cm.

Find the area of the triangle.

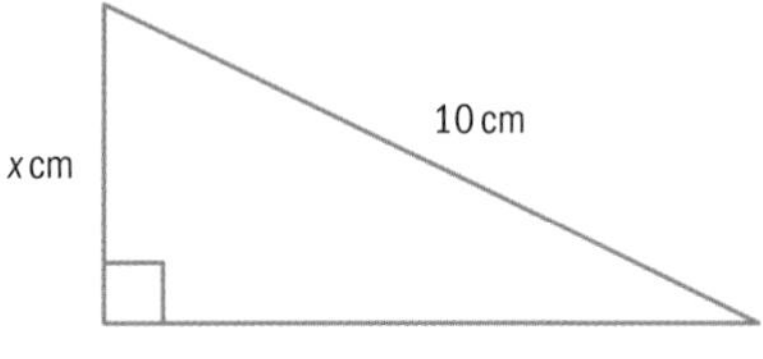

12. A rectangle has a width of x cm.
The perimeter of the rectangle is 32 cm.

Find the maximum area of the rectangle.

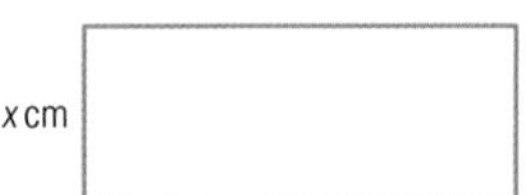

13. Find the maximum area of the triangle on the right.
State the value of x when this occurs.

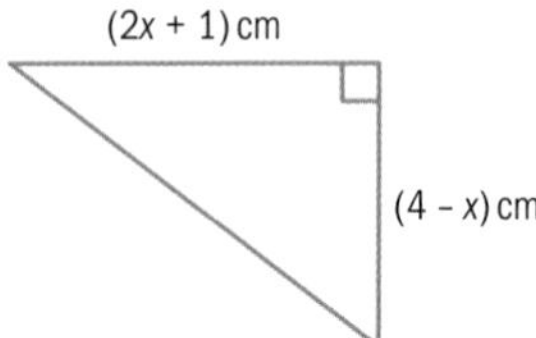

14. Faisal is x years old. Faisal has a brother called Omar.
The sum of the two boys' ages is 20 years.

a) Express the product of their ages in the form $y = a(x - b)^2 + c$.

b) How old must Faisal be to make the product of their ages a maximum?

Summary exercise 1

1. Solve $(6x - 5)(4 - 3x) = 0$.

2. a) Factorise $10x^2 - 29x - 21$.

   b) Hence or otherwise solve the equation $10x^2 - 29x = 21$.

3. Solve $y^6 - 9y^3 + 18 = 0$.

4. By using the substitution $y = \sqrt{x}$ or otherwise, solve the equation $2x - 5\sqrt{x} + 2 = 0$.

5. Solve the equation $2x^6 - 16x^3 - 3 = 0$ by completing the square. Write your answer correct to 3 significant figures.

6. Solve the equation $3x^4 + 5x^2 = 7$ by using the formula. Write your answer correct to 2 decimal places.

EXAM-STYLE QUESTION

7. A square piece of card with sides x cm has squares of sides 3 cm cut from the corners.
   The card is then folded to make an open box with volume 48 cm^2.
   Find the dimensions of the card.

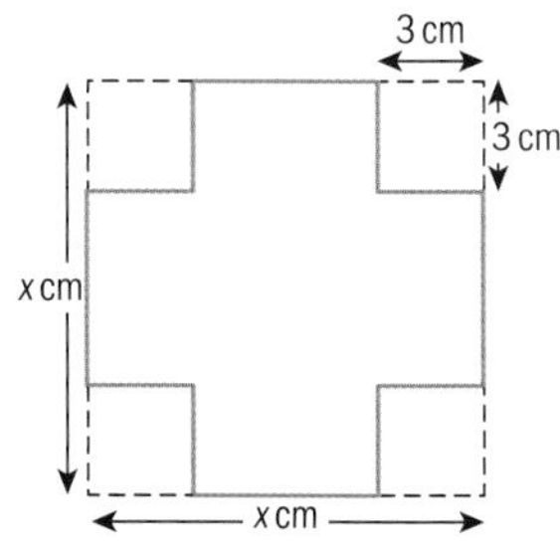

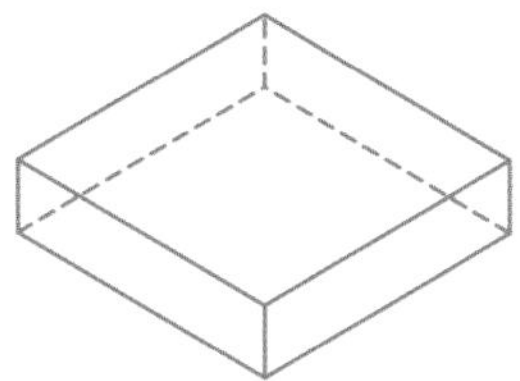

EXAM-STYLE QUESTIONS

8. Find the set of values for x for which $(x - 1)(x + 2) < 18$.

9. Find the set of values for x for which $20 + 7x - 6x^2 \geq 0$.

10. Solve the equation $2x^2 + 5x + 1 = 0$. Write your answer correct to 3 significant figures.

11. Use the formula to solve $7x^2 - 3x - 5 \leq 0$. Leave your answer in surd form.

EXAM-STYLE QUESTIONS

12. Solve $x(x - 1) - 3(x - 3) + 3(x - 2) = x + 6$.

13. Find the relationship between p and q, if the equation $px^2 + 3qx + 9 = 0$ has equal roots.

14. Work out whether each of these quadratic equations has two distinct roots, equal roots or no real roots.

    a) $x^2 + x + 1 = 0$

    b) $4x^2 + 12x + 9 = 0$

    c) $3x^2 + 4x + 1 = 0$

EXAM-STYLE QUESTIONS

15. The quadratic equation $x^2 + kx + 36 = 0$ has two different real roots.

    Find the set of possible values of k.

16. The equation $x^2 + px + q = 0$, where p and q are constants, has roots -1 and 4.

    a) Find the values of p and q.

    b) Using these values of p and q, find the value of r, where r is a constant, and the equation $x^2 + px + q + r = 0$ has equal roots.

EXAM-STYLE QUESTIONS

17. Solve the simultaneous equations $x + y = 1$ and $x^2 - y^2 = 5$.

18. Solve the simultaneous equations $x + y = 0$ and $2x^2 + y^2 = 6$.

19. Solve the simultaneous equations $x + y = 3$ and $x^2 + 2xy = 5$.

20. Solve the simultaneous equations $3x + y = 8$ and $3x^2 + y^2 = 28$.

21. a) Express $3x^2 + 12x + 5$ in the form $p(x + q)^2 + r$.

b) Find the minimum value of $3x^2 + 12x + 5$.

22. A square garden of side x metres is surrounded by a path of width 1 metre. The area of the garden is the same as the area of the path.

Find the value of x.

Leave your answer in surd form.

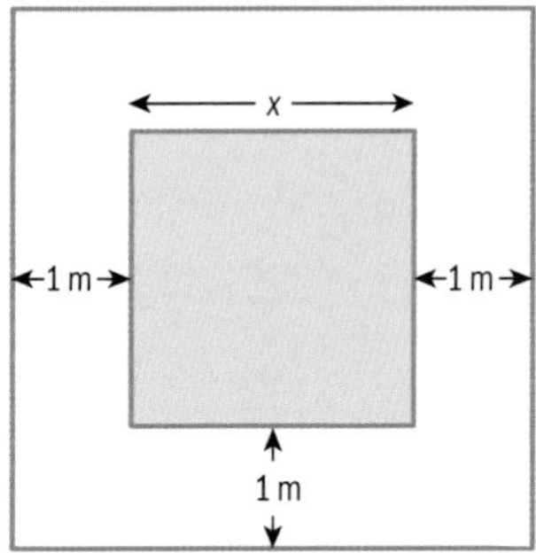

23. A curve has equation $y = 2x^2 + x + 11$.

i) Find the set of values of x for which $y \leq 17$.

24. Rectangle **A** has a height of $(x + 1)$ metres and an area of $4\,\text{m}^2$.

Rectangle **B** has a height of $(2x - 1)$ metres and an area of $6\,\text{m}^2$.

The sum of the widths of the two rectangles is less than 8 metres.

Find the set of values for x.

25. Find the set of values of k for which the line $y + 4 = kx$ intersects the curve $y = x^2$ at two distinct points.

26. The equation $x^2 + 8x - k(x + 8) = 0$ has equal roots. Find the value of k.

27. Find the real roots of the equation $\frac{8}{x^4} + \frac{5}{x^2} = 3$.

Chapter summary

To solve quadratic equations

- **Factorise** and solve.
- Or use the **method of completing the square**.
- Or use the quadratic formula $x = \frac{-b \pm \sqrt{b^2 - 4ac}}{2a}$.

To solve quadratic inequalities

- Solve the corresponding equation $ax^2 + bx + c = 0$ to find the values of x.
- Sketch the graph of $y = ax^2 + bx + c$.
- Write down the solutions using inequality signs.

The discriminant of the quadratic equation $ax^2 + bx + c = 0$

- Discriminant = $b^2 - 4ac$

  If $b^2 - 4ac > 0$ there are two distinct real roots.
  If $b^2 - 4ac = 0$ there are equal roots (one repeated root).
  If $b^2 - 4ac < 0$ there are no real roots.

$b^2 - 4ac > 0$

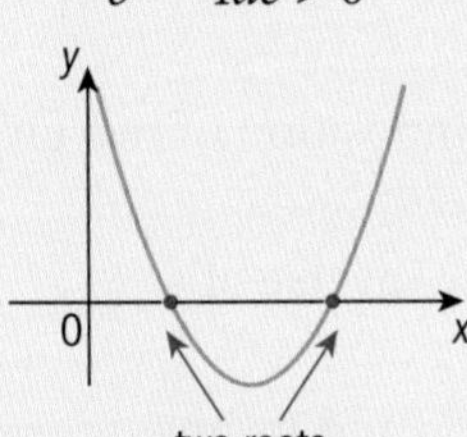

$b^2 - 4ac = 0$

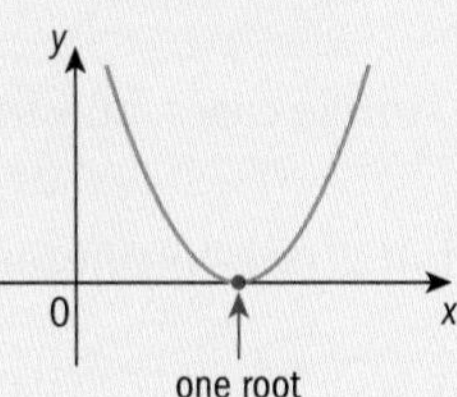

$b^2 - 4ac < 0$

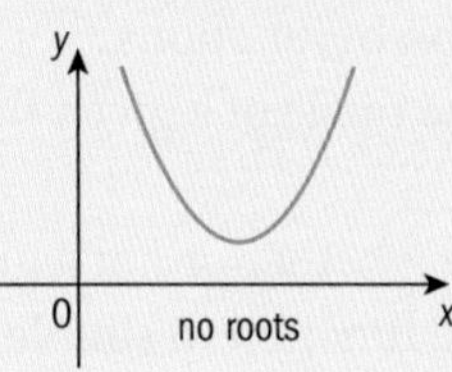

$ax^2 + bx + c = 0$, where a, b and c are constants, $a \neq 0$.

Graphs of quadratic equations $y = ax^2 + bx + c$ where a, b and c are constants, $a \neq 0$

- If $a > 0$, the curve is $\cup$ shaped and has a **minimum** value.
- If $a < 0$, the curve is $\cap$ shaped and has a **maximum** value.
- The **equation of the line of symmetry** is $x = \frac{-b}{2a}$.

To solve simultaneous equations where one is linear and one is quadratic

- Rearrange the linear equation to write x in terms of y or y in terms of x.
- Then substitute this into the quadratic equation.

Recognise and solve equations in x which are quadratic in some function of x

- e.g. To solve $x^6 - 2x^3 + 1 = 0$, let $y = x^3$ and solve $y^2 - 2y + 1 = 0$.

2 Functions and transformations

There are many situations in the real world where the application of functions is used. Galileo Galilei (1564–1642), who was an Italian astronomer, mathematician and physicist, invented the idea of a function to simplify equations. He grasped the concept of a mapping between two sets and used this to study the relationship between concentric circles. Functions can be used to describe many relationships, including the orbit of the planets around the Sun.

Objectives

- Understand the terms function, domain, range, one-to-one function, inverse function and composition of functions.
- Identify the range of a given function in simple cases, and find the composition of two given functions.
- Determine whether or not a given function is one-to-one, and find the inverse of a one-to-one function in simple cases.
- Illustrate in graphical terms the relation between a one-to-one function and its inverse.
- Understand and use the transformations of the graph of $y = f(x)$ given by $y = f(x) + a$, $y = f(x + a)$, $y = af(x)$, $y = f(ax)$ and simple combinations of these.

Before you start

You should know how to:

1. Substitute into f(x),

e.g. $f(x) = 3x - 2\sqrt{x}$

$f(1) = 3 - 2 = 1$

$f(100) = 300 - 20 = 280$

$f(4) = 12 - 4 = 8$.

2. Sketch the graph of $y = x^2$,

e.g.

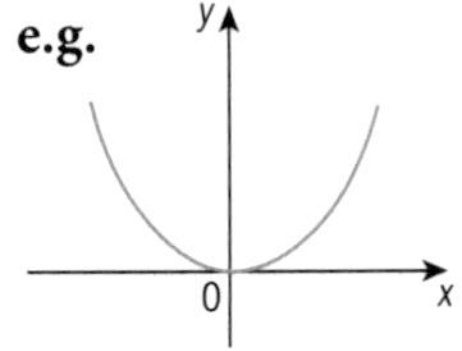

Skills check:

1. $f(x) = 5x + 2x^3$

Find

a) f(1) **b)** f(−2) **c)** f(−10)

d) f(8) **e)** f(−3) **f)** f(−1).

2. Sketch the graphs of

a) $y = x^3$ **b)** $y = (x - 2)(x + 1)$

c) $y = 2^x$ **d)** $y = x^2(x + 3)$

e) $y = \frac{1}{x}$ **f)** $y = (5 - x)(x + 4)$.

2.1 Mappings

A mapping looks at the relationship between two sets of numbers.

Consider the mapping $\times 2 + 1$

We can input a set of numbers, for example $\{-2, -1, 1, 3, 6\}$.

The input set is called the domain.

The output set can be obtained by applying the operation to each number in the input set, e.g. input $= -2$, output $= -2 \times 2 + 1 = -3$.

The output set is called the range.

The output set for a domain of $\{-2, -1, 1, 3, 6\}$ is $\{-3, -1, 3, 7, 13\}$.

We say that each number is a member ($\in$) of that set. $x \in \mathbb{R}$ means that x is a member of the set of all the real numbers.

A function is defined as a mapping where every element of the domain (x-values) is mapped onto exactly one element of the range (y-values). The diagram on the right shows an example of a function.

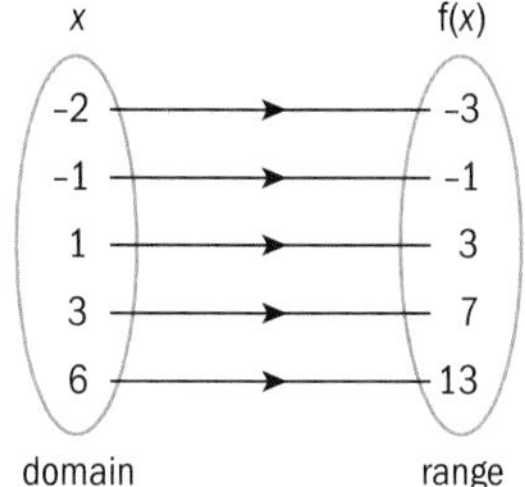

We can write this function in two different ways:

$$f(x) = 2x + 1 \qquad \text{or} \qquad f: x \mapsto 2x + 1$$

We say: f maps x onto $2x + 1$.

$f: x \mapsto 2x + 1$

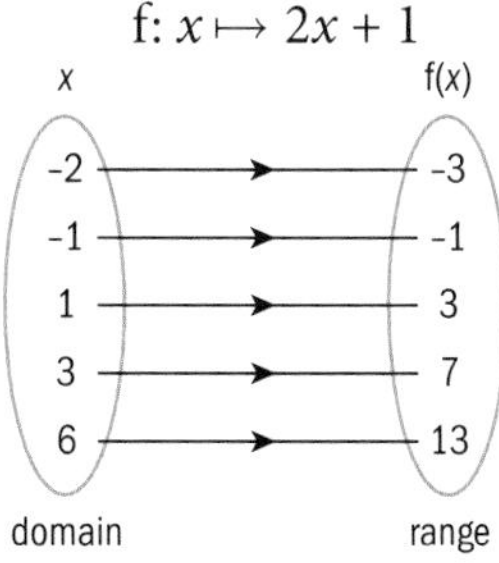

If every element in the domain is mapped onto exactly one element in the range and every element in the range is mapped onto exactly one element in the domain, we say the function is a one-to-one function.

$f: x \mapsto x^2$

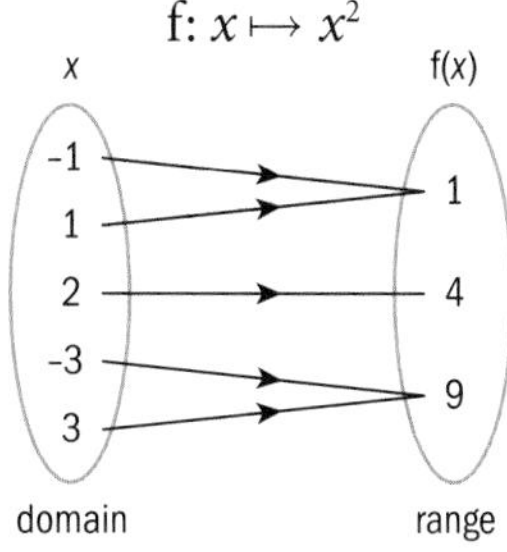

If every element in the domain is mapped onto exactly one element in the range, but some elements in the range arise from more than one element in the domain, we say the function is a many-to-one function.

$f: x \mapsto \pm\sqrt{x}, x \geq 0$

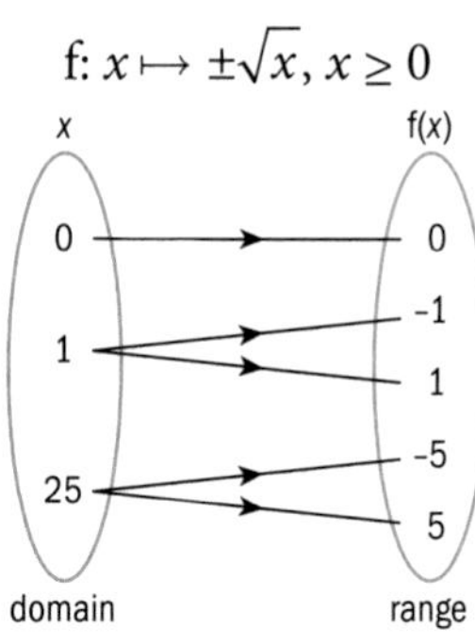

This is **not** a function, because it is one-to-many. Some elements in the domain are mapped onto more than one element in the range.

Example 1

Find the range of these functions.

a) $f(x) = x + 5, x \geq 5$ **b)** $g(x) = x^3, -2 \leq x < 4$

a) Given domain is $x \geq 5$
Range: $f(x) \geq 10$

The least value of $f(x)$ is $5 + 5 = 10$.

b) Given domain is $-2 \leq x < 4$
Range: $-8 \leq g(x) < 64$

$(-2)^3 = -8$ and $(4)^3 = 64$

Example 2

a) Sketch the graph of the function defined by

$$f(x) = \begin{cases} 4 - x & \text{when } -2 \leq x \leq 1 \\ 2x + 1 & \text{when } 1 < x \leq 3 \end{cases}$$

b) Find the range.

a)

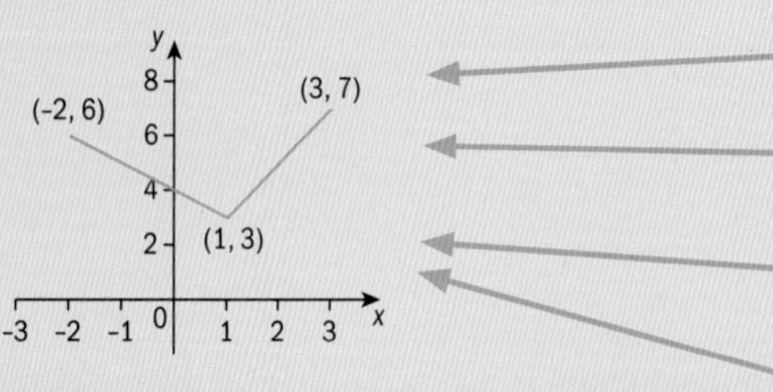

$f(x) = 2x + 1$ $x = 3$ $y = 2(3) + 1 = 7$

$f(x) = 4 - x$ $x = -2$ $y = 4 - (-2) = 6$

$f(x) = 2x + 1$ $x = 1$ $y = 2(1) + 1 = 3$

$f(x) = 4 - x$ $x = 1$ $y = 4 - 1 = 3$

b) Range: $3 \leq f(x) \leq 7$

From the graph, the least value of y is 3 and the greatest value is 7.

Example 3

$h(x) = 2x^2 - 12x + 22, x \in \mathbb{R}$

a) Express $h(x)$ in the form $a(x + b)^2 + c$.

b) Find the range.

a) $h(x) = 2(x^2 - 6x + 11)$

$= 2[(x - 3)^2 - 9 + 11]$

$= 2(x - 3)^2 + 4$

Use the method of completing the square seen in Chapter 1.

b) Least value of $h(x)$ is $0 + 4 = 4$

Range: $h(x) \geq 4$

$2(x - 3)^2$ cannot be less than 0.

Example 4

$f(x) = \dfrac{5}{(x-2)^2}, x \in \mathbb{R}, f(x) \geq 5$

Find the greatest possible domain.

When $f(x) = 5$, $\dfrac{5}{(x-2)^2} = 5$

$(x - 2)^2 = 1$

Since $5(x - 2)^2 = 5$.

$x - 2 = 1$ or -1

$x = 3$ or 1

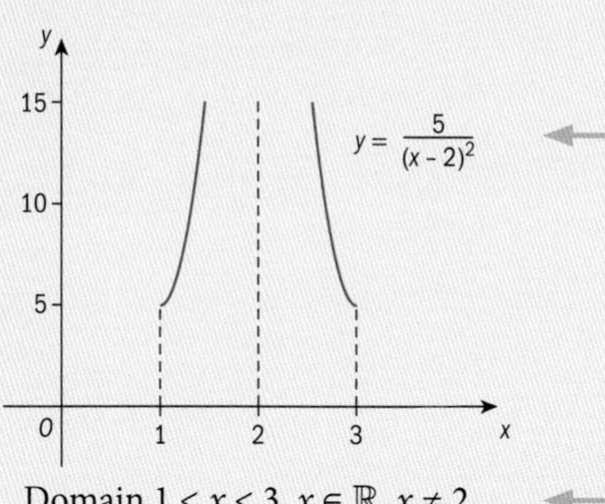

Draw a sketch of $f(x)$ so that you can see the domain.

Domain $1 < x < 3, x \in \mathbb{R}, x \neq 2$

Always include $x \in \mathbb{R}$ and $x \neq 2$ as the denominator cannot be 0.

Exercise 2.1

1. Find the range of these functions.

a) $f(x) = x - 2, x \leq 3$

b) $g(x) = 2x^2, -3 \leq x \leq 1$

c) $h(x) = x^3 + 4, x > -2$

d) $f(x) = \dfrac{1}{x}, x \geq 1$

e) $g(x) = 7x + 1, -1 \leq x \leq 3$

f) $h(x) = x^4, x \in \mathbb{R}$

2. By sketching their graphs or otherwise, find the range of these functions given the domain. State whether each function is a one-to-one function or a many-to-one function.

a) $f: x \mapsto 2x + 1, \quad x \in \mathbb{R}$

b) $g: x \mapsto x^2 - 4, \quad x \in \mathbb{R}, -3 < x < 3$

c) $h: x \mapsto -x^3, \quad x \in \mathbb{R}, -2 < x \leq 1$

d) $f: x \mapsto (x - 3)^2, \quad x \in \mathbb{R}, x \geq 4$

e) $g: x \mapsto 1 - 2x, \quad x \in \mathbb{R}, 1 \leq x \leq 3$

f) $h: x \mapsto (x + 1)^2, \quad x \in \mathbb{R}, x \geq -1$

3. a) Sketch the graph of the function defined by

$$f(x) = \begin{cases} 5 - 2x & \text{when } x < 2 \\ x^2 + 1 & \text{when } x \geq 2 \end{cases}$$

b) Find the range.

4. $f(x) = 3x^2 + 6x - 18, x \in \mathbb{R}$

a) Express $f(x)$ in the form $a(x + b)^2 + c$.

b) Find the range.

5. $f(x) = x^2 - 8x + 16, x \in \mathbb{R}, f(x) \geq 6$. Find the domain. Leave your answer in surd form.

6. Find the range of the function defined by $h(x) = 5 - 2x - x^2, x \in \mathbb{R}$ by the method of completing the square.

7. $$f(x) = \begin{cases} x^2 - 1 & \text{when } 0 \leq x \leq 2 \\ 2x - 1 & \text{when } 2 \leq x < 4 \end{cases}$$

$$g(x) = \begin{cases} x^2 + 2 & \text{when } 0 \leq x \leq 2 \\ 2x - 1 & \text{when } 2 \leq x < 4 \end{cases}$$

Explain why f is a function and g is not a function.

8. Find the range of the function defined by

$$f(x) = \begin{cases} 2x^2 & \text{when } -3 \leq x \leq 0 \\ x^3 + 1 & \text{when } 0 < x \leq 2 \end{cases}$$

9. For each of the following functions state

i) the greatest possible domain for which the function is defined

ii) the corresponding range of the function

iii) whether the function is one-to-one or many-to-one.

a) $f: x \mapsto 4x - 5$

b) $g: x \mapsto \sqrt{x}$

c) $h: x \mapsto \frac{1}{x^2}$

10. The function $g: x \mapsto x^2 + 10x + p$ is defined for the domain $x \in \mathbb{R}$, where p is a constant.

a) Express $g(x)$ in the form $(x + a)^2 + b + p$, where a and b are constants.

b) State the range of g in terms of p.

2.2 Composite functions

When you combine two or more functions you get a **composite function**.

Consider the functions $f(x) = x^2 + 1, x \in \mathbb{R}$ and $g(x) = x - 1, x \in \mathbb{R}$ and the subset of numbers $\{1, 2, 3, 4, 5\}$.

If we apply function f to this set of numbers we get $\{2, 5, 10, 17, 26\}$.
If we apply function g to this new set of numbers we get $\{1, 4, 9, 16, 25\}$.

If we apply f first and then g we write this combined function as gf.

> The composite function **gf(x)** means apply f first followed by g.

Consider the same set of numbers $\{1, 2, 3, 4, 5\}$ and the functions $f(x) = x^2 + 1$ and $g(x) = x - 1$.

If we apply function g to this set of numbers we get $\{0, 1, 2, 3, 4\}$.
If we apply function f to this new set of numbers we get $\{1, 2, 5, 10, 17\}$.

If we apply g first and then f, we write this combined function as fg.

> The composite function **fg(x)** means apply g first followed by f.

Note: fg will only exist if the range of g is contained within the domain of f. In general, $fg(x) \neq gf(x)$

Example 5

The functions f and g are defined by

$f: x \mapsto 3x + 2, x \in \mathbb{R}$ $\quad$ $g: x \mapsto 7 - x, x \in \mathbb{R}$

a) Find fg(x). **b)** Find ff(x). **c)** Solve the equation $gf(x) = 2x$.

a) $fg(x) = f(7 - x)$ ← Substitute $7 - x$ for g(x).

$= 3(7 - x) + 2$ ← f: × 3 + 2

$= 23 - 3x$

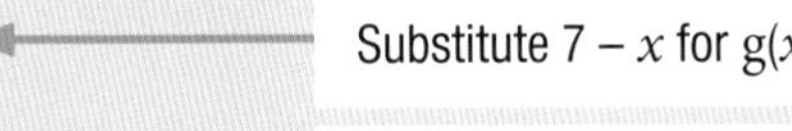

b) $ff(x) = f(3x + 2)$ ← Substitute $3x + 2$ for f(x).

$= 3(3x + 2) + 2$ ← Multiply $3x + 2$ by 3 and then add 2.

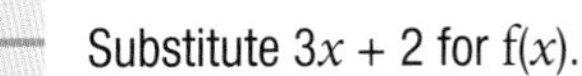

$= 9x + 8$

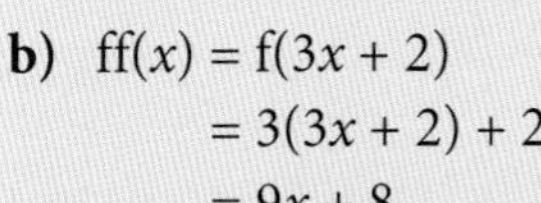

c) $gf(x) = g(3x + 2)$ ← Substitute $3x + 2$ for f(x).

$= 7 - (3x + 2)$

$= 5 - 3x$ ← g: subtract from 7.

but $gf(x) = 2x$ ← Put $gf(x) = 2x$ and solve.

$5 - 3x = 2x$

$5 = 5x$

$x = 1$

Example 6

The functions f, g, and h are defined by

f: $x \mapsto 4x - 1, x \in \mathbb{R}$ g: $x \mapsto \frac{1}{x+2}, x \neq -2$ h: $x \mapsto (2 - x)^2, x \in \mathbb{R}$

Find **a)** fg(x) **b)** hh(x)

a) $\text{fg}(x) = \text{f}\left(\frac{1}{x+2}\right)$ ← Substitute g(x) into fg(x).

$= 4\left(\frac{1}{x+2}\right) - 1$ ← f: × 4 – 1

$= \frac{4-1(x+2)}{x+2}$ ← Simplify using a common denominator of $x + 2$.

$= \frac{2-x}{x+2}, x \neq -2$

b) $\text{hh}(x) = \text{h}[(2 - x)^2]$

$= [2 - (2 - x)^2]^2$ ← h: subtract from 2 and then square.

$= [2 - (4 - 4x + x^2)]^2$

$= (-2 + 4x - x^2)^2$

Example 7

The functions f and g are defined by

f: $x \mapsto ax^2, x \in \mathbb{R}$ g: $x \mapsto 3x + b, x \in \mathbb{R}$ where a and b are constants.

Given that gf(1) = 5 and gg(2) = 14 find

a) the values of a and b **b)** the value of fg(–3).

a) $\text{gf}(1) = \text{g}[a(1)^2] = \text{g}(a)$ ← Substitute $x = 1$ in to f(x).

$= 3a + b$ ← g: multiply by 3 and then add b.

$\text{gf}(1) = 5$

So, $3a + b = 5$

$\text{gg}(2) = \text{g}(6 + b)$ ← Substitute $x = 2$ in to g(x).

$= 3(6 + b) + b$

$\text{gg}(2) = 14$

So, $18 + 4b = 14$ ← Solve for b.

$b = -1$

$3a - 1 = 5$ ← Substitute for $b = -1$ in $3a + b = 5$.

$a = 2$

b) $\text{f}(x) = 2x^2$ $\text{g}(x) = 3x - 1$

$\text{fg}(-3) = \text{f}(3 \times -3 - 1) = \text{f}(-10)$ ← First work out g(–3).

$\text{f}(-10) = 2(-10)^2 = 200$ ← Then substitute this value for x in to f.

Exercise 2.2

1. The functions f and g are defined for $x \in \mathbb{R}$ by
   f: $x \mapsto 5x - 1$ g: $x \mapsto x^2$.
   Find **a)** fg(x) **b)** ff(x) **c)** gf(-3) **d)** gg(1).

2. Given that f(x) = $2x^2 + 1$ and g(x) = $3 - x$, and $x \in \mathbb{R}$,
   find **a)** fg(-2) **b)** gf(x).

3. The functions f and g are defined for $x \in \mathbb{R}$ by
   f: $x \mapsto \dfrac{1}{x-3}$ g: $x \mapsto 1 + x$.
   Find **a)** fg(x) **b)** gf(x) **c)** ff(-2).
   Give your answer as a single fraction in its simplest form.

4. Given that f(x) = $2x - 1$, g(x) = $x^2 + 1$ and h(x) = $1 - x$ and $x \in \mathbb{R}$,
   find **a)** fg(x) **b)** hf(x) **c)** hg(-1) **d)** gf(2).

5. The functions f and g are defined for $x \in \mathbb{R}$ by
   f: $x \mapsto 4x - 1$ g: $x \mapsto 2x + k$.
   Find the value of k for which fg = gf.

6. Given that f(x) = $3\sqrt{x}$, for $x \in \mathbb{R}$, $x \geq 0$ and g(x) = $2x^2 - 1$ for $x \in \mathbb{R}$,
   solve the equation gf(x) = 8.

7. f(x) = $3 - x$, g(x) = $x^2 - 19$ and h(x) = $x - 2$ for $x \in \mathbb{R}$.
   Given that f(x) = gh(x), find the values of x.

8. Given that f(x) = $x^2 + 1$, g(x) = $x\sqrt{2}$ and fg(x) = 1.5, for $x \in \mathbb{R}$, find the value of x.

9. h(x) = $\dfrac{1}{x-1}$ for $x \neq 1$, $x \in \mathbb{R}$.
   Prove that hh(x) = $\dfrac{x-1}{2-x}$.

10. The functions f and g are defined for $x \in \mathbb{R}$, by f: $x \mapsto x + 4$, $x \in \mathbb{R}$ and g: $x \mapsto x^2 + 2$, $x \in \mathbb{R}$.
    Find the range of fg(x).

11. The functions f is defined by f: $x \mapsto \dfrac{x-1}{x}$, $x \in \mathbb{R}$, $x \neq 0$.
    Solve ff(x) = -2.

12. The functions f and g are defined for $x \in \mathbb{R}$ by
    f: $x \mapsto a - x$ g: $x \mapsto x^2 + ax + b$, where a and b are constants.
    Given that fg(-1) = -1 and gf(-1) = -1, find the values of a and b.

13. Given that f(x) = $3x - 1$, g(x) = $x^2 + 4$ and fg(x) = gf(x), where $x \in \mathbb{R}$, show that $x^2 - x - 1 = 0$.

14. Functions f and g are defined by f: $x \mapsto x - 2$, $x \in \mathbb{R}$ and g: $x \mapsto x^2 - x$, $x \in \mathbb{R}$.
    Find the set of values of x which satisfy fg(x) ≥ 0.

15. The functions f and g are defined for $x \in \mathbb{R}$ by
    f: $x \mapsto 2x - 1$ g: $x \mapsto x^2 + x$.
    Express gf(x) in the form $a(x + b)^2 + c$, where a, b and c are constants.

2.3 Inverse functions

Consider the function $f(x) = x^2 + 1$ with domain {1, 2, 3, 4, 5}.
The range of this set of numbers is {2, 5, 10, 17, 26}.

The **inverse function** provides structured evidence that leads to a given result of f. The inverse function is written as $f^{-1}(x)$ and maps the range {2, 5, 10, 17, 26} back onto the domain {1, 2, 3, 4, 5}.

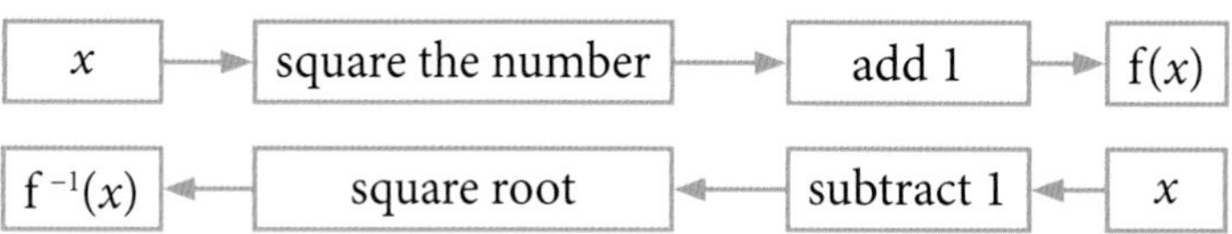

When $f(x) = x^2 + 1$, with domain {1, 2, 3, 4, 5},
$f^{-1}(x) = \sqrt{x-1}$
e.g. when $f(x) = 2$, $x = \sqrt{2-1} = 1$
e.g. when $f(x) = 26$, $x = \sqrt{26-1} = 5$

We can draw the graph of any inverse function $f^{-1}(x)$ by reflecting $f(x)$ in the line $y = x$.

Note:

- We do not write $f^{-1}(x) = \pm\sqrt{x-1}$ as all the numbers in the domain are positive.
- $f^{-1}(x) = \pm\sqrt{x-1}$ would not be a function as for every value of $x \geq 1$ we would get more than one value of $f^{-1}(x)$.
- We often have to restrict the domain so that the function is a one-to-one.
- We can only find the inverse function $f^{-1}(x)$ if $f(x)$ is a one-to-one function.

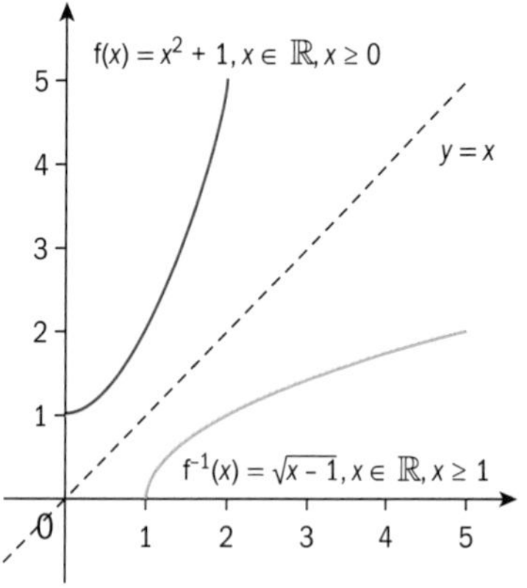

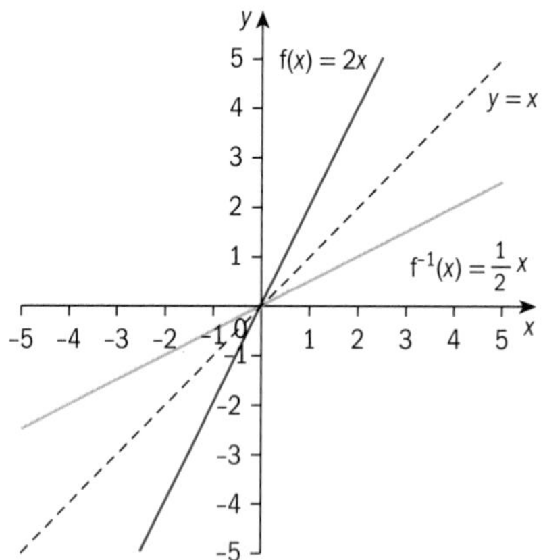

Here are some steps we will use to find the inverse function $f^{-1}(x)$ given $f(x)$.

1. Let $f(x) = y$.
2. Change x to y and y to x.
3. Rearrange to get y in terms of x.
4. Write in the correct form.

Example 8

Functions f and g are defined by

f: $x \mapsto 3x - 4, x \in \mathbb{R}$ g: $x \mapsto \frac{1}{x+1}, x \in \mathbb{R}, x \neq -1$

a) Express in terms of x **i)** $f^{-1}(x)$ **ii)** $g^{-1}(x)$

b) Sketch in a single diagram the graphs of $y = f(x)$ and $y = f^{-1}(x)$, making clear the relationship between the graphs.

a) i) Let $3x - 4 = y$ ← Let $f(x) = y$.

$3y - 4 = x$ ← Change x to y and y to x.

$y = \frac{x+4}{3}$ ← Rearrange to get y in terms of x.

$f^{-1}: x \mapsto \frac{x+4}{3}, x \in \mathbb{R}$ ← Write in the correct form.

Note: The domain of f^{-1} is the range of f.

ii) Let $\frac{1}{x+1} = y$ ← Let $g(x) = y$.

$\frac{1}{y+1} = x$ ← Change x to y and y to x.

$1 = xy + x$

$y = \frac{1-x}{x}$ ← Rearrange to get y in terms of x.

$g^{-1}: x \mapsto \frac{1-x}{x}, x \in \mathbb{R}, x \neq 0$ ← Write in the correct form.

b)

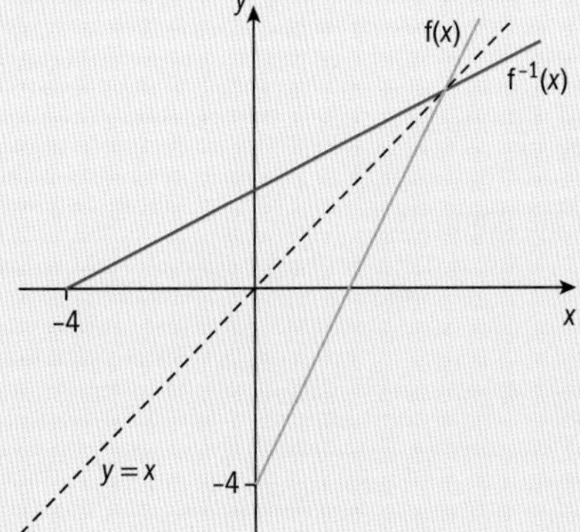

$f^{-1}(x)$ is the reflection of $f(x)$ in the line $y = x$.

Example 9

The functions f is defined by f: $x \mapsto x^2 - 6x + 2$ for $0 \le x \le 3$.

a) Express f(x) in the form $(x + a)^2 + b$, where a and b are constants.

b) State the range of f.

c) State the domain of $f^{-1}(x)$.

d) Find an expression for $f^{-1}(x)$.

e) Express $ff^{-1}(x)$ in terms of x.

f) Sketch on the same diagram the graphs of $y = f(x)$ and $y = f^{-1}(x)$, making clear the relationship between the graphs.

a) $x^2 - 6x + 2 = (x - 3)^2 - 9 + 2$ ← Use the method of completing the square.

$= (x - 3)^2 - 7$

b) Range of f: $-7 \le f(x) \le 2$ ← When $x = 0$, $f(x) = 2$; when $x = 3$, $f(x) = -7$ (which is the minimum value from **(a)** so f(x) cannot be less than this).

c) Domain of $f^{-1}(x)$: $-7 \le x \le 2$ ← The domain of f^{-1} is always the same as the range of f.

d) Let $y = (x - 3)^2 - 7$ ← Let $f(x) = y$.

$x = (y - 3)^2 - 7$ ← Change x to y and y to x.

Then, $x + 7 = (y - 3)^2$

Hence, $y = 3 \pm \sqrt{x+7}$ ← Rearrange to get y in terms of x.

$f^{-1}: x \mapsto 3 + \sqrt{x+7}$ ← However, f^{-1} must be one-to-one, so we must find whether a negative or positive square root is appropriate.

Since $f(0) = 2$, we know that $f^{-1}(2) = 0$.

This means that we need the negative square root giving

$f^{-1}: x \mapsto 3 - \sqrt{x+7}, -7 \le x \le 2$

e) $ff^{-1}(x) = f(3 + \sqrt{x+7})$

$= (3 + \sqrt{x+7} - 3)^2 - 7$

$= (\sqrt{x+7})^2 - 7$

$= x + 7 - 7$

$= x$ ← $ff^{-1}(x)$ is always the same as x.

f)

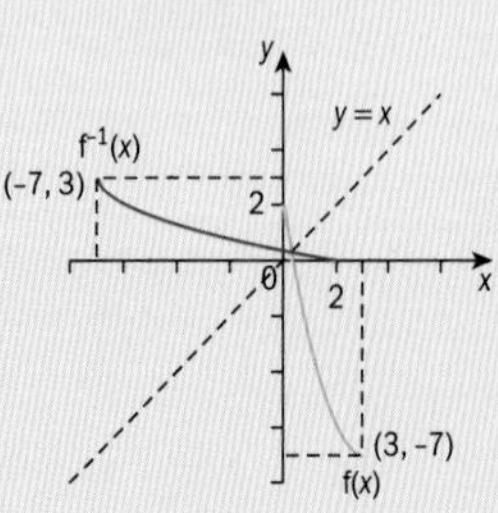

$f^{-1}(x)$ is the reflection of $f(x)$ in the line $y = x$.

Exercise 2.3

- The domain of f^{-1} is always the same as the range of f.
- The range of f^{-1} is always the same as the domain of f.
- $ff^{-1}(x) = f^{-1}f(x) = x$
- It is always best to draw a rough sketch of the function so that the domain and range are clear.

1. Find an expression for $f^{-1}(x)$ for each of the following functions.

a) $f(x) = 2x + 3, x \in \mathbb{R}$ **b)** $f(x) = x^2 - 2, x \in \mathbb{R}, x \geq 0$

c) $f(x) = 1 - 4x, x \in \mathbb{R}$ **d)** $f(x) = \frac{x-3}{2}, x \in \mathbb{R}$

e) $f(x) = \frac{1}{x}, x \in \mathbb{R}, x \neq 0$ **f)** $f(x) = 2(x - 1), x \in \mathbb{R}$

2. Find the inverse for each of these functions.

a) $f: x \mapsto 3 - x$ domain: $x \in \mathbb{R}$

b) $g: x \mapsto 6(4x - 1)$ domain: $x \in \mathbb{R}$

c) $h: x \mapsto 3x^2 + 2$ domain: $x \in \mathbb{R}, x \geq 0$

d) $f: x \mapsto \frac{2}{x+5}$ domain: $x \in \mathbb{R}, x \neq -5$

e) $g: x \mapsto \frac{2}{x+1}$ domain: $x \in \mathbb{R}, x \neq -1$

f) $h: x \mapsto \sqrt{x-3}$ domain: $x \in \mathbb{R}, x \geq 3$

3. The function f is defined by $f: x \mapsto \frac{1}{x+4}, x \in \mathbb{R}, x \neq -4$. Evaluate $f^{-1}(-3)$.

4. $f(x) = \frac{1+2x}{x-1}, x \in \mathbb{R}, x \neq 1$. Find $f^{-1}(x)$, stating its domain.

5. The function f is defined by $f: x \mapsto x^2 + 2x - 1$ for $0 \leq x \leq 2$.

a) Express $f(x)$ in the form $(x + a)^2 + b$, where a and b are constants.

b) State the range of f.

c) Find an expression for $f^{-1}(x)$, stating its domain.

d) Show that $ff^{-1} = x$.

e) Sketch on the same diagram the graphs of $y = f(x)$ and $y = f^{-1}(x)$, making clear the relationship between the graphs.

6. $f(x) = 2 + \frac{1}{1-x}, x \in \mathbb{R}, x \neq 1$. Find an expression for $f^{-1}(x)$, stating its domain.

7. $f(x) = \frac{2x-5}{7x+4}$. Show that $f^{-1}(-2) - f^{-1}(2) = \frac{43}{48}$.

8. $f(x) = \frac{x}{x+3}, x \in \mathbb{R}, x \neq -3$.

a) If $f^{-1}(x) = -5$, find the value of x.

b) Show that $ff^{-1}(x) = x$.

9. $f(x) = \frac{x}{x-1}, x \in \mathbb{R}, x \neq 1$. Show that $f(x)$ is a self-inverse function, that is $f(x) = f^{-1}(x)$.

10. Find if the following functions are self-inverse functions.

a) $f(x) = \frac{6}{x}, x \in \mathbb{R}, x \neq 0$

b) $g(x) = 2x - 1, x \in \mathbb{R}$

c) $h(x) = \frac{1}{1-\frac{1}{x}}, x \in \mathbb{R}, x \neq 0, x \neq 1$

11. a) The function $f: x \mapsto 2x^2 - 12x + 8$ is defined for $x \in \mathbb{R}$.

i) Express $f(x)$ in the form $a(x + b)^2 + c$, where a, b and c are constants.

ii) Find the range of $f(x)$.

b) The function $g: x \mapsto 2x^2 - 12x + 8$ is defined for $x \geq D$.

i) Find the smallest value of D for which g has an inverse.

ii) For this value of D, find an expression for $g^{-1}(x)$ in terms of x.

12. The function f is defined by $f: x \mapsto 4x - x^2$ for $x \geq 2$.

a) Express $4x - x^2$ in the form $a - (x - b)^2$, where a and b are positive constants.

b) Express $f^{-1}(x)$ in terms of x.

13. A function f is such that $f(x) = 2 + \sqrt{\frac{x-1}{3}}$, for $x \geq 1$.

a) Find $f^{-1}(x)$ in the form $ax^2 + bx + c$, where a, b and c are constants.

b) Find the domain of f^{-1}.

14. The functions f and g are defined for $x \in \mathbb{R}$ by

$f: x \mapsto 4x - a$ $\quad\quad$ $g: x \mapsto b + ax$

where a and b are constants.

Given that $f^{-1}(1) = 2$ and $g^{-1}(-1) = -3$, find the values of a and b.

15. The function f is defined by $f: x \mapsto \frac{x-1}{x+2}, x \in \mathbb{R}, x \neq -2$.

a) Find $f^{-1}(x)$.

b) Show that $f^{-1}(x) = -2$ has no solutions.

2.4 Transformations: translations

You can transform the graph of a function by moving it horizontally or vertically. This **transformation** is called a **translation**.

> $f(x + a)$ is a horizontal translation of $-a$.
> $f(x) + a$ is a vertical translation of a.

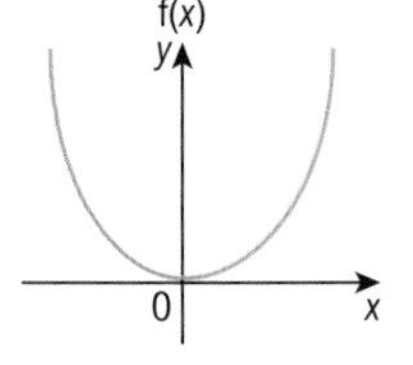

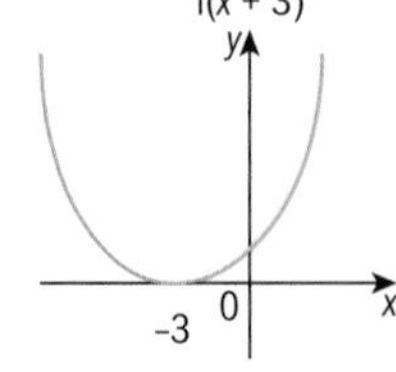

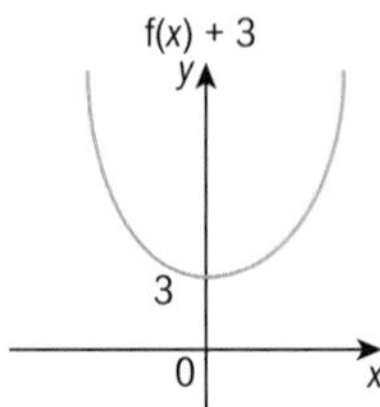

Note: We can write $y = f(x)$, $y = f(x + 3)$ and $y = f(x) + 3$ for these graphs.

Example 10

$f(x) = x^3$ $\qquad g(x) = \frac{1}{a}$

Sketch the graphs of the following functions, marking on each sketch where the curve cuts the axes and state the equations of any asymptotes:

a) $f(x - 2)$ **b)** $g(x) - 2$

a) The graph of $f(x) = x^3$ is

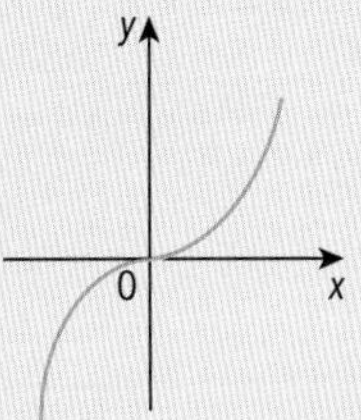

First sketch the graph of f(x), marking where it crosses the axes.

So the graph of $f(x - 2)$ is

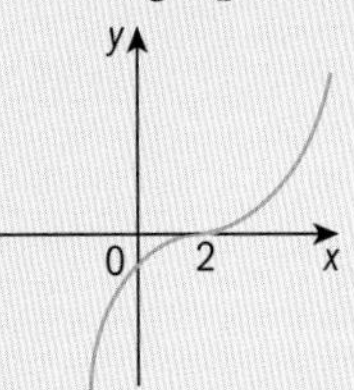

Translate the graph 2 units to the **right**, so it cuts the x-axis at 2, i.e. a translation by the vector $\begin{pmatrix} 2 \\ 0 \end{pmatrix}$.

b) The graph of $g(x) = \frac{1}{a}$ is

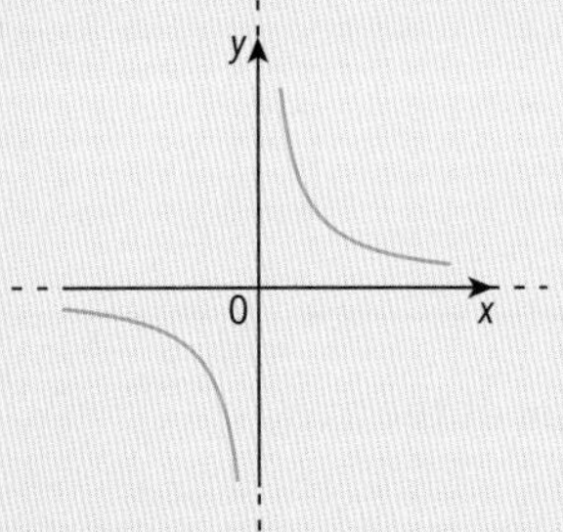

Sketch the graph of g(x), showing the asymptotes with dotted lines.

So the graph of $g(x) - 2$ is

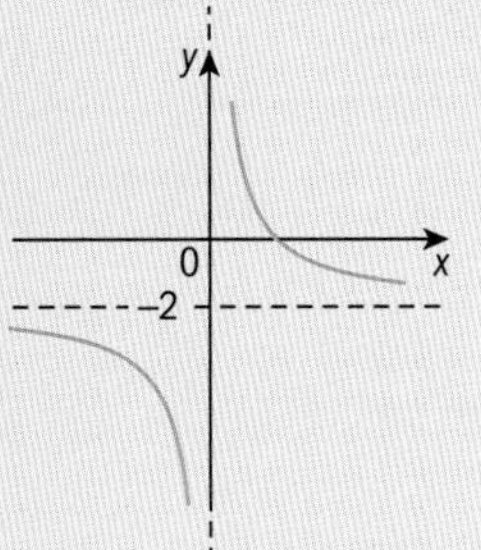

Translate the graph and the asymptotes 2 units **down**, i.e. a translation by the vector $\begin{pmatrix} 0 \\ -2 \end{pmatrix}$.

Write these equations on or below the graph.

The equations of the asymptotes are $x = 0$ and $y = -2$.

Exercise 2.4

Note: If the transformation is $f(x + a)$ we move the graph to the left ($a > 0$) or right ($a < 0$). If the transformation is $f(x) + a$ we move the graph up ($a > 0$) or down ($a < 0$).

1. a) Sketch the graph of $y = f(x)$, where $f(x) = (x + 1)(x - 3)$.
   b) On separate diagrams, sketch the graphs of
      i) $y = f(x) + 1$ ii) $y = f(x + 3)$ iii) $y = f(x - 1)$.

   Mark on each sketch, where possible, the coordinates of the points where the curve cuts the axes.
2. a) Sketch the graph of $y = g(x)$, where $g(x) = 2^x$.
   b) On separate diagrams, sketch the graphs of
      i) $y = g(x + 2)$ ii) $y = g(x) - 1$ iii) $y = g(x) + 4$.

   Mark on each sketch, where possible, the coordinates of the points where the curve cuts the axes and state the equations of any asymptotes.
3. The graph of $y = f(x)$ is transformed to the graph of $y = f(x + 5)$. Describe the transformation.
4. $f(x) = x(x + 3)$
   a) Find and simplify the equation of the curve $3 + f(x - 1)$.
   b) Describe the transformation.
5. The curve $y = x^2 - x + 1$ is translated by $\begin{pmatrix} 0 \\ 3 \end{pmatrix}$.
   Find and simplify the equation of the translated curve.
6. The curve $y = x^2 - 3$ is translated by $\begin{pmatrix} -1 \\ 0 \end{pmatrix}$.
   Find and simplify the equation of the translated curve.

2.5 Transformations: reflections

You can transform the graph of a function by reflecting the graph in one of the axes.

$-f(x)$ is a reflection in the x-axis.
$f(-x)$ is a reflection in the y-axis.

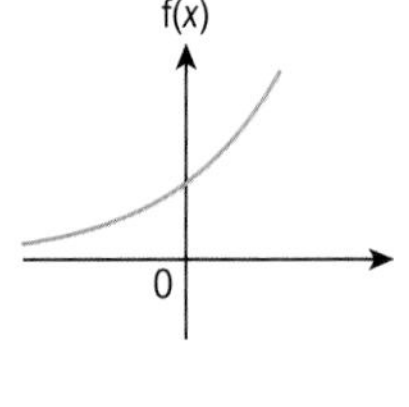

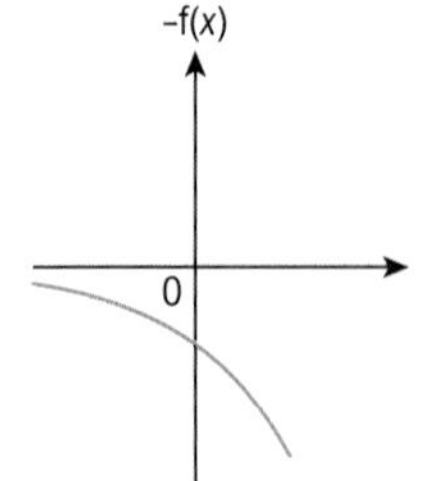

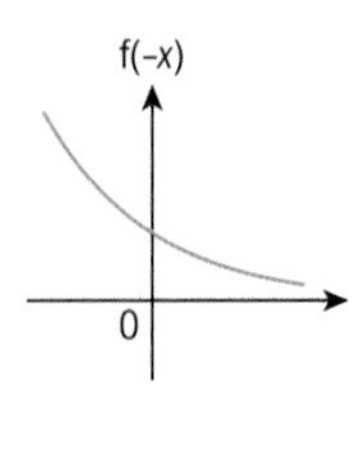

Example 11

$f(x) = x(x+1)(x-2)$

Sketch the graphs of the following functions

a) $1 + f(-x)$ **b)** $-f(x+3)$.

a) The graph of $f(x) = x(x+1)(x-2)$ is

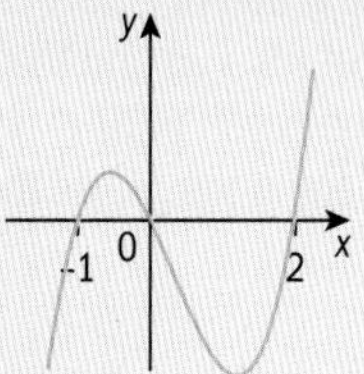

Sketch the graph of $f(x)$, marking where it cuts the axes.

The graph of $f(-x)$ is

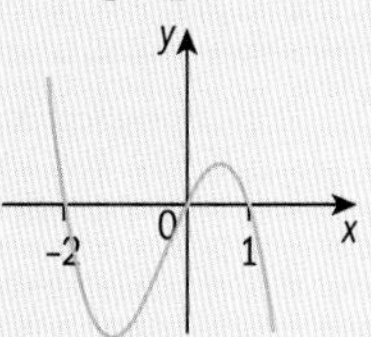

First reflect the graph in the y-axis.

The graph of $1 + f(-x)$ is

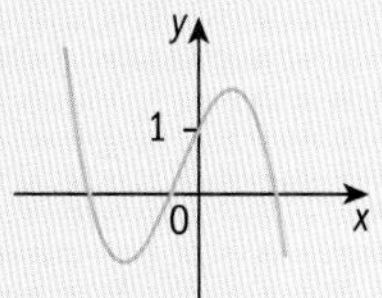

Then translate the graph 1 unit **up** and show where it cuts the y-axis.

b) The graph of $f(x) = x(x+1)(x-2)$ is

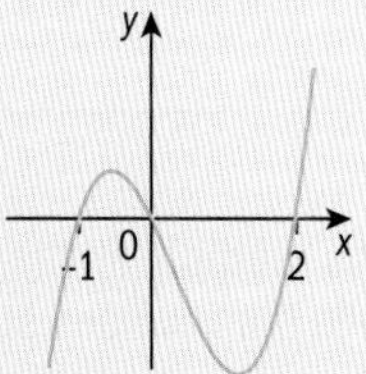

Sketch the graph of $f(x)$, marking where it crosses the x-axis.

The graph of $f(x+3)$ is

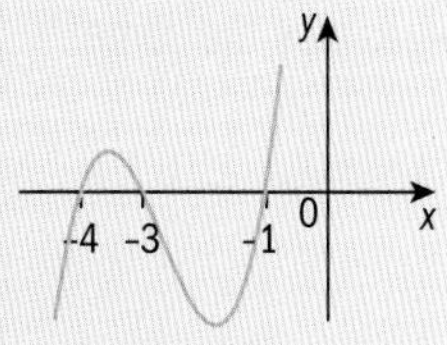

First translate the graph 3 units to the **left**, i.e. translation by the vector $\begin{pmatrix} -3 \\ 0 \end{pmatrix}$.

The graph of $-f(x+3)$ is

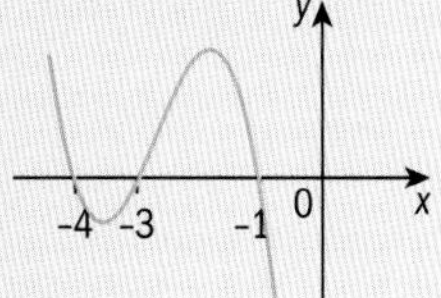

Then reflect the graph in the x-axis.

Exercise 2.5

> **Note:** When sketching the graph of a combined function, it is best to draw each stage separately to get to the final answer, as shown in the example.

1. a) Sketch the graph of $y = f(x)$, where $f(x) = (x + 2)(x - 2)$.

   b) On separate diagrams, sketch the graphs of

   i) $y = f(-x)$ ii) $y = -f(x)$.

   Mark on each sketch, where possible, the coordinates of the points where the curve cuts the axes.

2. The graph of $y = f(x)$ is transformed to the graph of $y = f(-x) + 7$.

   Describe the transformation.

> **Note:** You first transform what is **inside** the bracket and then what is **outside** the bracket. For example, to sketch the graph of $-f(9 + x)$, first apply the translation of $f(9 + x)$ and then the reflection. To sketch the graph of $f(4 - x)$, rewrite as $f(-(x - 4))$, reflect in the y-axis then translate 4 to the right.

3. The diagram shows a sketch of the curve $f(x)$, which passes through the points $A(0, 3)$ and $B(2, -1)$.

   Sketch the graphs of

   a) $y = -f(x)$ b) $y = f(-x)$ c) $y = -f(x + 4)$ d) $y = 3 - f(-x)$.

   In each case, mark the new position of the points A and B, writing down their coordinates.

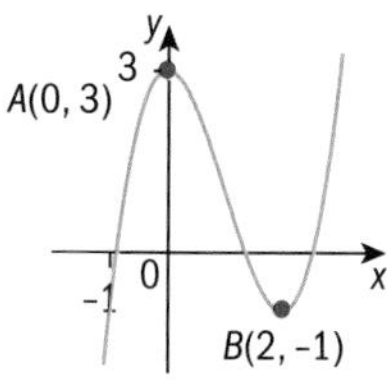

4. The points $P(-3, 5)$ and $Q(-2, -8)$ lie on the curve with equation $y = f(x)$.

   Find the coordinates of P and Q after the curve has been transformed by the following transformations:

   a) $f(-x)$ b) $-f(x)$ c) $f(-x + 1)$ d) $-f(x) - 5$

5. The curve $y = 3x^2 + 2x - 8$ is reflected in the x-axis.

   State the equation of the reflected curve in the form $y = ax^2 + bx + c$, where a, b and c are constants.

2.6 Transformations: stretches

You can transform the graph of a function by **stretching** (or compressing) the graph horizontally or vertically.

> $af(x)$ is a stretch with factor a in the y-direction.
>
> $f(ax)$ is a stretch with factor $\frac{1}{a}$ in the x-direction.

> **Note:** $af(x)$ means **multiply** all the y-values by a while the x-values stay the same.
>
> $f(ax)$ means **divide** all the x-values by a while the y-values stay the same.

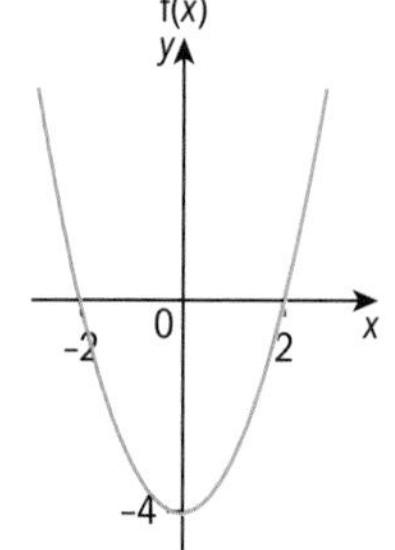

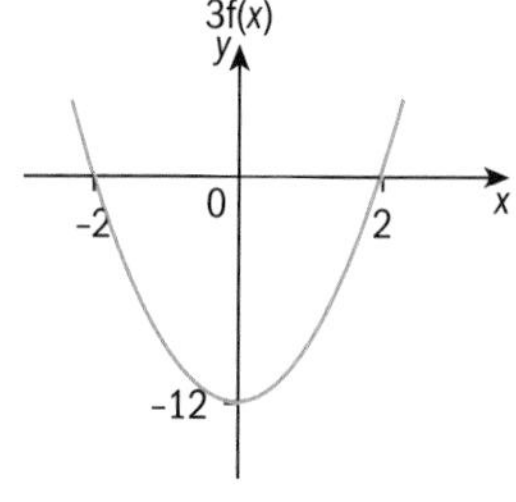

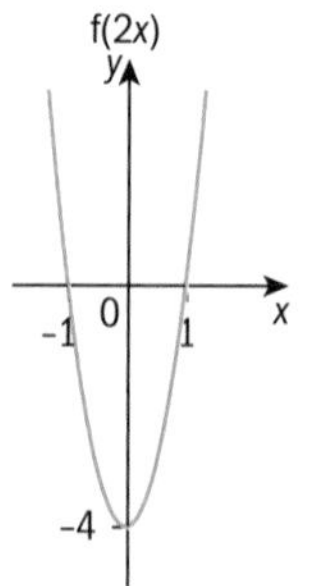

Example 12

The diagram shows a sketch of the curve f(x), which passes through the origin, O, and the points $A(-1, -2)$ and $B(3, -18)$.
Sketch the graphs of

a) $-2f(x)$ **b)** $f(-3x)$

In each case, mark the new position of the points O, A and B, writing down their coordinates.

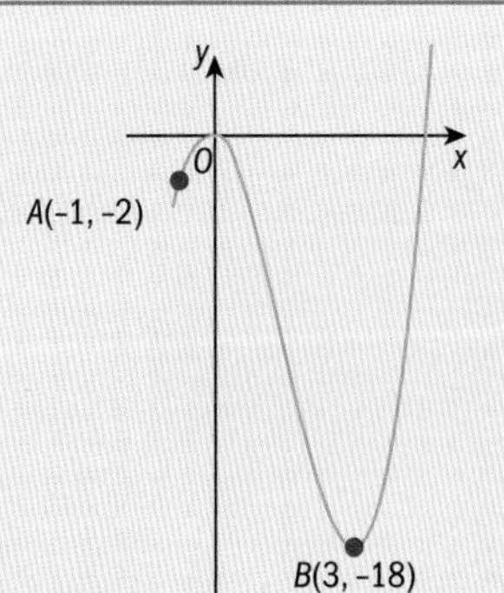

a) The graph of $2f(x)$ is

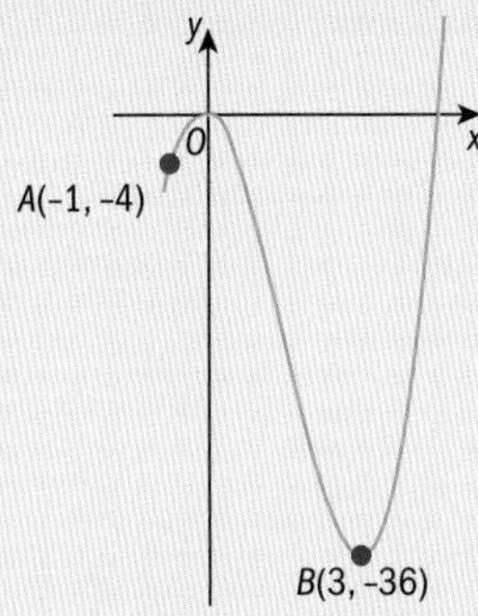

Apply a stretch with factor 2 in the y-direction, i.e. keep the x-values the same, double the y-values.

The graph of $-2f(x)$ is

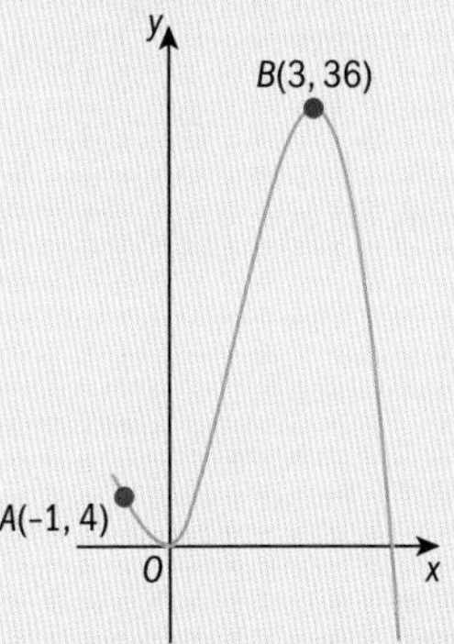

Then reflect the graph in the x-axis.

b) The graph of $f(3x)$ is

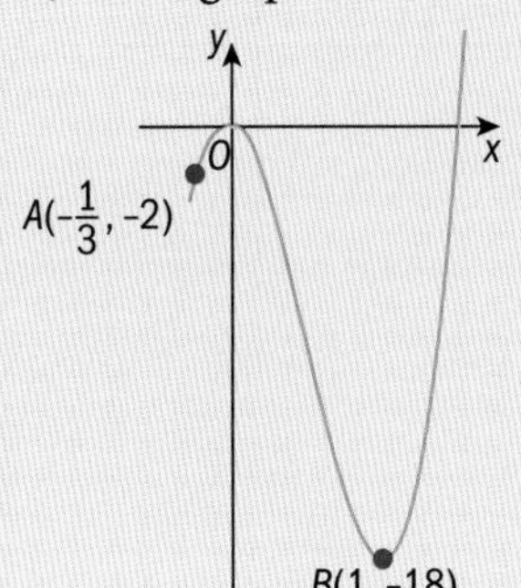

Apply a stretch with factor $\frac{1}{3}$ in the x-direction, i.e. divide the x-values by 3, keep the y-values the same.

▶ Continued on the next page

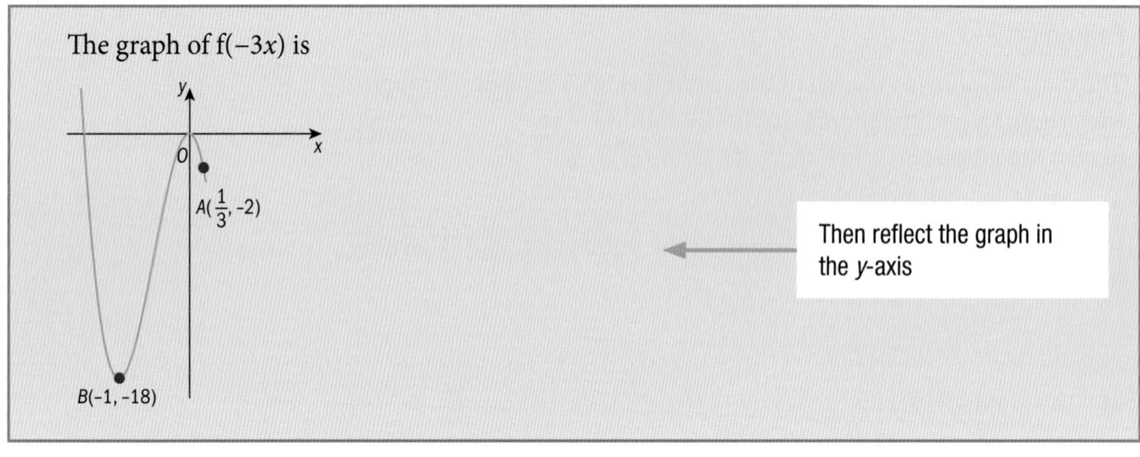

Exercise 2.6

1. a) Sketch the graph of $y = \mathrm{f}(x)$ where $\mathrm{f}(x) = 16 - x^2$.

   b) On separate diagrams, sketch the graphs of

      i) $y = \mathrm{f}(2x)$ ii) $y = 4\mathrm{f}(x)$ iii) $y = \mathrm{f}(-4x)$.

   Mark on each sketch the coordinates of the points where the curve cuts the axes.

2. The diagram shows a sketch of the curve $\mathrm{f}(x)$, which passes through the origin, O, and the points $A(-2, 8)$ and $B(1, -1)$.

   On separate diagrams, sketch the graphs of

   a) $y = 4\mathrm{f}(x)$ b) $y = \mathrm{f}(-\frac{1}{2}x)$ c) $y = -2\mathrm{f}(x + 1)$ d) $y = 1 + \mathrm{f}(-2x)$.

   Mark the new position of the points O, A and B on each transformation, stating their coordinates.

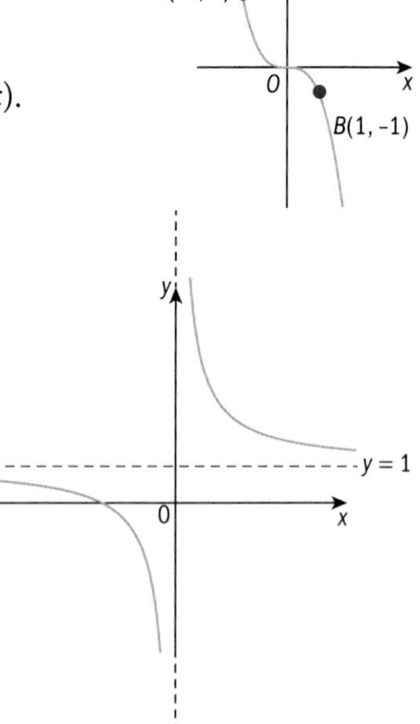

3. The diagram shows a sketch of the curve $\mathrm{f}(x)$.

   The curve has a horizontal asymptote with equation $y = 1$ and a vertical asymptote with equation $x = 0$.

   On separate diagrams, sketch the graphs of

   a) $y = \mathrm{f}(2x)$ b) $y = -3\mathrm{f}(x)$ c) $y = 2\mathrm{f}(x - 3)$

   d) $y = 4 - \mathrm{f}(\frac{1}{2}x)$ e) $y = 3 + 4\mathrm{f}(x)$ f) $y = 4\mathrm{f}(-\frac{1}{2}x)$.

   Mark on each sketch the equations of the asymptotes and, if possible, any points where the curve cuts the axes.

4. The graph of $y = \mathrm{f}(x)$ is transformed to the graph of $y = -\mathrm{f}(2x)$.

   Describe fully the two single transformations that have been combined to give the resulting transformation.

5. The graph of $y = \mathrm{f}(x)$ is transformed to the graph of $y = 4 + 3\mathrm{f}(x - 2)$.

   Describe fully the three single transformations that have been combined to give the resulting transformation.

Summary exercise 2

1. Find the range of these functions.

   a) $f(x) = x + 3, -2 < x \le 5$

   b) $g(x) = 7\sqrt{x+1}, 3 < x \le 8$

   c) $h(x) = \frac{1}{x-4}, \frac{1}{4} \le x \le 1$

   d) $f(x) = \frac{1}{x^2}, -4 \le x < -1$

EXAM-STYLE QUESTION

2. $f(x) = \begin{cases} 3x^2 + 2 & \text{when } -3 \le x \le -1 \\ 5x - 1 & \text{when } -1 \le x < 4 \end{cases}$

   $g(x) = \begin{cases} x^3 - 1 & \text{when } -3 \le x \le -1 \\ 2x & \text{when } -1 \le x < 4 \end{cases}$

   a) Explain why g is a function and f is not a function.

   b) Determine whether the function g is one-to-one or many-to-one.

3. $f: x \mapsto \frac{1}{1-x}$. Find the domain and range for which this function is defined.

EXAM-STYLE QUESTION

4. The functions f and g are defined by

   $f: x \mapsto 4x - 2$ $\qquad$ $g: x \mapsto (x + 1)^2$

   a) Find gf(x).

   b) Solve the equation fg(x) = 14.

5. Given that $f(x) = \sqrt{3x}$, $g(x) = x - 5$ and $h(x) = x^2$, find

   a) gh(x) $\qquad$ b) fg(x) $\qquad$ c) hg(−2).

EXAM-STYLE QUESTIONS

6. The function f is defined by $f: x \mapsto ax^2 + b$, where a and b are constants.

   Given that f(−2) = 2 and f(4) = 14, find the values of a and b.

7. The functions f and g are defined by

   $f: x \mapsto \frac{1}{2x+1}$ $\qquad$ $g: x \mapsto 8 - x$

   Solve the equation $gf(x) = \frac{20}{3}$.

8. The functions f and g are defined for $x \in \mathbb{R}$ by

   $f: x \mapsto 5 - 3x$ $\qquad$ $g: x \mapsto (x - 1)^2$

   a) Find the set of values which satisfy $gf(x) \le 25$.

   b) Sketch on the same diagram the graphs of $y = f(x)$ and $y = f^{-1}(x)$, showing the coordinates of their point of intersection and making clear the relationship between the graphs.

9. The functions f and g are defined for $x \in \mathbb{R}$ by

   $f: x \mapsto x^2 - 1$ $\qquad$ $g: x \mapsto 3x + 4$

   a) Find and simplify expressions for fg(x) and gf(x).

   b) Hence find the values of a for which fg(a) = 4a + gf(a).

10. The functions f and g are defined for $x \in \mathbb{R}$ by

    $f: x \mapsto 10x + x^2$ $\qquad$ $g: x \mapsto 2x - 1$.

    Express fg(x) in the form $a(x + b)^2 + c$, where a, b and c are constants.

11. The functions f, g, and h are defined by

    $f: x \mapsto 3x + 5 \qquad -1 \le x \le 5$

    $g: x \mapsto 4 - 2x^2 \qquad x \ge 0$

    $h: x \mapsto 4 + 2x^2 \qquad x \ge -1$

    For each of these functions **determine** whether or not the inverse function exists. If it does, write down the inverse in its simplest form. If it does not, explain why.

12. $f(x) = \frac{x^2+1}{2x^2+1}, x \in \mathbb{R}, x \ge 2$

    a) Find an expression for $f^{-1}(x)$.

    b) Find the domain for which this inverse function is defined.

    c) Show that $ff^{-1}(x) = x$.

13. The functions f and g are defined for $x \in \mathbb{R}$ by

$f: x \mapsto 3x + 2$ $\qquad$ $g: x \mapsto x^2 + 1$

a) Find the values of x for which $fg(x) = gf(x)$.

b) Find an expression for $(fg)^{-1}(x)$, stating its domain.

c) Explain why $(fg)^{-1}(x)$ is not a function.

14. $f(x) = \sqrt{2x + a}$, $x \in \mathbb{R}$, $2x \geq -a$.

a) Find an expression for $f^{-1}(x)$, stating its domain.

b) Find an expression for a in terms of x, $a \neq 0$, when $f(x) = f(2x + a)$.

15. The graph shows a sketch of f(x).

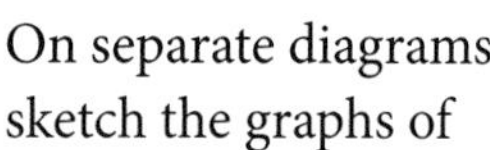

On separate diagrams, sketch the graphs of

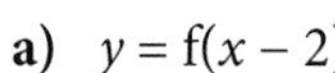

a) $y = f(x - 2)$

b) $y = -f(x)$

c) $y = f(2x)$

d) $y = -f(-x)$

e) $y = 3 + f(1 - x)$.

Mark the new position of the points O, A and B on each transformation, stating their coordinates.

16. The diagram shows the graph of $y = f(x)$.

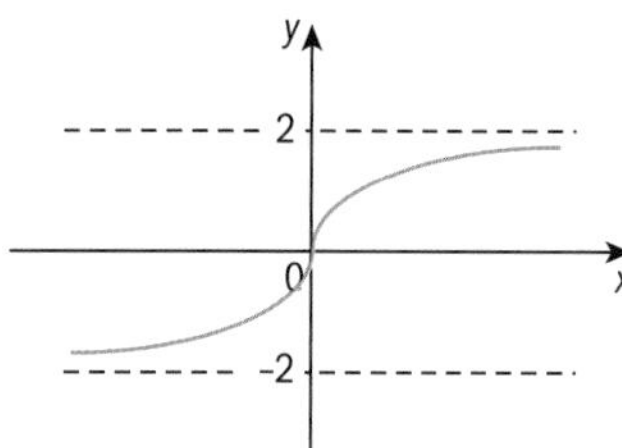

On separate diagrams, sketch the graphs of

a) $y = f(-x)$

b) $y = f(x - 1)$

c) $y = -f(2x)$

d) $y = 3f(4 - x)$.

On each sketch, write the equation of any asymptotes and, where possible, write the coordinates where the graph cuts the axes.

17. The diagram shows the graph of $f(x)$.

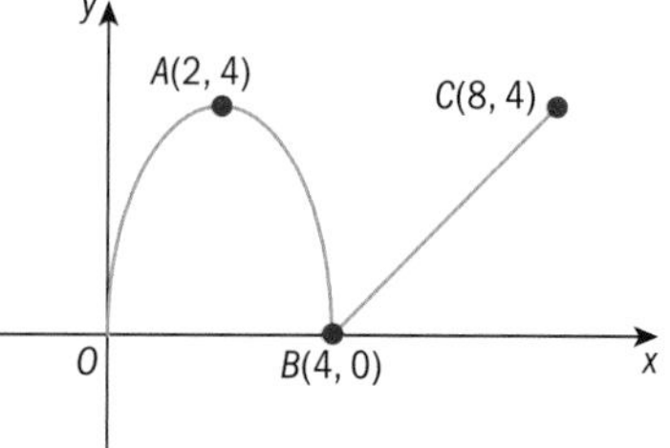

On separate diagrams, sketch the graphs of

a) $y = f(4x)$

b) $y = f(3 - x)$

c) $y = -f(x + 1)$

d) $y = 2f(-x)$.

On each sketch, mark the new position of the points O, A, B and C, writing down their coordinates.

18. The graph of a function is reflected in the x-axis and then translated by the vector $\begin{pmatrix} 4 \\ 3 \end{pmatrix}$.

After the transformation its equation is $y = 3 + \dfrac{1}{(x-4)^2}$.

Determine the equation of the original function.

19. The points $P(-1, -1)$ and $Q(2, \frac{1}{2})$ lie on the graph of $f(x) = \dfrac{1}{x}$.

Sketch the graph of $y = 2f(x - 1) + 2$.

Mark the images of the points P and Q, stating their coordinates, and write down the equations of any asymptotes.

EXAM-STYLE QUESTIONS

20.

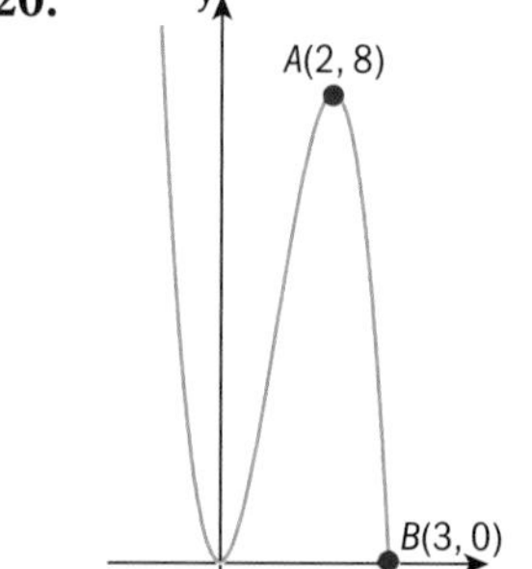

The diagram shows a sketch of the graph with equation $f(x) = 6x^2 - 2x^3$.

There is a minimum at the origin, a maximum at the point $A(2, 8)$ and it cuts the x-axis at $B(3, 0)$.

a) On separate diagrams, sketch the graphs with equations

i) $y = f(2x)$

ii) $y = f(x + 1)$.

On each sketch, show where the curve cuts the x-axis and mark the coordinates of A and B.

b) The curve with equation $y = f(x) + c$ has a maximum point at $(2, 3)$.

State the value of c.

c) The curve with equation $y = df(x + e)$ has a maximum point at $(4, 4)$.

State the value of d and the value of e.

21. The graph of $y = f(x)$ is transformed to the graph of $y = 5 - f(x)$.

Describe fully the two single transformations that have been combined to give the resulting transformation.

22.

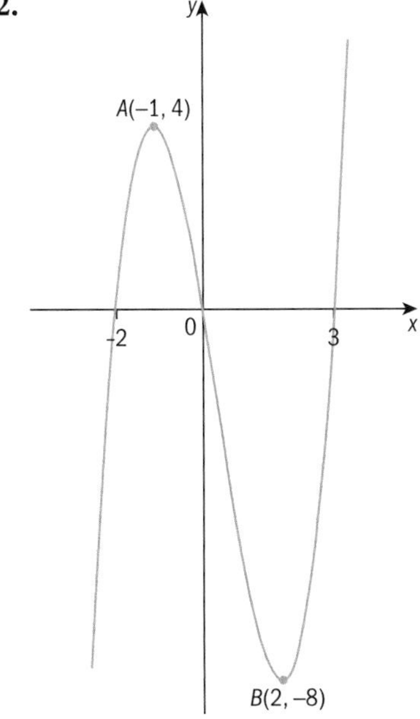

The curve shown has a maximum at $A(-1, 4)$ and a minimum at $B(2, -8)$.

On separate diagrams sketch

a) $y = f(x) + 4$

b) $y = \frac{1}{2}f(x)$

Show clearly the coordinates of the maximum and minimum points and where the curve cuts the y-axis.

23. The graph of $y = f(x)$ is shown. It cuts the axes at $(0, 4)$ and $(4, 0)$.

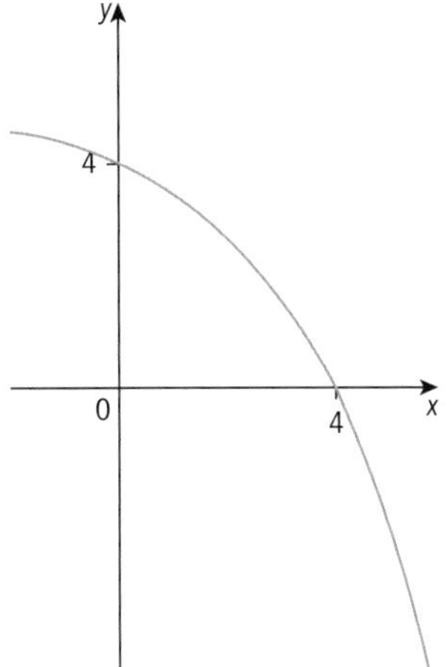

a) Sketch $y = -3f(x)$ and give the coordinates where the graph cuts the axes.

b) The graph of $y = f(x)$ is transformed to the graph of $y = f(-\frac{1}{2}x)$. Describe fully the two single transformations that have been combined to give the resulting transformation.

Chapter summary

Domain and range

- If a function maps x onto $f(x)$, then the values of x is the **domain** of the function and the corresponding values of $f(x)$ is the **range** of the function.

Functions

- A **function** is defined as a mapping where every element of the domain (set x) is mapped onto exactly one element of the range (set y).
- If every element in the domain is mapped onto exactly one element in the range we say the function is a **one-to-one** function.
- If every element in the domain is mapped onto exactly one element in the range, but some elements in the range arise from more than one element in the domain we say the function is a **many-to-one** function.
- We say that each member of a set is an **element** of that set.
- $x \in \mathbb{R}$ means that x is a member of all the real numbers.

Composite functions

- The composite function **fg(*x*)** means apply g first followed by f.
- The composite function **gf(*x*)** means apply f first followed by g.

Inverse functions

- The **inverse function** of f, written as $\mathbf{f^{-1}(x)}$, maps the range back onto the domain.
- The steps for finding the inverse function $f^{-1}(x)$ given $f(x)$ are:
  1. Let $f(x) = y$.
  2. Change x to y and y to x.
  3. Rearrange to get y in terms of x.
  4. Write in the correct form.
- The domain of f^{-1} is always the same as the range of f.
- The range of f^{-1} is always the same as the domain of f.
- $ff^{-1}(x)$ and $f^{-1}f(x)$ are always the same as x.
- The graph of $f^{-1}(x)$ is the reflection of the graph of $f(x)$ in the line $y = x$.
- If $f(x) = f^{-1}(x)$, then we say $f(x)$ is a **self-inverse** function. If f is self-inverse, then $ff(x) = x$.

Transformations

- We can transform the graph of a function by moving it horizontally or vertically. This transformation is called a **translation**.
- $\mathbf{f(x + a)}$ is a horizontal translation of $-a$.
- $\mathbf{f(x) + a}$ is a vertical translation of a.

- We can transform the graph of a function by **reflecting** the graph in one of the axes.
- $-\mathbf{f}(\boldsymbol{x})$ is a reflection in the x-axis.
- $\mathbf{f}(-\boldsymbol{x})$ is a reflection in the y-axis.
- We can transform the graph of a function by **stretching** the graph horizontally or vertically.
- $\boldsymbol{a}\mathbf{f}(\boldsymbol{x})$ is a stretch with factor $\boldsymbol{a}$ in the y-direction.
- $\mathbf{f}(\boldsymbol{ax})$ is a stretch with factor $\frac{1}{a}$ in the x-direction.

3 Coordinate geometry

Since the first plane took paying passengers in the 1920s, airlines have been trying hard to maximise their profits by selling some tickets at a discounted rate so they could fill seats that would otherwise not have been sold. But how many seats should they sell at the discounted price? This revenue management problem of how much to charge for each seat so that profit is maximised can be solved by using a technique called linear programming which includes formulating and using the equation of straight lines and graphing them on coordinate axes.

Objectives

- Find the equation of a straight line given sufficient information.
- Interpret and use any of the forms $y = mx + c$, $y - y_1 = m(x - x_1)$ and $ax + by + c = 0$ in solving problems.
- Understand that the equation $(x - a)^2 + (y - b)^2 = r^2$ represents the circle with centre (a, b) and radius r.
- Use algebraic methods to solve problems involving lines and circles.
- Understand the relationship between a graph and its associated algebraic equation, and use the relationship between points of intersection of graphs and solutions of equations.

Before you start

You should know how to:

1. Find the length of the third side of a right-angled triangle given two of the sides,

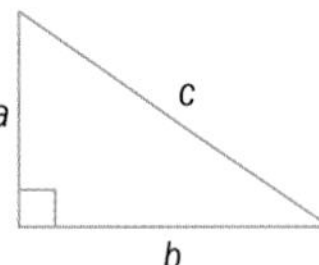

e.g. $a = 2$ cm, $b = 3$ cm. Find c.

$c = \sqrt{2^2 + 3^2} = \sqrt{13} = 3.61$ cm.

e.g. $a = 7$ cm, $c = 10$ cm. Find b.

$b = \sqrt{10^2 - 7^2} = \sqrt{51} = 7.14$ cm.

2. Find the gradient (m) of the following lines,

e.g. $y = 3 - 2x$
$m = -2$

e.g. $x - y = 4$
$x - 4 = y$; $m = 1$

Skills check:

1. Find the length of the third side of this triangle given two sides.

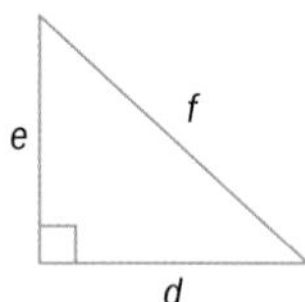

a) $d = 3$ cm, $e = 5$ cm. Find f.

b) $d = 1$ cm, $f = 4$ cm. Find e.

c) $e = 2.3$ cm, $f = 5.6$ cm. Find d.

d) $e = \sqrt{17}$ cm, $d = \sqrt{19}$ cm. Find f.

2. Find the gradient (m) of the following lines

a) $y = 2x + 7$ **b)** $x + y = 5$

c) $y = 8 - 3x$ **d)** $3x - 6y = 1$

e) $2y = 9 + x$ **f)** $4x + 2y = 3$.

3.1 Line segments

If you have the coordinates of the end-points of a line segment we can find the mid-point of the line segment.

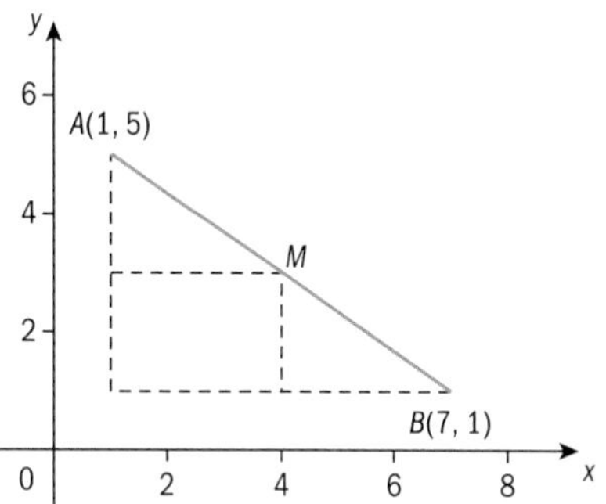

A has coordinates (1, 5).
B has coordinates (7, 1).

We can see from the diagram that the coordinates of the mid-point of the line *AB* is (4, 3).

The *x*-coordinate is halfway between $x = 1$ and $x = 7$.
The *y*-coordinate is halfway between $y = 1$ and $y = 5$.

> The **mid-point** of the line joining (x_1, y_1) and (x_2, y_2) is given by
> $$\left(\frac{x_1 + x_2}{2}, \frac{y_1 + y_2}{2}\right)$$

In the above example $\left(\frac{x_1 + x_2}{2}, \frac{y_1 + y_2}{2}\right) = \left(\frac{1+7}{2}, \frac{5+1}{2}\right) = (4, 3)$

> The **gradient** of the line joining (x_1, y_1) and (x_2, y_2) is given by $\frac{y_2 - y_1}{x_2 - x_1}$.

The gradient **of the line** $AB = \frac{\text{change in } y}{\text{change in } x} = \frac{-4}{6} = -\frac{2}{3}$

> The **length of the line segment** between (x_1, y_1) and (x_2, y_2) is given by
> $$\sqrt{(x_2 - x_1)^2 + (y_2 - y_1)^2}$$

Using Pythagoras' theorem the **length of the line segment**

$AB = \sqrt{6^2 + 4^2} = \sqrt{52} = 7.211\ldots$

Example 1

P has coordinates $(-5, 4)$ and Q has coordinates $(-3, -1)$.

Find

a) the gradient of PQ

b) the length of PQ

c) the coordinates of the mid-point of PQ.

a) gradient $= \dfrac{-1-4}{-3-(-5)}$

$= \dfrac{-5}{2} = -\dfrac{5}{2}$

If you start with -1 in the numerator you **must** start with -3 in the denominator.

b) length $PQ = \sqrt{(-3-(-5))^2 + (-1-4)^2}$

$= \sqrt{(2)^2 + (-5)^2}$

$= \sqrt{29} = 5.385$

c) mid-point of $PQ = \left(\dfrac{-5+-3}{2}, \dfrac{-1+4}{2}\right)$

$= (-4, 1.5)$

Check: -4 is halfway between -3 and -5. 1.5 is halfway between 4 and -1.

Example 2

The mid-point M of the line AB has coordinates $(-3, 1)$. B has coordinates $(-1, -2)$. Given that a point C has coordinates $(2, 5)$, find the exact length of the line AC.

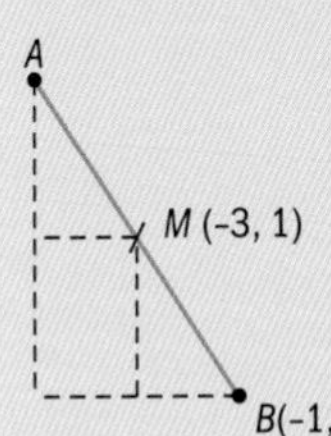

It is useful to draw a diagram.

Let A be the point (x, y).

Work out the coordinates of A.

Then $\dfrac{x+-1}{2} = -3$, so $x = -5$

Also $\dfrac{y+-2}{2} = 1$, so $y = 4$

$A = (-5, 4)$ $C = (2, 5)$

length $AC = \sqrt{(2-(-5))^2 + (5-4)^2}$

$= \sqrt{1+49} = \sqrt{50}$

We are asked for an *exact* answer so do not work out answer as a decimal.

Example 3

The point P has coordinates $(-2, 4)$. The point Q has coordinates $(6, -4)$. The point R has coordinates $(3, -1)$. **Show that** the three points are **collinear**.

Hint: Collinear means the three points are all on the same straight line.

Gradient of $PQ = \frac{-4-4}{6-(-2)} = -1$ ← Find the gradient of the line joining any two points.

Gradient of $QR = \frac{-1-(-4)}{3-6} = -1$ ← Find the gradient of the line joining another two points.

Since the gradients are the same, and Q is on both lines, the three points are collinear.

Note: We could also show the gradient of $PR = -1$.

Note: In examination questions, 'Find' will usually mean 'Find by calculation'. Examiners will be looking for evidence of the method used.
You need to ensure you show your method clearly including labelling of calculations,
e.g. *gradient of PQ* = ...
e.g. *length of PQ* = ...

Exercise 3.1

1. Find the mid-points of the straight lines joining each of the following sets of points.

   a) $(1, 3)$ and $(7, 13)$ **b)** $(2, -1)$ and $(-4, -5)$
   c) $(-6, -3)$ and $(-2, 7)$ **d)** $(-8, 5)$ and $(3, -2)$
   e) $(-k, -9)$ and $(-3k, 5)$ **f)** $(3p, -4p)$ and $(-p, 6p)$

2. Find the gradients of the straight lines joining each of the following sets of points.

   a) $(-3, -6)$ and $(2, -1)$ **b)** $(4, -6)$ and $(1, 0)$
   c) $(5, -7)$ and $(-3, 9)$ **d)** $(-8, -3)$ and $(2, -7)$
   e) $(-3k, -4k)$ and $(-k, 6k)$ **f)** $(4q, -4q)$ and $(7q, 8q)$

3. Find the lengths of the straight lines joining each of the following sets of points.

   a) $(-8, 4)$ and $(-2, -4)$ **b)** $(-1, 1)$ and $(-5, -2)$
   c) $(6, -1)$ and $(9, -4)$ **d)** $(-7, -3)$ and $(-6, -8)$
   e) $(4q, -4q)$ and $(7q, -8q)$ **f)** $(2p, -6p)$ and $(-3p, 6p)$

4. The mid-point of the line joining the points $A(p, -2)$ and $B(6, -8)$ is $(1, q)$. Work out the values of p and q.

5. The straight line joining the points $A(3, -5)$ and $B(6, k)$ has a gradient of 4. Work out the value of k.

6. The straight line joining the points $A(p, 3)$ and $B(1, -1)$ has a length of 5. Work out the values of p.

7. M is the mid-point of the line joining the points $A(-5, -2)$ and $B(3, -8)$. Given that a point C has the coordinates $(1, 4)$, find

   a) the gradient of the line MC

   b) the length of the line MC.

8. The gradient of the line joining the points $A(1, k)$ and $B(5, 1)$ is 1. Find the mid-point of the line AB.

9. The mid-point of the line joining the points $P(b, -1)$ and $Q(-2, 7)$ is $(-2a, a)$. Work out the values of a and b.

10. $ABCD$ is a square with vertices $A(-1, -2)$, $B(1, 2)$, $C(5, 0)$ and $D(3, -4)$.

    a) Show that the diagonals intersect at the point $(2, -1)$.

    b) Find the area of $ABCD$.

11. ABC is a triangle with angle $ABC = 90°$. The points A, B and C have coordinates $(2, 5)$, $(1, 3)$ and $(5, 1)$ respectively. Find the area of triangle ABC.

12. PQR is a triangle with vertices $P(4, -2)$, $Q(-1, 7)$ and $R(k, 0)$. $PR = QR$. Find the value of k.

13. The points A, B and C have coordinates $(4, 0)$, $(p, 6)$ and $(7, 1)$ respectively. The length of AB is twice the length of AC. Find the possible values of p.

14. ABC is a triangle with vertices $A(-2, -3)$, $B(6, -3)$ and $C(2, 3)$. Show that triangle ABC is an isosceles triangle.

15.

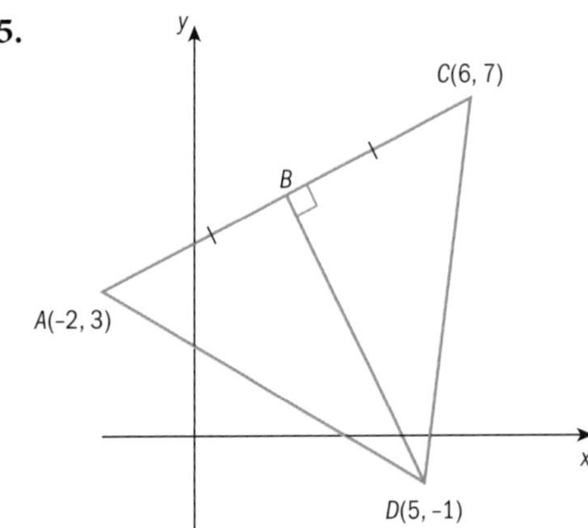

The diagram above shows a straight line AC with the mid-point B. BD is perpendicular to AC. Find the area of triangle ACD.

3.2 Parallel and perpendicular lines

Two straight lines are **parallel** if their gradients are the same.

$$m_1 = m_2$$

We can see from the diagram that

the gradient of $AB = \dfrac{\text{change in } y}{\text{change in } x} = \dfrac{-4}{6} = -\dfrac{2}{3}$

the gradient of $CD = \dfrac{\text{change in } y}{\text{change in } x} = -\dfrac{2}{3}$

The lines are parallel as the two gradients are the same.

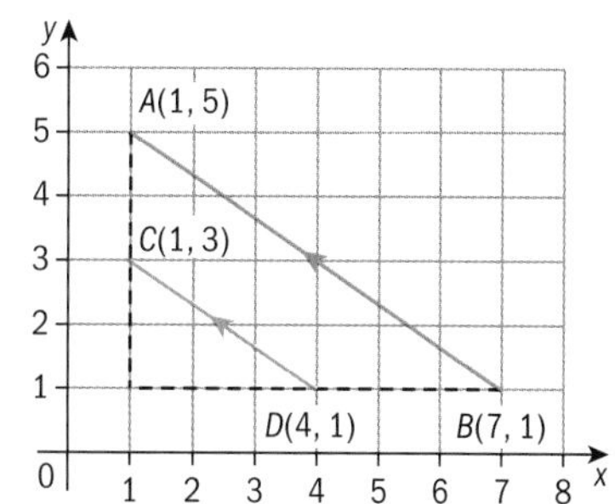

Two straight lines are **perpendicular** if the product of their gradients = −1.

$$m_1 \times m_2 = -1$$

We can see from the diagram that

the gradient of $AB = \dfrac{\text{change in } y}{\text{change in } x} = \dfrac{-4}{6} = -\dfrac{2}{3}$

the gradient of $CD = \dfrac{\text{change in } y}{\text{change in } x} = \dfrac{3}{2}$

The lines are perpendicular as $m_1 \times m_2 = -\dfrac{2}{3} \times \dfrac{3}{2} = -1$.

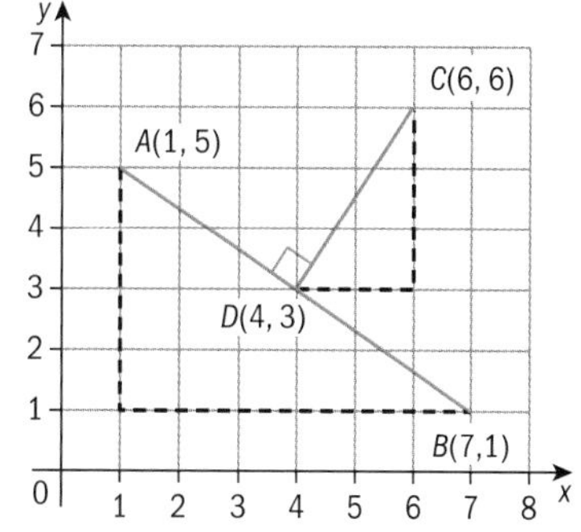

Example 4

The points P, Q, R and S have coordinates (−5, 5), (−2, −1), (1, 3) and (−3, 1) respectively.

Show that PQ is perpendicular to RS.

Gradient of $PQ = \dfrac{-1-5}{-2-(-5)} = -2$

> If you start with −1 in the numerator you must start with −2 in the denominator.

Gradient of $RS = \dfrac{1-3}{-3-1} = \dfrac{1}{2}$

$m_{PQ} \times m_{RS} = -2 \times \dfrac{1}{2} = -1$

> m_{PQ} can be used to denote the gradient of PQ.

Therefore the lines are perpendicular.

> Always finish a proof with a statement.

Example 5

Find the gradient of the line that is perpendicular to the lines with these gradients.

a) $\frac{7}{3}$ **b)** $-\frac{1}{5}$ **c)** 1 **d)** −8

Note: Since $m_1 \times m_2 = -1$, $m_2 = -\frac{1}{m_1}$

a) $-\frac{3}{7}$ ← Change the sign and invert the fraction.

b) $\frac{5}{1} = 5$ ← Change the sign and invert the fraction.

c) $-\frac{1}{1} = -1$

d) $\frac{1}{8}$

← $1 = \frac{1}{1}$ and $-8 = -\frac{8}{1}$

Example 6

Find in each case whether the lines are parallel to each other, perpendicular to each other, or neither.

a) $y = 5 - 3x$, $y = 3x + 4$ **b)** $x - 2y = 4$, $6y = 3x - 1$ **c)** $3y = 5x - 2$, $10y = 3 - 6x$

a) $m_1 = -3$ $m_2 = 3$
The lines are not parallel or perpendicular.

← The gradient is equal to the coefficient of x.

b) $m_1 = \frac{1}{2}$ $m_2 = \frac{3}{6} = \frac{1}{2}$
The lines are parallel, as $m_1 = m_2$.

← $2y = x - 4$, $y = \frac{1}{2}x - 2$ and $y = \frac{3}{6}x - \frac{1}{6}$

c) $m_1 = \frac{5}{3}$ $m_2 = -\frac{6}{10} = -\frac{3}{5}$
The lines are perpendicular, as $m_1 m_2 = -1$.

← $y = \frac{5}{3}x - \frac{2}{3}$ and $y = \frac{3}{10} - \frac{6}{10}x$

Exercise 3.2

1. Find the gradient of a line that is **(i)** parallel, **(ii)** perpendicular to the lines with these gradients.

a) −2 **b)** $-\frac{2}{7}$ **c)** 9

d) $3\frac{1}{3}$ **e)** $-3k$ **f)** $-2\frac{1}{4}$

2. Find the gradient of a line that is **(i)** parallel, **(ii)** perpendicular to the lines with these equations.

a) $y = 7x - 3$ **b)** $x + y = 1$ **c)** $5y = 2 - 15x$

d) $4x - 3y = 6$ **e)** $3x = 4 + 3y$ **f)** $y = -4x$

3. Find in each case whether the lines are parallel to each other, perpendicular to each other, or neither.

a) $y = 1 - x$
$y = x + 4$

b) $x - 2y = 4$
$6y = 3x - 1$

c) $3y = 9x + 1$
$x + 3y = 4$

d) $4y = 8x + 1$
$2y = 3 - 4x$

4. The line PQ has a gradient of $\frac{1}{3}$. PQ is parallel to the line joining the points $(3, k)$ and $(-6, 5)$.

Find k.

5. The line joining points $A(x, 3)$ and $B(2, -3)$ is perpendicular to BC where C is the point $(10, 1)$.
Find the value of x.

6. The line that passes through the points $(3, 0)$ and $(-4, y)$ is perpendicular to the line with equation $7x + 4y = 5$. Find the value of y.

7. ABC is a right-angled triangle with AB perpendicular to BC.
The coordinates of the points A, B and C are $(-5, 0)$, $(x, -5)$ and $(3, -2)$ respectively. Find the possible values of x.

8. Show that $A(4, 3)$, $B(-3, 2)$, $C(-5, -2)$ and $D(2, -1)$ are the vertices of a parallelogram.

9. The diagram shows triangle ABC. The coordinates of A, B and C are $(0, 3)$, $(-4, -5)$ and $(6, -1)$ respectively. M is the mid-point of AB, and N is the mid-point of AC.

Show that MN is parallel to BC.

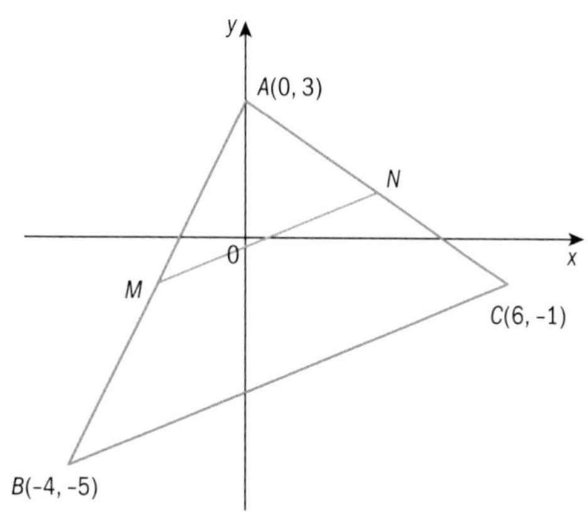

10. The diagram shows the straight line AB that cuts the x-axis at P. PC is perpendicular to AB.

The point A is $(-3, -4)$, the point B is $(6, 2)$ and the point C is $(5, -3)$.
Show that point P lies on the x-axis and find its coordinates.

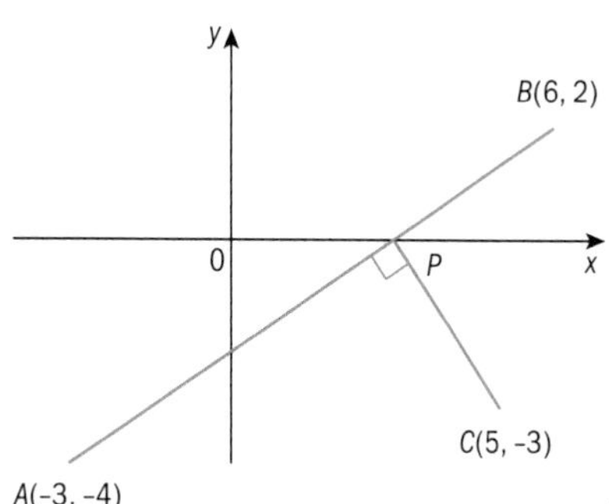

3.3 Equation of a straight line

We know that the equation of a straight line is given by $y = mx + c$ where m is the gradient and c is the **y-intercept**.

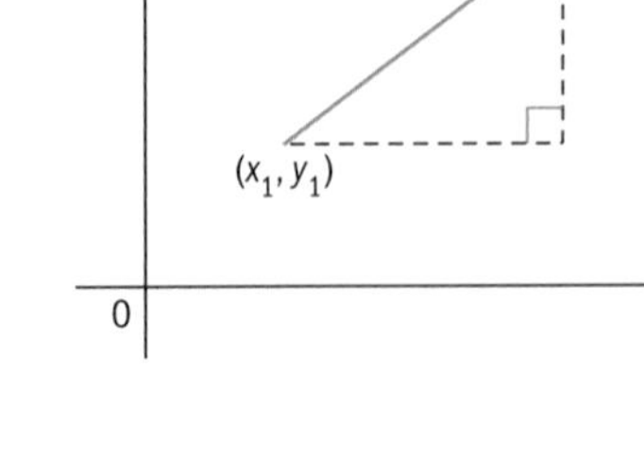

We can find the equation of a straight line when given the gradient and the coordinates of a point on the line.

We can also find the equation of a straight line when given the coordinates of two points on the line.

Consider the line passing through (x_1, y_1) with gradient of m.

$$\text{Gradient} = \frac{y - y_1}{x - x_1} = m$$

$$(y - y_1) = m(x - x_1)$$

The equation of the straight line through the point (x_1, y_1) and with gradient m is: $(y - y_1) = m(x - x_1)$ (given point and gradient)

Consider the line passing through (x_1, y_1) and (x_2, y_2).

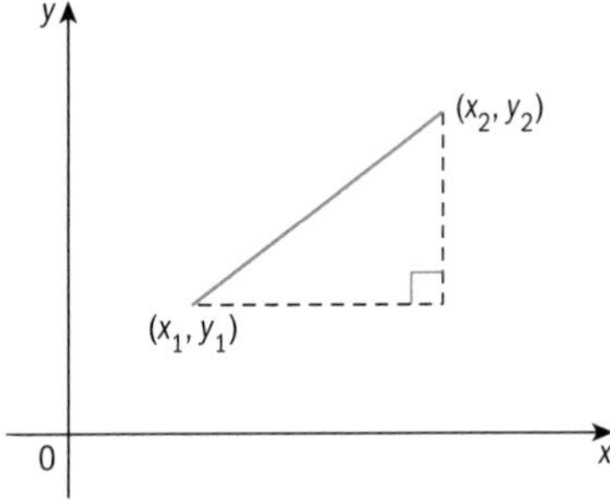

$$\text{Gradient} = \frac{y_2 - y_1}{x_2 - x_1} = m$$

But $(y - y_1) = m(x - x_1)$

therefore $(y - y_1) = \dfrac{y_2 - y_1}{x_2 - x_1}(x - x_1)$

and $\dfrac{y - y_1}{x - x_1} = \dfrac{y_2 - y_1}{x_2 - x_1}$

The equation of the straight line through the points (x_1, y_1) and (x_2, y_2) is:
$\dfrac{y - y_1}{x - x_1} = \dfrac{y_2 - y_1}{x_2 - x_1}$ (given two points)

Example 7

Find the equation of the line through the points $(-2, -7)$ and $(-4, 1)$.

Give your answer in the form $ax + by + c = 0$, where a, b and c are integers.

Let $(-2, -7) = (x_1, y_1)$ and $(-4, 1) = (x_2, y_2)$. — First, choose either point for (x_1, y_1).

The equation is $\dfrac{y - y_1}{x - x_1} = \dfrac{y_2 - y_1}{x_2 - x_1}$. — This is the equation when we are given two points.

$$\frac{y - (-7)}{x - (-2)} = \frac{1 - (-7)}{-4 - (-2)}$$

We then substitute in the formula.

$$\frac{y + 7}{x + 2} = \frac{8}{-2} = -4$$

$$y + 7 = -4(x + 2)$$

$$4x + y + 15 = 0$$

Ensure the answer is in the form as specified in the question.

Example 8

Find the equation of the line that passes through the point (4, −3) and is perpendicular to the line with equation $2y = 5x$.

$2y = 5x \qquad m = \frac{5}{2}$

The gradient of the line is equal to the coefficient of x.

Gradient of the perpendicular line $= -\frac{2}{5}$

Change the sign and inverse the fraction.

Equation: $y + 3 = -\frac{2}{5}(x - 4)$

$5y + 15 = -2x + 8$

$2x + 5y + 7 = 0$

Multiply $(y + 3)$ by 5 and multiply $(x - 4)$ by −2.

Example 9

Find the equation of the **perpendicular bisector** of the line PQ, where $P(3, -1)$ and $Q(7, 3)$.

Mid-point of $PQ = \left(\frac{3+7}{2}, \frac{-1+3}{2}\right) = (5, 1)$

Gradient $PQ = \frac{3-(-1)}{7-3} = 1$

Gradient of perpendicular $= -1$

$m_1 \times m_2 = -1$

Equation of perpendicular:

$y - 1 = -1(x - 5)$

$x + y - 6 = 0$

Example 10

$A(4, 3)$, $B(9, 13)$, $C(5, 11)$, and $D(12, 4)$ are four points.
The lines AB and CD meet at P.
Find the equations of AB and CD and hence the coordinates of P.

Gradient of $AB = \frac{13-3}{9-4} = 2$

Equation AB: $y - 3 = 2(x - 4)$

$y = 2x - 5$

Use the coordinates of A or B.

Gradient of $CD = \frac{4-11}{12-5} = -1$

Equation CD: $y - 11 = -1(x - 5)$

$y = -x + 16$

Use the coordinates of C or D.

At P: $2x - 5 = -x + 16$

Both equal y.

$3x = 21$

$x = 7$

When $x = 7$, $y = 2(7) - 5 = 9$

Use equation of AB or CD.

Coordinates of $P = (7, 9)$

Exercise 3.3

1. Find the equation of the lines with the given gradient (m) passing through the given point P.
   a) $m = 2 \quad P(5, 7)$ **b)** $m = -1 \quad P(-1, -2)$ **c)** $m = -3 \quad P(-4, 1)$
   d) $m = 4 \quad P(3, -2)$ **e)** $m = \frac{4}{3} \quad P(-6, -1)$ **f)** $m = -\frac{1}{5} \quad P(-4, 0)$

2. Find the equation of the line AB with the given pairs of coordinates.
   a) $A(1, 5)$ and $B(3, 7)$ **b)** $A(-1, -2)$ and $B(-3, 4)$
   c) $A(5, -3)$ and $B(-2, 4)$ **d)** $A(-2, 4)$ and $B(-6, 0)$
   e) $A(-8, -1)$ and $B(2, -6)$ **f)** $A(9, -3)$ and $B(-6, -2)$

3. Find the equation of the line parallel to the line $y = 4 - 3x$ and passing through the point $(5, 7)$.

4. Find the equation of the line perpendicular to the line $y = 5x$ and passing through the point $(2, -1)$.

5. Find the equation of the line through the point $(-1, 3)$ parallel to the line with equation $2x - 5y = 10$.

6. Find the equation of the line through the point $(-1, 3)$ perpendicular to the line with equation $6x + 9y - 7 = 0$.

7. Find the equation of the line through the point $(1, 8)$ perpendicular to the line with equation $3x - 4y + 4 = 0$.

8. A is the point $(3, -4)$ and B is the point $(-5, -10)$. Find the equation of the perpendicular bisector of AB.

9. $A(9, 0)$, $B(5, 8)$, $C(1, 1)$ and $D(3, 2)$ are four points.
   The lines AB and CD meet at P.
   Find the equations of AB and CD and hence the coordinates of P.

10. $A(1, -2)$, $B(-1, -3)$, $C(-2, 14)$, and $D(2, 2)$ are four points.
    The lines AB and CD meet at P. Find the coordinates of P.

11. 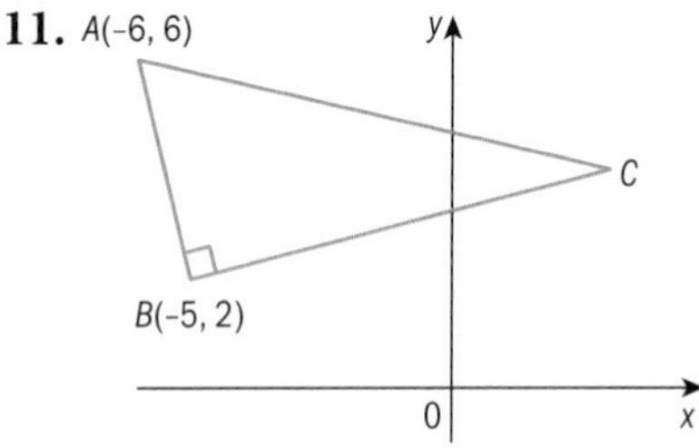

    The diagram shows triangle ABC with AB perpendicular to BC. A is the point $(-6, 6)$ and B is the point $(-5, 2)$. The gradient of AC is $-\frac{2}{9}$. Find the coordinates of C.

12. The straight line with equation $3x - 2y = 6$ cuts the x-axis at P and the y-axis at Q.
    a) Find the coordinates of P and Q.
    b) Find the equation of the line that is perpendicular to PQ and passes through the point $(4, 3)$.

13. $A(5, -5)$, $B(3, -7)$, $C(12, 3)$, and $D(-3, -6)$ are four points.
    The lines AB and CD meet at P. Find the coordinates of P.

3.4 Circles

If we take any point $P(x, y)$ on the circumference of a circle, centre O, we can work out the equation of the circle using Pythagoras' theorem.

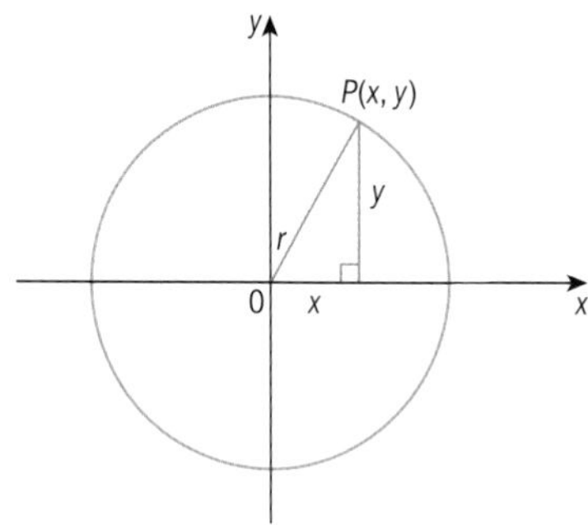

The equation of a circle, centre (0, 0), radius r, is given by $x^2 + y^2 = r^2$.

If we translate this circle by the vector $\begin{pmatrix} a \\ b \end{pmatrix}$,

we can get the equation of any circle, centre (a, b), radius r.

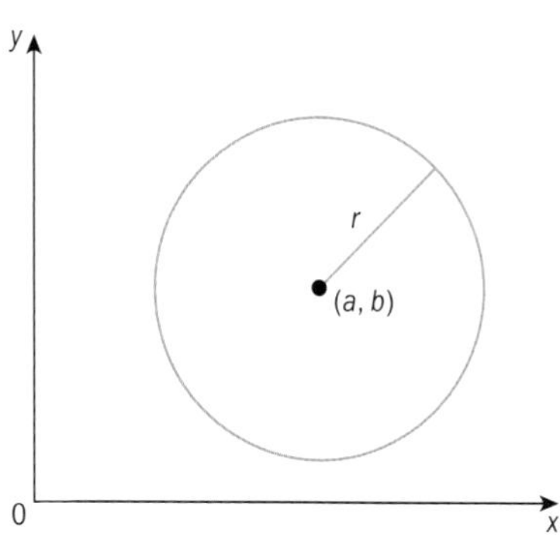

The equation of a circle, centre (a, b), radius r is given by

$$(x - a)^2 + (y - b)^2 = r^2.$$

If we expand this equation, we get $x^2 - 2ax + a^2 + y^2 - 2by + b^2 = r^2$. When this is simplified, we get the expanded form of the equation of a circle:

$x^2 + y^2 - 2ax - 2by + c = 0 \quad$ where $c = a^2 + b^2 - r^2$

Note: When the circle is given in this expanded form, we can use the method of completing the square to write the equation in the form $(x - a)^2 + (y - b)^2 = r^2$.

The expanded form of the equation of a circle is given by

$$x^2 + y^2 + 2gx + 2fy + c = 0, \text{ where } c = f^2 + g^2 - r^2.$$

Example 11

Find the equation of the circle with centre (−5, 7) and radius 6.

$(x + 5)^2 + (y - 7)^2 = 36$ ← Substitute into $(x - a)^2 + (y - b)^2 = r^2$, where $a = -5$, $b = 7$ and $r = 6$.

Example 12

Find the centre and radius of the circle with equation $x^2 + y^2 - 10x + 12y + 12 = 0$.

$x^2 - 10x + y^2 + 12y + 12 = 0$ — First rearrange the equation to this form.

$(x - 5)^2 - 25 + (y + 6)^2 - 36 + 12 = 0$ — Complete the square for x and y.

$(x - 5)^2 + (y + 6)^2 = 49$ — Rearrange to this form.

Centre is $(5, -6)$, radius $= 7$. — $a = 5, b = -6, r = \sqrt{49}$

Example 13

Find the equation of the circle with centre $(4, -3)$ that passes through the point $(-2, 5)$.

$(x - 4)^2 + (y + 3)^2 = r^2$ — Write the equation of circle centre $(4, -3)$.

$(-2 - 4)^2 + (5 + 3)^2 = r^2$ — Substitute for $x = -2$ and $y = 5$.

$36 + 64 = r^2$ — Simplify

$r^2 = 100$ — Work out r^2.

Equation of the circle is

$(x - 4)^2 + (y + 3)^2 = 100$ — Write down the equation of the circle.

Exercise 3.4

1. Find the equation of each of these circles.

a) centre $(9, 1)$, radius 4

b) centre $(-5, 3)$, radius 7

c) centre $(-4, -7)$, radius 5

d) centre $(6, -2)$, radius 3

2. **Determine** the centre and radius of each of these circles.

a) $(x + 2)^2 + (y - 1)^2 = 9$

b) $(x - 3)^2 + (y - 8)^2 = 36$

c) $(x - 5)^2 + (y + 9)^2 = 20$

d) $(x + 6)^2 + (y + 7)^2 = 1$

e) $x^2 + y^2 - 6x + 8y + 10 = 0$

f) $x^2 + y^2 + 4x + 2y - 1 = 0$

g) $x^2 + y^2 - x - 10y + 5 = 0$

h) $x^2 + y^2 + 2x - 3y - 7 = 0$

i) $2x^2 + 2y^2 - 8x + 2y + 2 = 0$

j) $3x^2 + 3y^2 - 42x + 6y - 4 = 0$

3. Find the equation of these circles.
   a) centre $(-2, -1)$ and passes through the point $(2, 0)$
   b) centre $(-3, 7)$ and passes through the point $(-2, 4)$
   c) centre $(5, -4)$ and passes through the point $(0, 8)$
   d) centre $(6, 0)$ and passes through the point $(-1, -3)$
4. The circle with equation $(x - 2)^2 + (y + a)^2 = 26$ passes through the point $(3, -1)$. Find the value of a where $a < 0$.
5. PQ is a diameter of the circle, where P is $(-8, 3)$ and Q is $(2, -5)$. Find the equation of the circle.
6. a) Show that $(x - 5)(x - 2) + (y - 7)(y - 1) = 0$ represents a circle.
   b) Find the centre and radius of this circle.
7. A circle with equation $x^2 + y^2 + 4x - 6y = 12$ has centre C. The circle cuts the x-axis at the points A and B. Calculate the area of the triangle ABC.

3.5 Points of intersection and circle properties

You know that the equation of a straight line is given by: $\boldsymbol{y = mx + c}$, where m is the gradient and c is the y-intercept. In Chapter 1 you learnt that you can find the points of intersection of a line and a curve by solving their equations simultaneously.

When a line touches a curve in exactly one place we call that line a tangent to the curve.

This also means that there will only be one solution when solving the equations simultaneously.

If you form a quadratic equation there will be only **one (repeated)** root,

i.e. $b^2 - 4ac = 0$

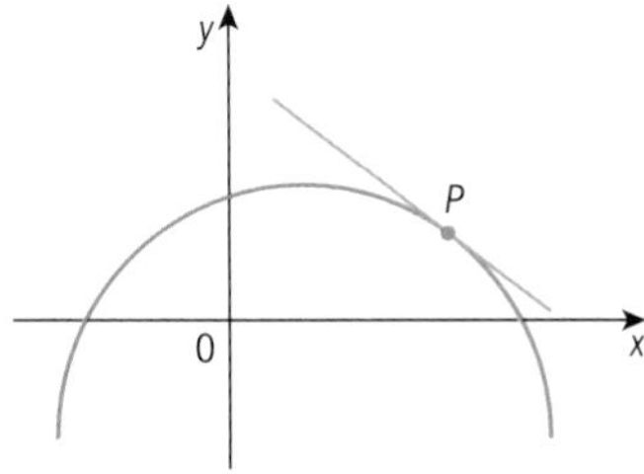

Example 14

Find the set of values of k for which the line $x = y + 2$ intersects the curve $y = x^2 + 3x + k$ at two distinct points.

They meet when $x^2 + 3x + k = x - 2$.

$x^2 + 2x + (k + 2) = 0$ ← Rearrange as a quadratic equation.

There are two distinct roots when $b^2 - 4ac > 0$.

$(2)^2 - 4(1)(k + 2) > 0$ ← $a = 1$, $b = 2$ and $c = k + 2$

$-4k - 4 > 0$

$-4k > 4$ ← Use inequality signs throughout.

$k < -1$

Example 15

A line has equation $y = 4k - x$ and a curve has equation $y = 3x - kx^2$, where k is a constant.

a) Find the two values of k for which the line is a tangent to the curve.

b) Hence find the coordinates of the points where the line touches the curve.

a) The line and the curve meet when

$4k - x = 3x - kx^2$ ← Both equal y.

$kx^2 - 4x + 4k = 0$ ← Rearrange in the form $ax^2 + bx + c = 0$.

The line is a tangent so there is one repeated root.

$(-4)^2 - 4(k)(4k) = 0$ ← $b^2 - 4ac = 0$

$16 = 16k^2$, $k^2 = 1$

$k = 1$ or $k = -1$

b) When $k = 1$:

$x^2 - 4x + 4 = 0$ ← Substitute in $kx^2 - 4x + 4k = 0$.

$(x - 2)(x - 2) = 0$

$x = 2$, $y = 4k - x = 4 - 2 = 2$

When $k = -1$:

$-x^2 - 4x - 4 = 0$

$x^2 + 4x + 4 = 0$

$(x + 2)(x + 2) = 0$

$x = -2$, $y = 4k - x = -4 + 2 = -2$

Line meets curve at (2, 2) and (−2, −2). ← The question asked for coordinates.

Example 16

Determine the equation of the tangent to the circle, centre C, with equation $x^2 + y^2 - 2x - 4y - 5 = 0$ at the point $A(2, 5)$. Give your answer in the form $ax + by + c = 0$, where a, b and c are integers.

$(x - 1)^2 - 1 + (y - 2)^2 - 4 - 5 = 0$ ← Complete the square to find the centre.

C is $(1, 2)$. ← We do not need to find the radius.

Gradient of $AC = \dfrac{5-2}{2-1} = 3$ ← Use the points (2, 5) and (1, 2).

Gradient of the tangent $= -\dfrac{1}{3}$ ← The tangent is perpendicular to the radius.

Equation of the tangent: $y - 5 = -\dfrac{1}{3}(x - 2)$ ← The gradient is $-\dfrac{1}{3}$ through the point (2, 5).

$3y - 15 = -x + 2$

$x + 3y - 17 = 0$ ← Simplify and put in the required form.

Example 17

Determine the shortest distance from the point $P(1, 1)$ to the circle, centre C, with equation $x^2 + y^2 + 6x - 8y + 21 = 0$.

$(x + 3)^2 - 9 + (y - 4)^2 - 16 + 21 = 0$ ← Complete the square to find C and radius r.

C is $(-3, 4)$, $r = \sqrt{4} = 2$. ← Find the centre and the radius.

$PC = \sqrt{(1 - -3)^2 + (1 - 4)^2} = 5$ ← Find the distance between P and C.

Shortest distance $= 5 - 2 = 3$ ← See diagram below.

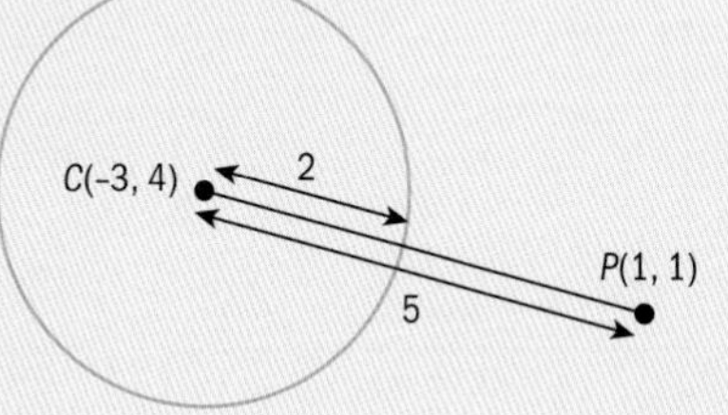

Remember that the tangent to a point on a circle is always perpendicular to the radius at that point and that angles in a semicircle are always right angles. The diagram reminds you why $\angle PRQ = 90°$ if PQ is a diameter of a circle with centre C. Drawing the radius CR creates two isosceles triangles, with one base angle from each making up $\angle PRQ$.

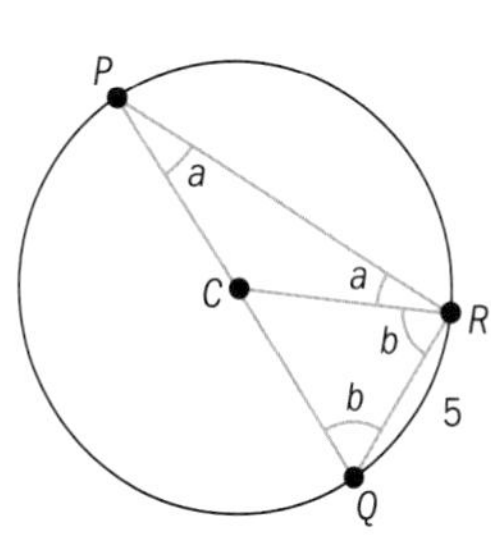

You can be asked to use these elementary geometrical properties of circles when working with coordinate geometry.

Example 18

The diagram shows a circle, centre C, with equation $(x-7)^2+(y-6)^2=25$. PQ is a diameter, where P is $(3, 9)$ and Q is $(11, 3)$. $R(12, 6)$ lies on the circle.

Show that the line perpendicular to QR and passing through R goes through P.

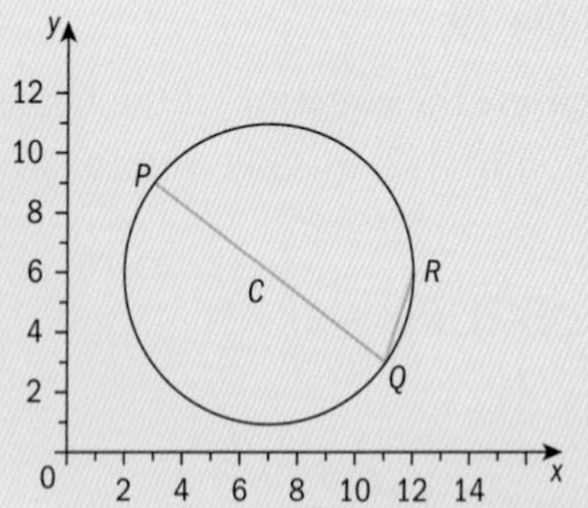

QR has gradient $\frac{6-3}{12-11}=3$, so the gradient of the perpendicular is $-\frac{1}{3}$.

Gradients of perpendicular lines multiply to –1.

The equation of the perpendicular line through (12, 6) is given by

$y-6=-\frac{1}{3}(x-12)$

Substitute the gradient and known point to get the equation of the line.

$y=-\frac{1}{3}x+10$

Rearrange to standard form.

When $x=3$, $y=-\frac{1}{3}\times 3+10=9$

so $P(3, 9)$ lies on the line.

Always finish the answer to a 'show that' question with a statement.

Example 19

AB is the diameter of a circle, where A is $(2, 6)$ and B is $(8, 2)$.

The tangent to the circle at B meets the x-axis at P and the y-axis at Q. Find the coordinates of P and Q.

A tangent is perpendicular to the diameter and we have two points defining the diameter.

AB has gradient $\frac{2-6}{8-2}=-\frac{2}{3}$, so the gradient of the tangent at B is $\frac{3}{2}$.

The equation of the perpendicular line through (8, 2) is given by

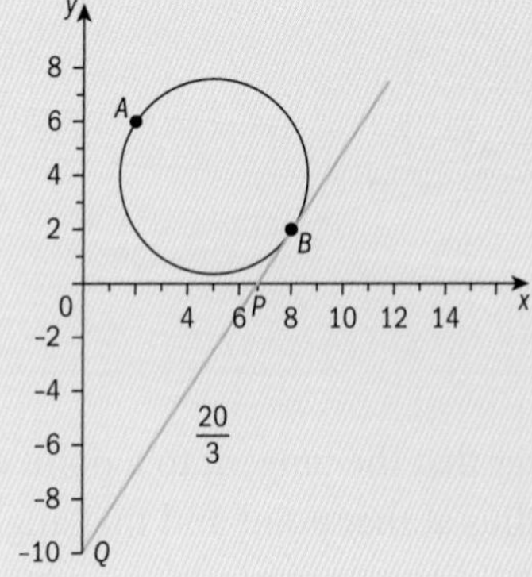

$y-2=\frac{3}{2}(x-8)$.

Substitute the gradient and known point to get the equation of the line.

$y=\frac{3}{2}x-10$

Rearrange to standard form.

Continued on the next page

When $y = 0$, $0 = \frac{3}{2}x - 10 \Rightarrow x = \frac{20}{3}$ ← Substitute $x = 0$ and $y = 0$ to find coordinates of intercepts.

so P is $\left(\frac{20}{3}, 0\right)$.

When $x = 0$, $y = -10$ so Q is $(0, -10)$.

Example 20

The points P, Q and R have coordinates $(0, 2)$, $(4, 0)$ and $(8, 8)$ respectively.

Use Pythagoras' theorem to show that PQR is a right-angled triangle and hence find the equation of the circle that passes through P, Q and R.

$PQ^2 = 2^2 + 4^2 = 20 \quad QR^2 = 4^2 + 8^2 = 80$ ← Calculate the squares of sides of triangle.

$PR^2 = 8^2 + 6^2 = 100$

so $PQ^2 + QR^2 = PR^2$ and PQR is a right-angled triangle with PR as the hypotenuse with the right angle at Q. ← Use Pythagoras' theorem as required.

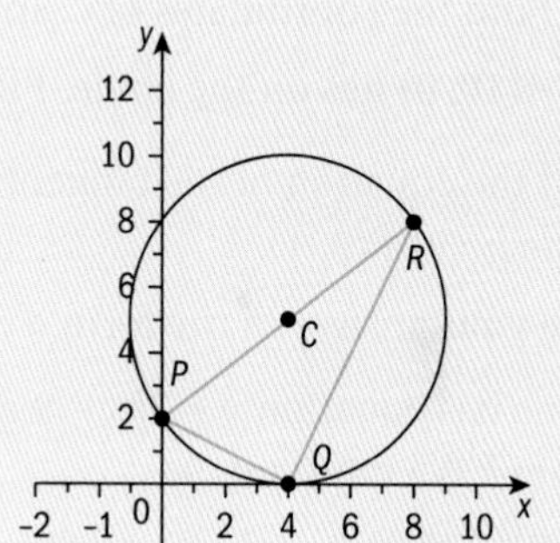

The circle through P, Q and R therefore has PR as diameter.

If C is the centre of the circle, then C is the mid-point of PR and the radius of the circle is half PR.

C is $(4, 5)$, radius $= 5$.

The equation of the circle is $(x - 4)^2 + (y - 5)^2 = 25$. ← Substitute centre coordinates and radius to get the equation of the circle.

Exercise 3.5

1. Find the value of k for which the line $y = 2kx + 7$ is a tangent to the curve $y = 3 + kx^2$.

2. A circle with centre C has equation $x^2 + y^2 - 6x + 4y + 8 = 0$.

   a) Express the equation in the form $(x - a)^2 + (y - b)^2 = r^2$.

   b) Find the coordinates of C and the radius of the circle.

3. Find the set of values of k for which the line $y = kx - 4$ intersects the curve $y = x^2$ at two distinct points.

4. The line $y = 5 - kx$, where k is an integer, is a tangent to the curve $y = 2k - x^2$.

   a) Find the possible values of k.

   When $k = 2$, the line $y = 5 - kx$ is a tangent to the curve $y = 2k - x^2$ at point A.

   b) Find the coordinates of A.

5. Determine the shortest distance from the point $A(1, 3)$ to the circle with equation $x^2 + y^2 - 10x - 12y + 45 = 0$.

6. A curve has equation $y = 2x^2 + kx - 1$ and a line has equation $x + y + k = 0$, where k is a constant.

   a) State the value of k for which the line is a tangent to the curve.

   b) For this value of k find the coordinates of the point where the line touches the curve.

7. The equation of a line is $y = x - k$, where k is a constant, and the equation of a curve is $x^2 + 2y = k$.

   a) When $k = 1$, the line intersects the curve at the points A and B. Find the coordinates of A and the coordinates of B.

   b) Find the value of k for which the line $y = x - k$ is a tangent to the curve $x^2 + 2y = k$.

8. A circle with centre C has equation $x^2 + y^2 - 8x - 6y - 20 = 0$.

   a) Find the coordinates of C and the radius of the circle.

   b) $A(10, 0)$ lies on the circle. Find the equation of the tangent to the circle at A.

9. Find the equation of the tangent to the circle $x^2 + y^2 - 12x + 26 = 0$ at the point $P(3, 1)$.

   Give your answer in the form $ax + by + c = 0$, where a, b and c are integers.

10. A line has equation $y = 2kx - 9$ and a curve has equation $y = x^2 - kx$, where k is a constant.

    a) Find the two values of k for which the line is a tangent to the curve.

    b) For each value of k, find the coordinates of the point where the line is a tangent to the curve, and find the equation of the line that joins these two points.

11. $P(2, 1)$ is a point on the circumference of the circle $x^2 + y^2 - 10x + 2y + 13 = 0$. PQ is a diameter of the circle. Find the equation of the line through P and Q.

12. Show that the line with equation $y = 2x - 3$ cannot be a tangent to the curve with equation $y = x^2 - 3x + 5$.

13. A line has equation $y = 2kx - 7$ and a curve has equation $y = x^2 + kx - 3$, where k is a positive integer.

    a) Find the value of k for which the line is a tangent to the curve.

    b) For this value of k, find the coordinates of the point where the line touches the curve.

14. Show that when $k > 2$, the line $2x + y = 1$ does not intersect the curve with equation $y = x^2 + k$.

15. Find the equation of the line joining the point (1, 4) to the centre of the circle with equation $x^2 + y^2 - 8x + 4y + 11 = 0$.

16. A straight line L passes through the point (1, 0) and has gradient m.

    a) Write down the equation of the line.

    The line L is a tangent to the curve $y = x^2 + 2x - 3$.

    b) Find the value of m.

    The line L passes through the point $(-2, k)$.

    c) Find the value of k.

17. A circle with equation $x^2 + (y - k)^2 = 10$ passes through the point (3, 5). Determine the two possible values of k.

18. Show that the line $y = 2x$ does not intersect the circle $(x - 5)^2 + (y - 2)^2 = 10$.

19. AB is a diameter of the circle $(x - 3)^2 + (y - 2)^2 = 5$, where A is (2, 0). $D(5, 3)$ lies on the circle. Show that triangle ABD is right angled at D.

Summary exercise 3

EXAM-STYLE QUESTIONS

1. The straight line joining the points $P(9, -1)$ and $Q(k, 5)$ has a gradient of -3. Work out the value of k.

2. The mid-point M of the line PQ has coordinates $(-2, -1)$. P has coordinates $(-3, 4)$. R has coordinates $(5, -2)$. Find the length of the line QR.

3. The point $(-4, 8)$ is the mid-point of the line joining the points $(a + b, 3b)$ and $(-2a, 2b - 3a)$. Find the values of a and b.

4. The point P has coordinates $(-5, 10)$. The point Q has coordinates (1, 6). The point R has coordinates (3, 5). Show that the three points are not collinear.

5.

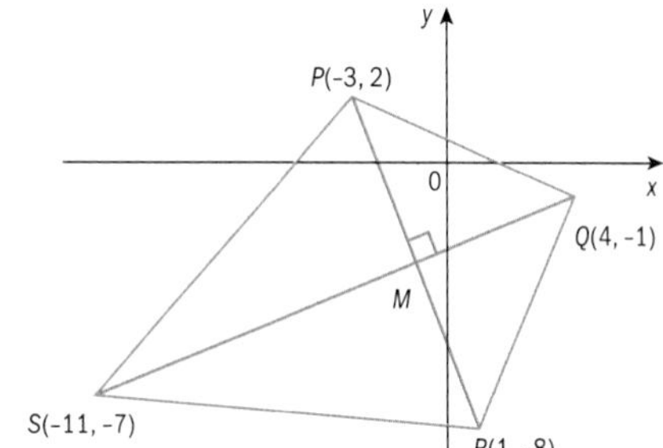

The diagram shows a kite $PQRS$.
The point P is $(-3, 2)$, the point Q is $(4, -1)$, point R is $(1, -8)$ and the point S is $(-11, -7)$.
The diagonal SQ bisects PR at M.
Show that the ratio $SM : MQ = 2 : 1$

6. Find if the lines with equations $9x - 6y = 8$ and $4y - 6x = 7$ are parallel.

7. **a)** Show that the lines with equations $2y = x + 3$ and $2x + y = 9$ are perpendicular.

   b) Find their point of intersection.

EXAM-STYLE QUESTIONS

8.

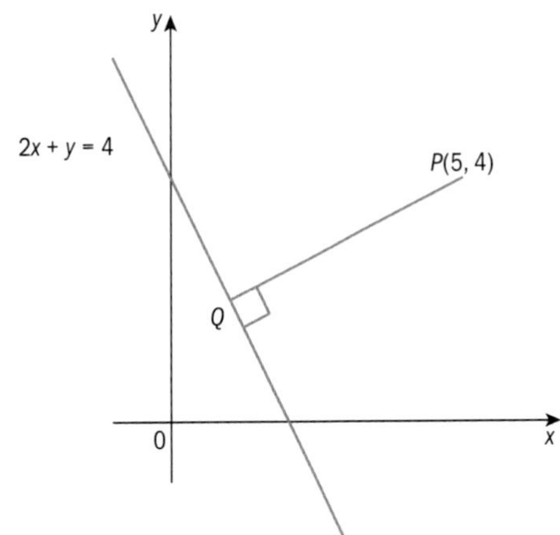

The line PQ is perpendicular to the line with equation $2x + y = 4$. P has coordinates $(5, 4)$. Find the coordinates of Q.

9. Find the equation of the straight line through the point $P(7, 1)$ and parallel to the line with equation $2x - 5y + 3 = 0$.

10. ABC is an isosceles triangle with $AB = BC$. A has coordinates $(3, 2)$, B has coordinates $(2, -5)$ and C has coordinates $(9, -4)$. Find the equation of the line of symmetry of the triangle.

11. P is the point with coordinates $(-2, 5)$ and Q is the point with coordinates $(-6, -3)$.

a) Find the gradient of the line PQ.

b) Find the equation of the line PQ. Give your answer in the form $ax + by + c = 0$.

c) Find the equation of the line that is parallel to PQ and passes through the point R with coordinates $(1, -2)$.

d) Find the exact length of QR.

12. The line with equation $x + y = 2$ meets the line with equation $2y - x = 7$ at the point P. Find the equation of the straight line that passes through P and is perpendicular to the line with equation $y = 2x$. Give your answer in the form $ax + by + c = 0$.

13. P is the point $(k, 3k)$ and Q is the point $(5k, -5k)$. Find the equation of the perpendicular bisector of PQ.

14.

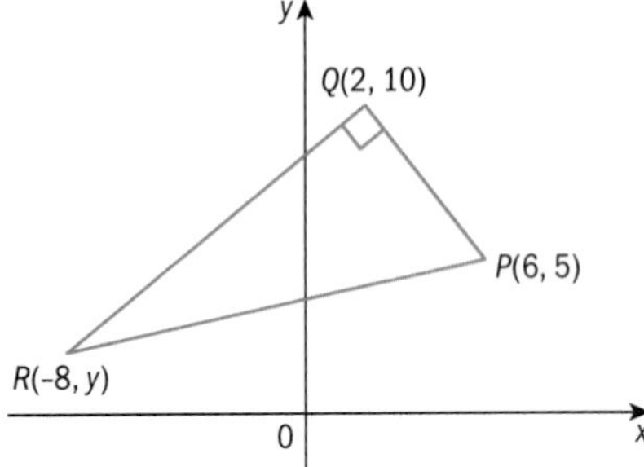

The diagram shows triangle PQR with PQ perpendicular to QR. P is the point $(6, 5)$, Q is the point $(2, 10)$, and R is the point $(-8, y)$. Find the value of y.

15. The line PQ has equation $3x - 2y = 12$. P is the point with coordinates $(6, 3)$ and Q is the point with coordinates $(-2, k)$.

a) Find the value of k.

b) Find an equation of the line through P that is perpendicular to PQ.

The point R has coordinates $(2, -5)$.

c) Find the exact length of PR.

16. The perpendicular bisector of the line joining the points $(3, 2)$ and $(7, 4)$ cuts the y-axis at the point $(0, k)$. Find the value of k.

17. The line $3x + y = 17$ intersects the curve $y = x^2 + 2x - 7$ at the point $P(3, 8)$ and the point Q.

a) Find the coordinates of Q.

b) Find the exact length of the line PQ.

18. Find the equation of the tangent to the circle $x^2 + y^2 - 4x + 2y + 3 = 0$ at the point $(3, -2)$.

19. Find the equation of the circle whose diameter is the line joining the points $P(-1, 5)$ and $Q(3, -1)$.

20. The line $y = kx - 6$ is a tangent to the curve $y = x^2 + x - 2$.

 a) Find the two possible values of k.

 b) For each of these values of k, find the coordinates of the point where the line touches the curve.

21. Determine the set of values of k for which the line $3x + y = k$ does not intersect the curve with equation $y = x^2 - 2x + 1$.

22. Calculate the *exact* distance of the point $P(-1, 1)$ to the point on the circle $x^2 + y^2 - 6x - 10y + 30 = 0$ that is closest to P.

23. Show that that the straight line with equation $3x - y + 8 = 0$ is a tangent to the circle with equation $x^2 + y^2 - 18x - 10y + 16 = 0$.

24. A line has equation $4x + y + k = 0$ and a curve has equation $y = kx^2 + 3$, where k is a constant.

 a) Find the two values of k for which the line is a tangent to the curve.

 b) Find the equation of the straight line joining the points on the curve for these two values of k.

25. The circles with equations $x^2 + y^2 + 6x - 10y = 26$ and $x^2 + y^2 - 12x + 4y = k$ are congruent.

 a) Determine the value of k.

 b) Calculate the distance between their centres.

EXAM-STYLE QUESTIONS

26. A circle with centre C has equation $x^2 + y^2 - 6x + 2y + 5 = 0$.

 a) Find the coordinates of C and the radius of the circle.

 b) $A(5, -2)$ lies on the circle. Find the equation of the tangent to the circle at A.

27. The circle $x^2 + y^2 + 2x + 4y - 35 = 0$ has centre C and passes through points P and Q.

 a) Find the coordinates of C.

 $M(3, 2)$ is the mid-point of PQ.

 b) Find the equation of the line PQ. Give your answer in the form $ax + by + c = 0$.

 c) Find the coordinates of P and Q.

28. PQ is a tangent to the circle, centre C, with equation $x^2 + y^2 + 6x - 8y = 0$.

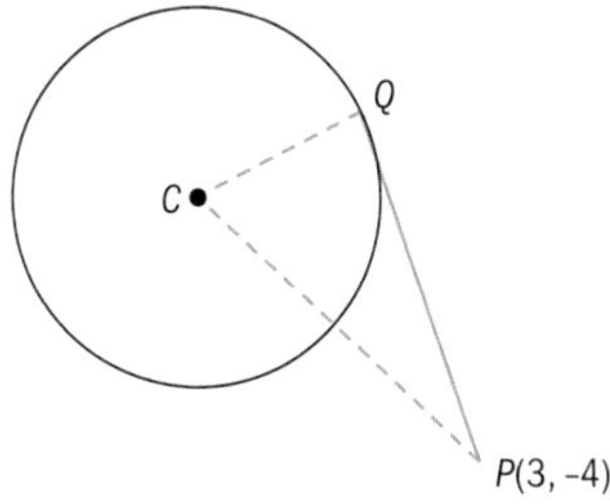

 a) Find the length of CQ.

 P has coordinates $(3, -4)$.

 b) Find the length of PQ.

 c) Calculate the length of the shortest distance from P to the circle.

29. The points P, Q and R have coordinates $(2, 1)$, $(6, 2)$ and $(2, 18)$ respectively.

 a) Use Pythagoras' theorem to show that PQR is a right-angled triangle.

 b) Hence find the equation of the circle that passes through P, Q and R.

30. The points A and B have coordinates $(5, -1)$ and $(-1, 7)$ respectively. C is the mid-point of AB.

 a) Find the coordinates of C.

 b) Find the equation of the circle with AB as diameter.

 c) Find the equation of the tangent to the circle at A.

31. A circle with centre C has equation $x^2 + y^2 + 8x + 4y - 5 = 0$.

   a) Find the coordinates of C and the radius of the circle.

   b) $A(0, -5)$ and $B(0,1)$ lie on the circle. Find the equation of the line perpendicular to AB which goes through B.

   c) Show that this line goes through the other end of the diameter through A.

32. The points K, L and M have coordinates $(10, 8)$, $(-2, -10)$ and $(-14, -2)$ respectively.

   a) Use Pythagoras' theorem to show that KLM is a right-angled triangle.

   b) Hence find the equation of the circle that passes through K, L and M.

33. $f(x) = -\frac{1}{x}$ $(x \neq 0)$, $g(x) = f(x) + 2$ and $h(x) = f(x + 2)$

   a) Sketch the graph of $y = g(x)$ and state the domain and range of $g(x)$.

   b) Give the coordinates where $y = g(x)$ crosses a coordinate axis.

   c) Sketch the graph of $y = h(x)$ and state the domain and range of $h(x)$.

   d) Give the coordinates where $y = h(x)$ crosses a coordinate axis.

34. A circle with centre C has equation $x^2 + y^2 + 6x + 2y - 8 = 0$.

   For the line $x + y = k$, determine the values of k for which the line and the circle

   a) meet at two distinct points

   b) meet at a single point (the line is a tangent to the circle)

   c) do not meet.

35. A circle with centre $C(2, -1)$ passes through the point $A(4, 2)$.

   a) Find the equation of the circle in the form $x^2 + y^2 + ax + by + k = 0$, where a, b and k are integers.

   b) Find the equation of the tangent to the circle at A.

   c) Find the coordinates of the points at which the tangent cuts the coordinate axes.

36. A circle with centre C has equation $(x - 7)^2 + (y - 4)^2 = 16$.

   The point $P(1, 7)$ lies outside the circle.

   Find the length of the two tangents to the circle from P.

Chapter summary

Mid-point of a line segment

- The mid-point of the line joining (x_1, y_1) and (x_2, y_2) is given by $\left(\frac{x_1+x_2}{2}, \frac{y_1+y_2}{2}\right)$.

Gradient of a line

- The gradient of the line joining (x_1, y_1) and (x_2, y_2) is given by $\frac{y_2-y_1}{x_2-x_1}$.

Length of a line segment

- The length of the line segment between (x_1, y_1) and (x_2, y_2) is given by $\sqrt{(x_2-x_1)^2+(y_2-y_1)^2}$.

Parallel and perpendicular lines

- Two straight lines are parallel if their gradients are the same $(m_1 = m_2)$.
- Two straight lines are perpendicular if the product of their gradients $= -1$ $(m_1 \times m_2 = -1)$.

Equation of a line

- The equation of the straight line through the point (x_1, y_1) and with gradient m is: $(y - y_1) = m(x - x_1)$ (given point and gradient).
- The equation of the straight line through the points (x_1, y_1) and (x_2, y_2) is: $\frac{y-y_1}{x-x_1} = \frac{y_2-y_1}{x_2-x_1}$ (given two points).

Equation of a circle

- The equation of a circle, centre (a, b), radius r, is given by $(x - a)^2 + (y - b)^2 = r^2$.
- The expanded form of the equation of a circle is given by $x^2 + y^2 + 2gx + 2fy + c = 0$, where $c = f^2 + g^2 - r^2$.

Tangent to a curve

- When a line touches a curve in exactly one place we call that line a tangent to the curve.
- When you form a quadratic equation in order to solve the equation of the tangent and the equation of the quadratic simultaneously, there will be only one repeated root, i.e. use $b^2 - 4ac = 0$.

Maths in real-life

Parabolic reflectors

In Chapter 1 we studied quadratic functions and how we can use their equations to solve problems. This principle is used in parabolic reflectors where the properties of the parabola – which in two dimensions is the shape of a simple quadratic function – is used to receive and transmit energy, such as light, sound and radio waves.

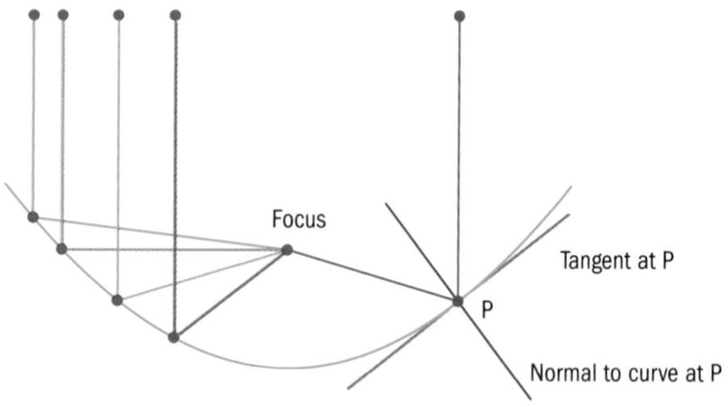

▲ The pink ray hits the curve at P. The angle between the path that it comes in on and the tangent at P is the same as the angle the reflected ray makes with the tangent.

In a parabola, any ray entering parallel to the axis of symmetry is reflected through a single point, known as the **focus**. This allows signals to be greatly amplified.

The properties of the parabola have been known for thousands of years. Greek mythology describes Archimedes destroying enemy ships by setting them on fire using a giant mirror or series of mirrors during the siege of Syracuse in the early third century.

Nowadays, historians and scientists debate the credibility of the story. There is no doubt that the Greeks knew of the practical principle of parabolas nearly 2000 years ago, but it is perhaps more likely that the story is an exaggeration of a more modest example of the phenomena.

The same principle which Archimedes uses in the Greek myth is traditionally used at the start of each Olympic Games, where a parabolic reflector ignites a fire and lights the Olympic torch in Olympia.

► A rehearsal of the lighting of the Olympic Flame at Olympia, Greece in 2008.

In the modern world, parabolic reflectors are used in satellite dishes and telescopes to focus incoming rays onto a receiver.

The radio telescope in the Stanford foothills has a 46 m diameter dish. It was built in 1966 by the Stanford Research Institute and is owned by the U.S. government. The original purpose was to study the chemical composition of the atmosphere. Since then the dish has been used to communicate with satellites and spacecraft.

The parabolic reflector of this radio telescope concentrates electromagnetic waves (radio waves) into an aluminium collector dish. In order for scientists to read the data, electromagnetic waves from space need to be manipulated by transducers and fed into computers.

The Odeillo-Font-Romeau Solar Furnace in France consists of a total of 10 000 mirrors that automatically track the sun and concentrate the light on a parabolic reflector. The reflector then concentrates the rays onto a single point to produce a temperature of 3000 degrees centigrade. This energy is used in numerous ways, including generating electricity via a steam turbine, making hydrogen fuel, testing re-entry materials for space vehicles and performing high-temperature metallurgic experiments.

4 Circular measure

The Wiener Riesenrad (which in German means 'Viennese giant wheel'), originally constructed in 1897, is one of the most frequently visited attractions in Vienna, Austria's capital city. The wheel is 64.75 metres tall and was the world's tallest Ferris wheel from 1920 until 1985. The wheel turns through an angle of approximately 85° or 1.5 radians every minute. It is driven by a circumferential cable which leaves the wheel and passes through the drive mechanism under the base.

Objectives

- Understand the definition of a radian and use the relationship between radians and degrees.
- Use the formulae $s = r\theta$ and $A = \frac{1}{2}r^2\theta$ (where θ is measured in radians) in solving problems concerning the arc length, s, and sector area, A, of a circle.

Before you start

You should know how to:

1. Find the perimeter and area of shapes,

e.g. Perimeter $= 10 + 4 + 4 + \frac{\pi \times 10}{2}$
$= 33.7\,\text{cm}$ (3 s.f.)

Area $= 10 \times 4 + \frac{\pi \times 5^2}{2}$
$= 79.3\,\text{cm}^2$ (3 s.f.)

4 cm
10 cm

2. Find arc length and area of a sector of a circle using angles measured in degrees by using:

$$\text{Arc length} = \frac{\theta°}{360°} \times \pi d$$

$$\text{Area of sector} = \frac{\theta°}{360°} \times \pi r^2,$$

e.g. Arc length $= \frac{110}{360} \times \pi \times 10$
$= 9.60\,\text{cm}$ (3 s.f.)

110°
5 cm

Area of sector $= \frac{110}{360} \times \pi \times 5^2$
$= 24.0\,\text{cm}^2$ (3 s.f.)

Skills check:

1. Find the perimeter and area of the shaded shape made from a square and a quarter circle.

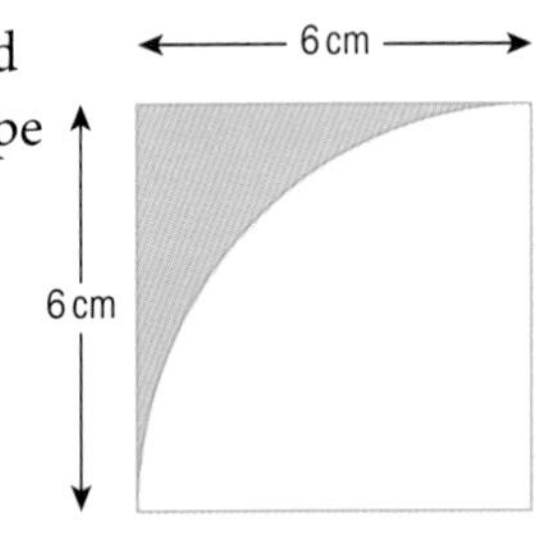

2. The diagram shows the sector of a circle. Find the arc length AB and the area of the sector AOB.

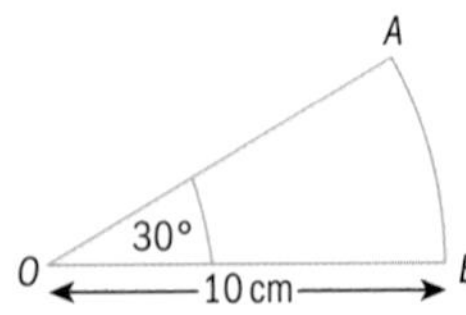

Note: You will also need to know and be able to use trigonometry and Pythagoras' theorem.

4.1 Radians

There is more than one way to measure angles. So far you have probably used degrees. This chapter introduces you to radians.

One **radian** is the angle subtended at the centre of a circle radius r by an arc of length r.

That is, when a sector is drawn with the arc length equal to the radius, then the sector angle is 1 radian.

For the circle shown, circumference $= 2\pi r$

$$= 2\pi \times \text{arc length } PQ$$

Therefore the total angle subtended by the circle $= 2\pi \times 1$ radian.

This means

$360° = 2\pi$ radians

So $1° = \dfrac{2\pi}{360}$ radians $= 0.0175$ rad (3 s.f.) and 1 radian $= \dfrac{360°}{2\pi} = 57.3°$ (1 d.p.)

To convert from degrees to radians, multiply by $\dfrac{2\pi}{360}$ or $\dfrac{\pi}{180}$.

To convert from radians to degrees, multiply by $\dfrac{360}{2\pi}$ or $\dfrac{180}{\pi}$.

Some angles measured in radians can be written as simple fractions of π.

You should learn these.

Degrees	0°	30°	45°	60°	90°	180°	270°	360°
Radians	0	$\frac{\pi}{6}$	$\frac{\pi}{4}$	$\frac{\pi}{3}$	$\frac{\pi}{2}$	π	$\frac{3\pi}{2}$	2π

Example 1

Convert 165° to radians.

$165 \times \dfrac{\pi}{180} = \dfrac{11\pi}{12} = 2.88$ radians (3 s.f.)

Multiply by $\dfrac{\pi}{180}$.

Example 2

Convert $\dfrac{2\pi}{3}$ radians to degrees.

$\dfrac{2\pi}{3} \times \dfrac{180}{\pi} = 120°$

Multiply by $\dfrac{180}{\pi}$.

Or

$\dfrac{2\pi}{3} = 2 \times \dfrac{\pi}{3} = 2 \times 60°$

Note: It is a useful skill to be able to find multiples of known equivalents quickly in your head.

Exercise 4.1

1. Convert each angle to radians, giving your answer in terms of π.

   a) 15° **b)** 225° **c)** 135°

   d) 315° **e)** 63° **f)** 72°

2. Convert each angle to radians, giving your answer to 3 significant figures.

   a) 25° **b)** 100° **c)** 250°

   d) 80° **e)** 137° **f)** 318°

3. Convert each angle to degrees.

   a) $\frac{\pi}{10}$ **b)** $\frac{7\pi}{180}$ **c)** $\frac{3\pi}{8}$

   d) $\frac{5\pi}{6}$ **e)** $\frac{4\pi}{3}$ **f)** $\frac{2\pi}{9}$

4.2 Arc length and sector area

Using radians as the angle of measure gives us simple formulae for the length of an arc and the area of a sector.

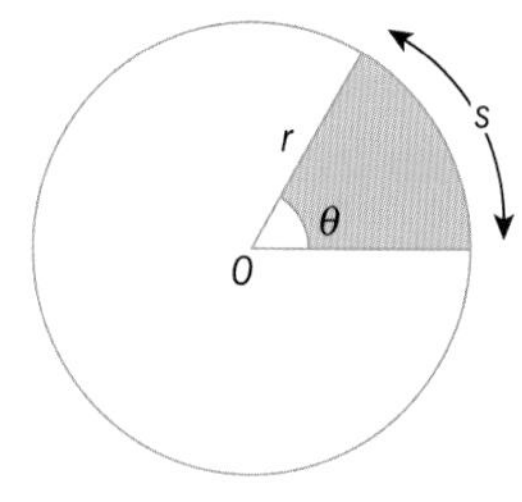

From the diagram we can see that, if θ is measured in radians,

Arc length, $s = \frac{\theta}{2\pi} \times 2\pi r$ (where $2\pi r$ circumference of circle with radius r)

$= r\theta$

Remember the angle for a whole circle = 2π rad.

Also,

area of sector, $A = \frac{\theta}{2\pi} \times \pi r^2$ (area of circle with radius, r)

$= \frac{1}{2}r^2\theta$

Arc length, $s = r\theta$ Sector area, $A = \frac{1}{2}r^2\theta$
where θ is measured in radians and r is the radius of the circle.

Example 3

OAB is a sector of a circle with centre O and radius 8 cm.

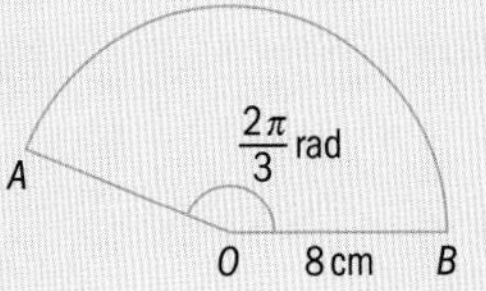

a) Find the area of the sector in terms of π.

b) Find the perimeter of the sector in terms of π.

a) $A = \frac{1}{2}r^2\theta = \frac{1}{2} \times 8^2 \times \frac{2\pi}{3} = \frac{64\pi}{3}\text{ cm}^2$

b) $s = r\theta = 8 \times \frac{2\pi}{3} = \frac{16\pi}{3}$

Perimeter = arc length + 8 + 8 = $\frac{16\pi}{3} + 16$ cm

Substitute $r = 8$ and $\theta = \frac{2\pi}{3}$ into the formulae.

Example 4

OAB is a sector of a circle with centre O and radius 25 cm.
The line AB is a chord of the circle.

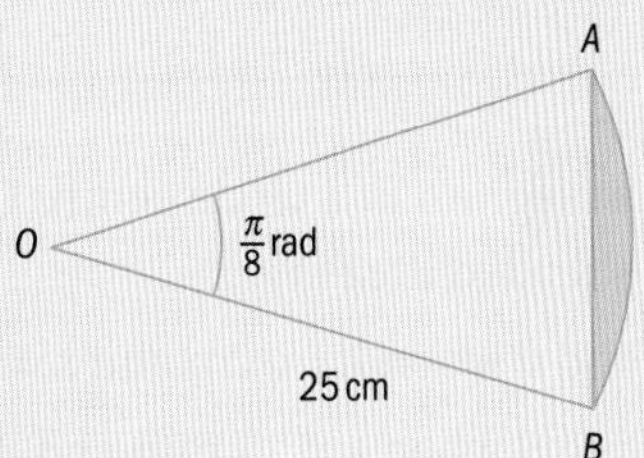

Find the area of the shaded region.
(You are given that the area of triangle OAB may be found by using the formula $\frac{1}{2} \times OA \times OB \times \sin AOB$.)

Area of shaded region = Area of sector OAB – Area of triangle OAB

Area of sector OAB $= \frac{1}{2}r^2\theta = \frac{1}{2} \times 25^2 \times \frac{\pi}{8}$ ← Find the area of the sector.

Area of triangle OAB $= \frac{1}{2} \times 25 \times 25 \times \sin\frac{\pi}{8}$ ← Find the area of the triangle.

Area of the shaded region $= \frac{1}{2} \times 25^2 \times \frac{\pi}{8} - \frac{1}{2} \times 25^2 \times \sin\frac{\pi}{8}$

$= 3.13\text{ cm}^2$ (3 s.f.)

You should use the radian mode in your scientific calculator.

Exercise 4.2

1. Find the length of each arc of a circle.

a)

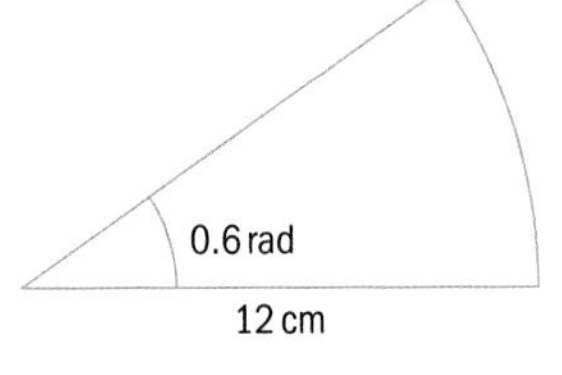

b)

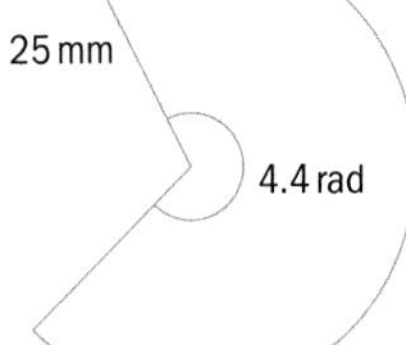

c)

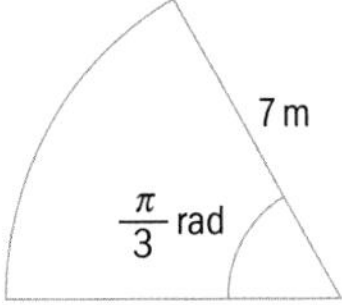

d)

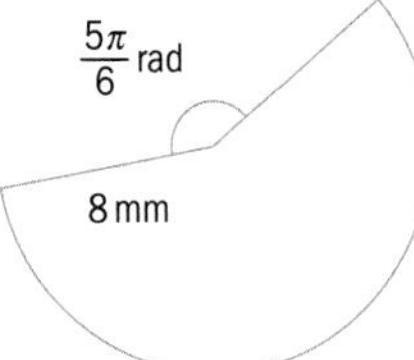

2. Find the area of the shaded sector in each circle.

a)

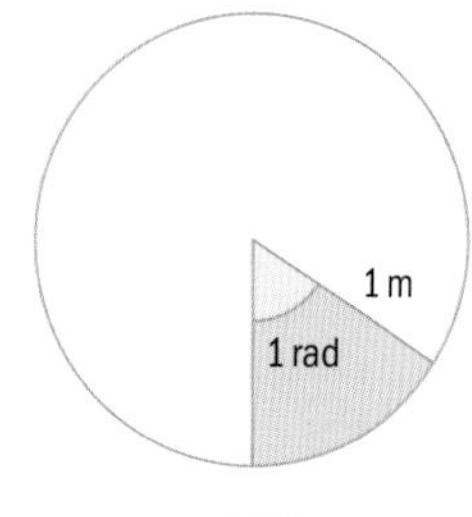

b)

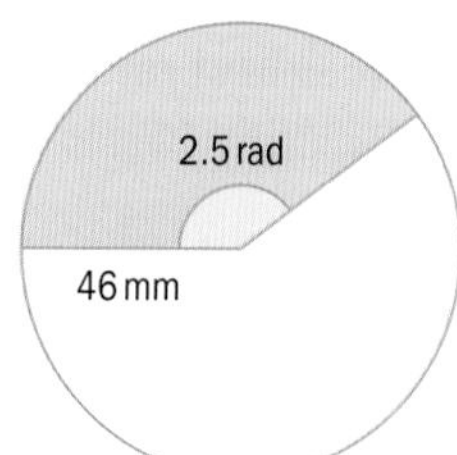

c)

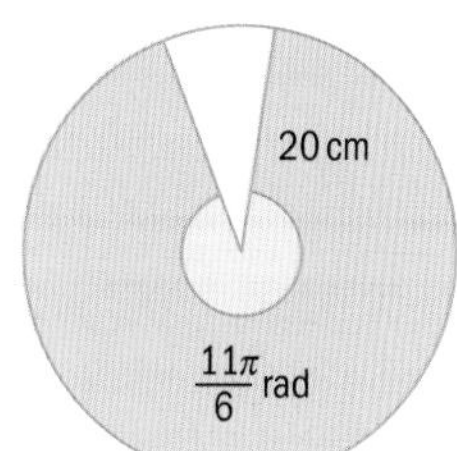

d)

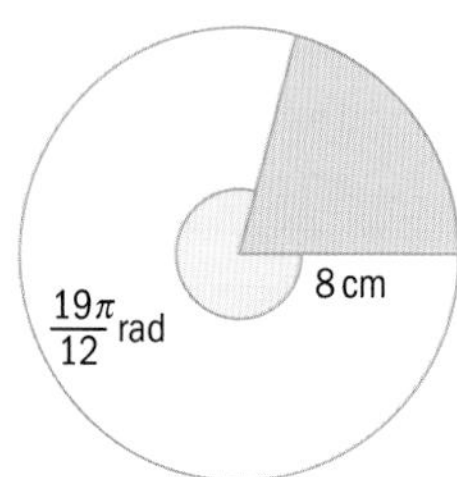

3. In the diagram of a circle centre O, the arc length is s and the area of the shaded area is A.

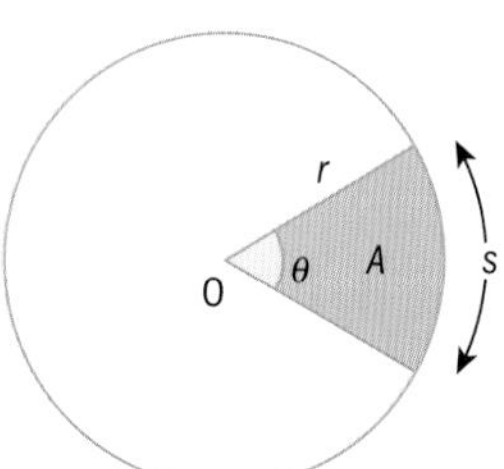

a) Find θ when $r = 6$ cm and $s = 10$ cm.

b) Find θ when $r = 25$ mm and $A = 1000\text{ mm}^2$.

c) Find r when $\theta = 0.8$ rad and $s = 20$ m.

d) Find r when $\theta = \frac{3\pi}{4}$ rad and $A = 50\text{ cm}^2$.

4. The diagram has three sectors with $OA = 2$ cm, $OB = 4$ cm and $OC = 6$ cm.

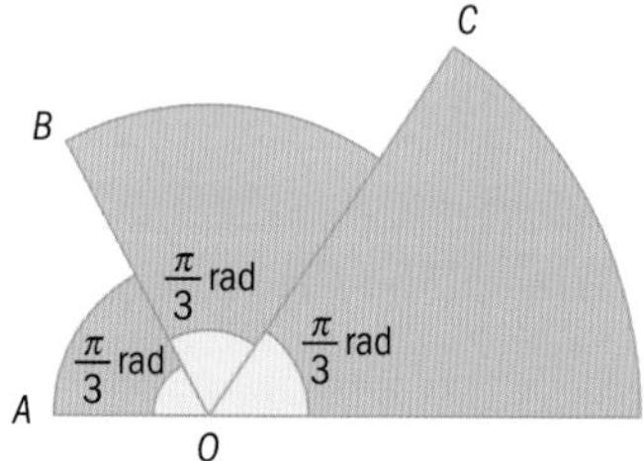

a) Find the total area of the shape in terms of π.

b) Find the total perimeter of the shape in terms of π.

5. OAB is a sector of a circle, with centre O and a radius of 12 cm.

Find the perimeter of the shaded segment.

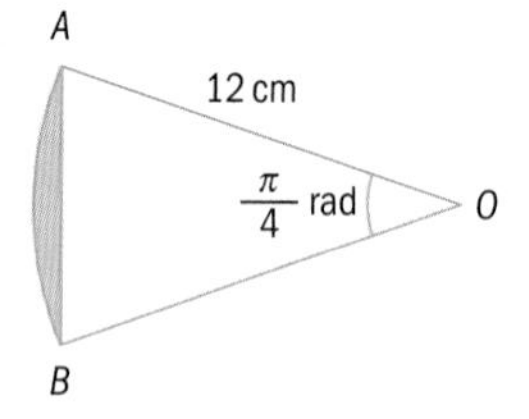

Hint: You will need to use trigonometry.

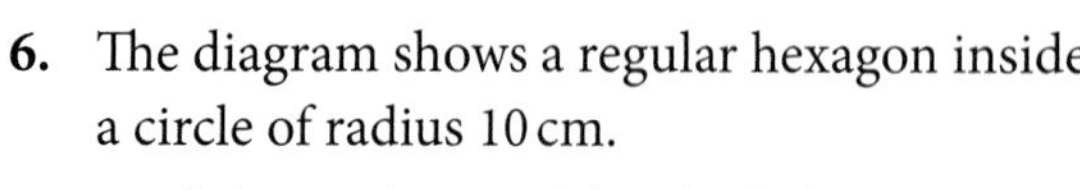

6. The diagram shows a regular hexagon inside a circle of radius 10 cm.

Find the total area of the shaded regions.

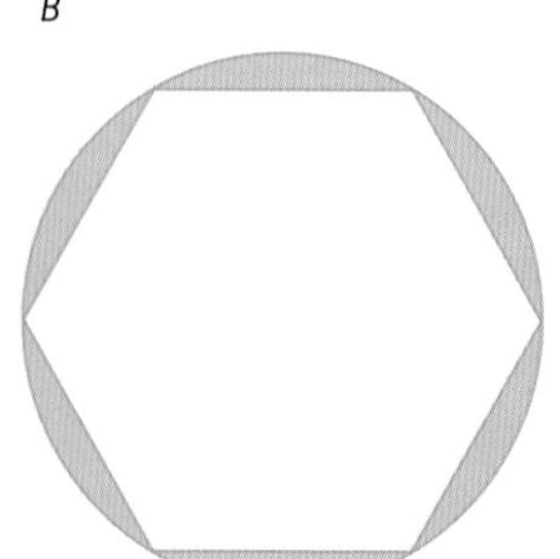

4.3 Further problems involving arcs and sectors

When solving problems involving compound shapes you may also need to know and be able to use other results you have already met. These will include:

Pythagoras' theorem	$a^2 + b^2 = c^2$
Right-angled triangle trigonometry	$\sin x = \frac{O}{H}$, $\cos x = \frac{A}{H}$, $\tan x = \frac{O}{A}$
Sine rule	$\frac{a}{\sin A} = \frac{b}{\sin B} = \frac{c}{\sin C}$
Cosine rule	$a^2 = b^2 + c^2 - 2bc\cos A$
Area of a triangle rule	$\text{Area} = \frac{1}{2}ab\sin C$

Example 5

In the diagram, BD is the arc of a circle with centre C and radius 20 m.

Angle $BCD = \frac{2\pi}{3}$ radians.

AB and AD are tangents to the circle at B and D, respectively.

a) Show that $AB = 20\sqrt{3}$ cm.

b) Show that the area of the shaded region is $400\left(\sqrt{3} - \frac{\pi}{3}\right)$ cm².

c) Find the perimeter of the shaded region in terms of π and $\sqrt{3}$.

Note: You may assume that $\tan\frac{\pi}{3} = \sqrt{3}$ in this question. Exact trigonometric ratios will be studied in Chapter 5.

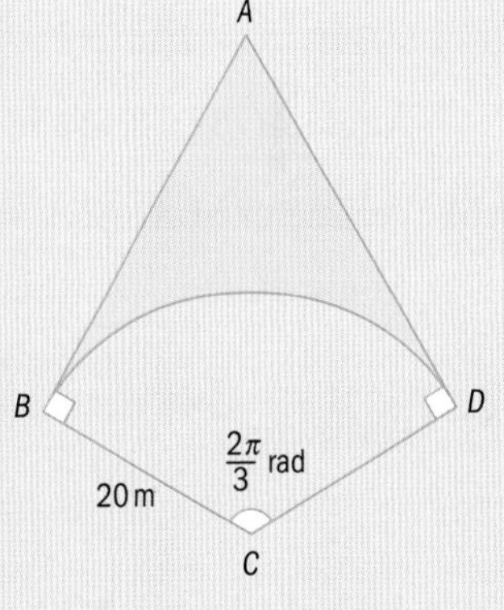

a) Angle $BCA = \frac{\pi}{3}$ — Use the right-angled triangle ABC.

$\tan\frac{\pi}{3} = \frac{AB}{20}$

$AB = 20\tan\frac{\pi}{3}$ — Find AB.

$AB = 20\sqrt{3}$ cm

b) Shaded area = 2 × area of triangle ABC – area of sector DCB

$= 2 \times \frac{1}{2} \times 20\sqrt{3} \times 20 - \frac{1}{2} \times 20^2 \times \frac{2\pi}{3}$

$= 400\sqrt{3} - \frac{400\pi}{3}$ — Use the formulae: area of a triangle $= \frac{1}{2}bh$ and area of a sector $= \frac{1}{2}r^2\theta$.

$= 400\left(\sqrt{3} - \frac{\pi}{3}\right)$ cm²

c) Perimeter = AB + arc length BD + AD

$= 20\sqrt{3} + 20 \times \frac{2\pi}{3} + 20\sqrt{3}$ — Use arc length = $r\theta$ and $AD = AB$.

$= 40\sqrt{3} + \frac{40\pi}{3}$ cm — Leave your answer in exact form.

Example 6

The diagram shows a shape $ABCD$ made by removing the two shaded regions from the circle of radius 8 cm and with centre O. AB is parallel to DC. $AB = DC = 14$ cm.

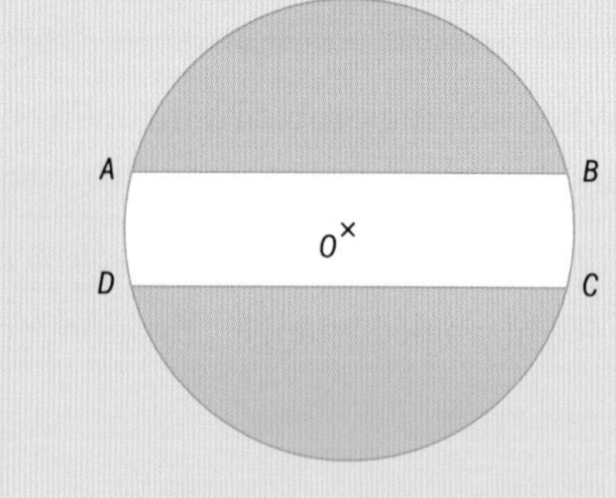

a) Find angle AOB. Give your answer in radians correct to 4 significant figures.

b) Find the perimeter of $ABCD$.

c) Find the area of $ABCD$.

---

a) Using the cosine rule:

$$\cos AOB = \frac{8^2 + 8^2 - 14^2}{2 \times 8 \times 8} = \frac{-17}{32}$$

Angle $AOB = 2.131$ rad (4 s.f.)

Recall the cosine rule:

$a^2 = b^2 + c^2 - 2bc \cos A$

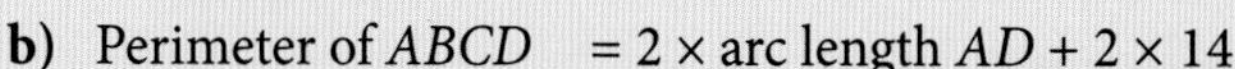

b) Perimeter of $ABCD = 2 \times$ arc length $AD + 2 \times 14$

Angle $AOD = \pi - 2.131$

$= 1.011$ rad

Perimeter $= (2 \times 8 \times 1.011) + (2 \times 14)$

$= 44.2$ cm (3 s.f.)

Give your answer to 3 significant figures.

c) Area of triangle $AOB = \frac{1}{2} \times 8 \times 8 \times \sin 2.131 = 27.11$

Area of sector $OAD = \frac{1}{2} \times 8^2 \times 1.011 = 32.35$

Area of $ABCD = 2 \times 27.11 + 2 \times 32.35$

$= 119$ cm^2 (3 s.f.)

Use $\frac{1}{2}ab \sin C$ and $\frac{1}{2}r^2\theta$.

2 × area of triangle + 2 × area of sector

Exercise 4.3

1. The diagram shows a square $ABCD$.

   The two arcs are arcs of circles, with centres A and C and radius r cm.

   Find the area of the shaded region in terms of r and π.

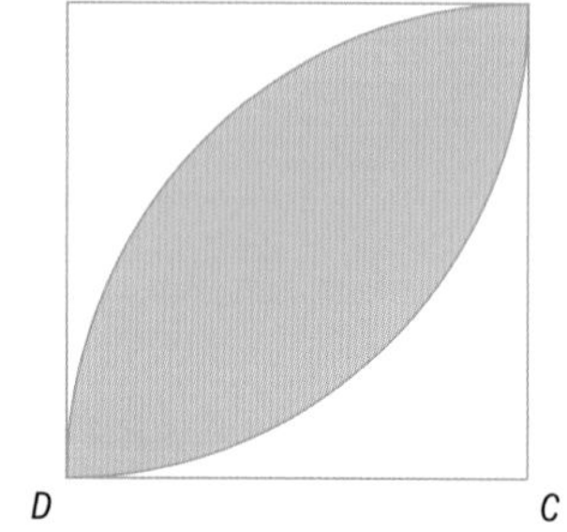

2. The diagram shows part of a circle centre O, with radius 12 cm, and a straight line AB. The length of AB is 12 cm.

   Find the perimeter of the shape.

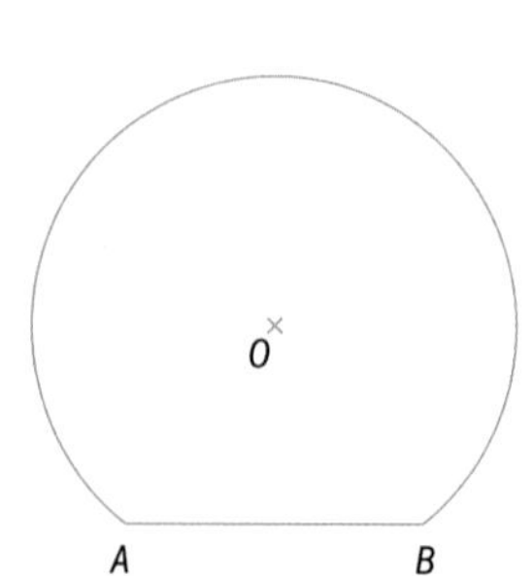

3. The diagram shows the area used for a sports event. The area is made from the sectors of two circles, centre O. Calculate the area of the shaded region.

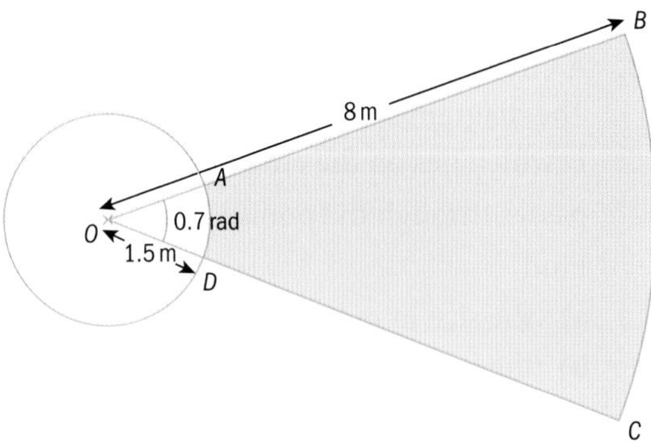

4. The diagram shows a shape ABC.

The boundary of the shape is made from three arcs.

The arcs AB, BC and AC have centres C, A, B, respectively.

a) Find the perimeter of the shape in terms of r and π.

b) Find the area of the shape when $r = 15$ cm.

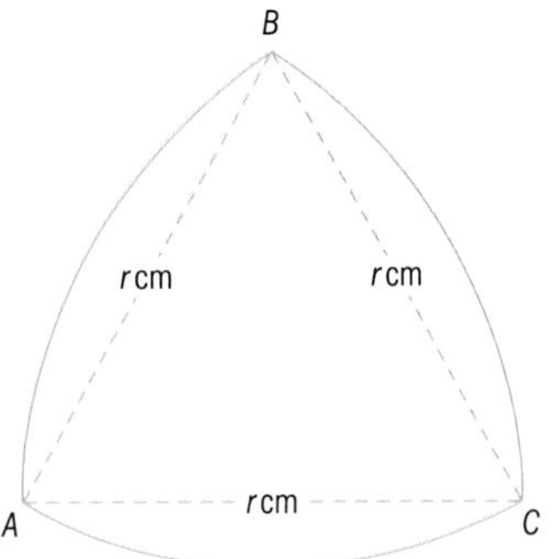

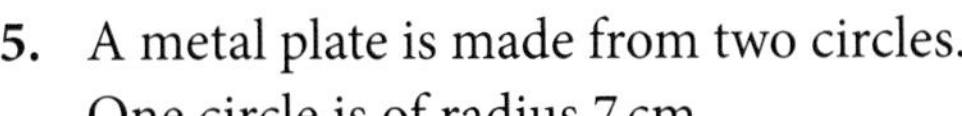

5. A metal plate is made from two circles.
One circle is of radius 7 cm.

The other circle is of radius 5 cm.

Their centres C_1 and C_2 are 9 cm apart.

a) Find, in radians, angles AC_1B and AC_2B.

b) Find the area of the metal plate.

c) Find the perimeter of the metal plate.

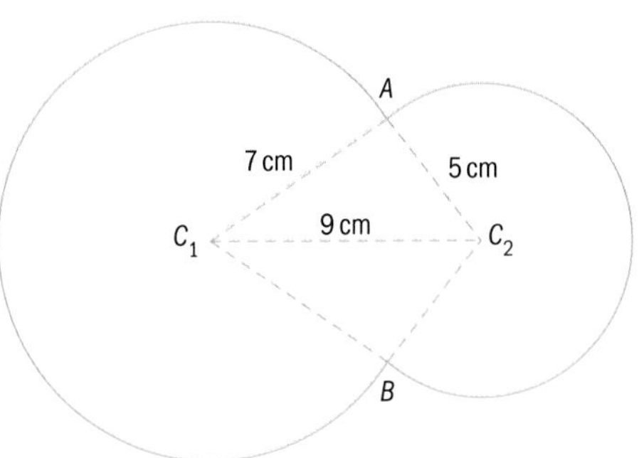

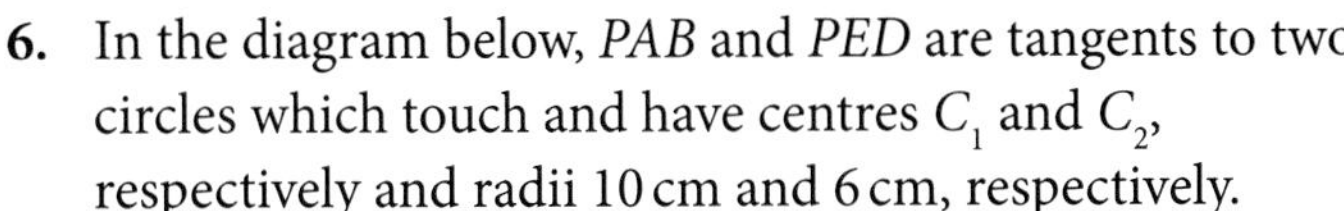

6. In the diagram below, PAB and PED are tangents to two circles which touch and have centres C_1 and C_2, respectively and radii 10 cm and 6 cm, respectively.

a) Calculate the angle APE in radians.

b) Calculate the the total area of the shaded regions.

c) Calculate the total of the perimeters of the shaded regions.

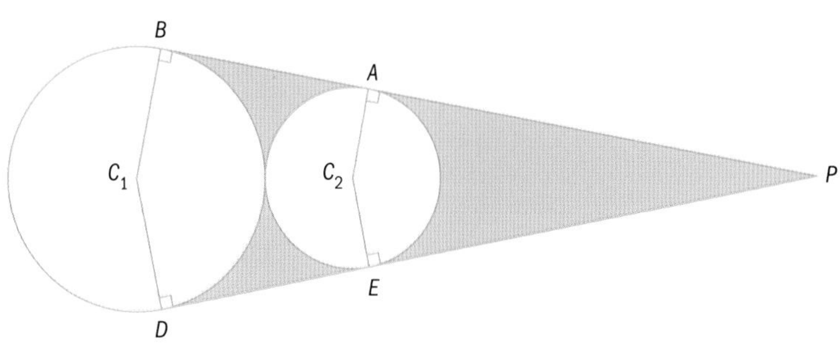

Summary exercise 4

EXAM-STYLE QUESTIONS

1. In the diagram, ABC is a sector of a circle with centre A and radius 8 m. The size of angle BAC is 0.65 rad and D is the mid-point of AC. Find
   a) the length of the arc BC
   b) the area of the shaded region BCD.

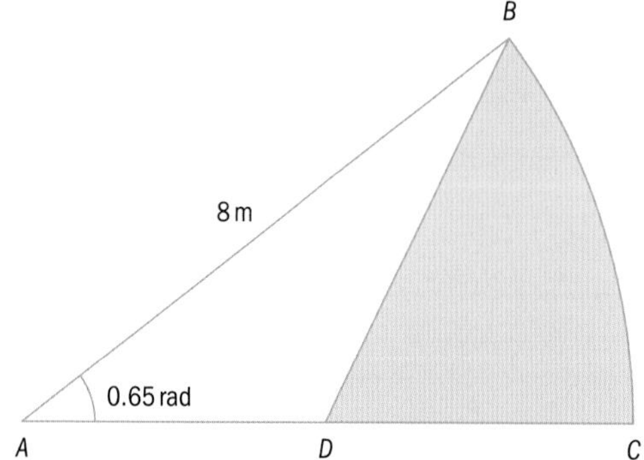

2. In the diagram below, $ABCD$ is a sector of a circle with centre B and radius 5 m.
   Given that the length of the arc ADC is $\frac{15\pi}{4}$ m, find
   a) the exact size of angle ABC in radians
   b) the exact area of the sector $ABCD$.

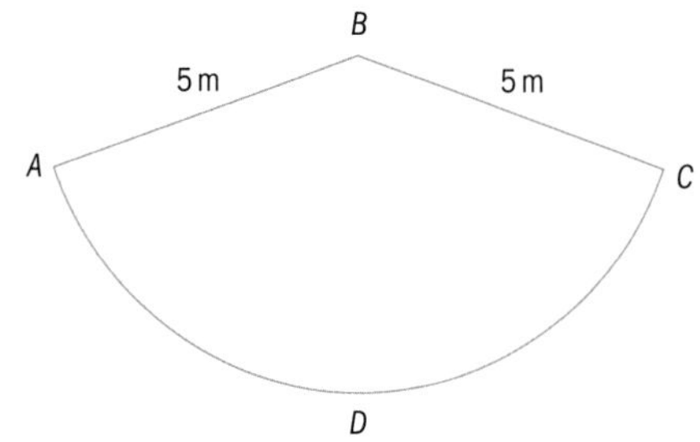

3. A circle has centre C and radius 4 cm. The diagram shows the circle with a point D which lies on the circle. The tangent at D passes through the point E. $EC = \sqrt{65}$ cm. Find
   a) the size in radians of angle DCE
   b) the area of the shaded region DEH.

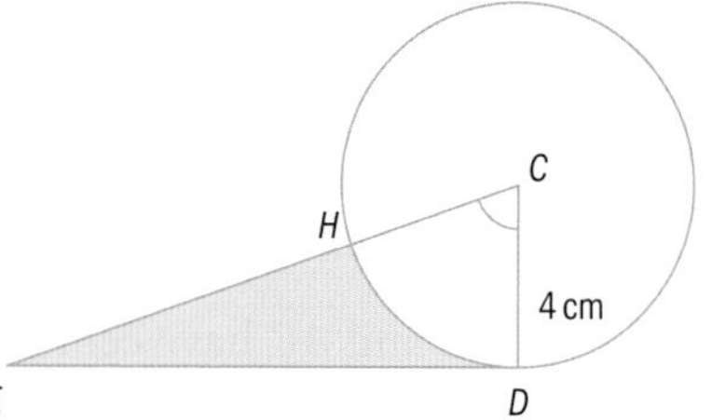

4. In the diagram below, OPQ is a sector of a circle, radius 6 m. The straight line PQ is 8 m long. Calculate
   a) the area of the sector OPQ
   b) the shaded area.

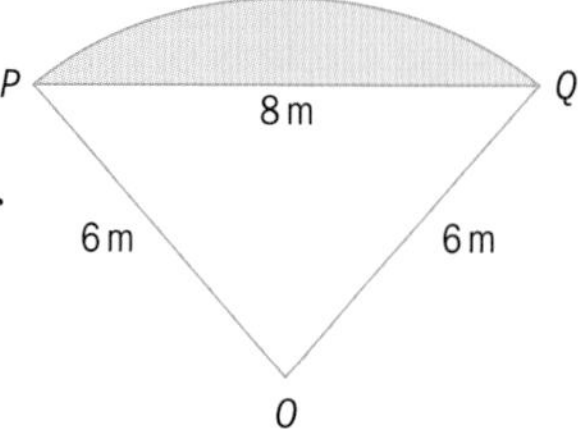

5. The diagram below shows a triangle ABC with $AB = 10$ cm, $AC = 13$ cm and angle $BAC = 0.5$ rad. BD is an arc of a circle with centre A and radius 10 cm. Find
   a) the length of arc BD
   b) the perimeter of the shaded region
   c) the area of the shaded region.

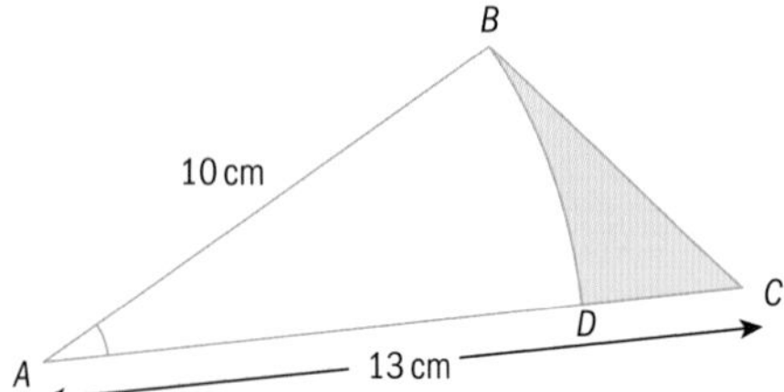

6. The diagram shows a shape $ABCD$. The curve AB is an arc of a circle with centre D. Angle ADB is 0.75 rad. Find

a) the length of the arc AB

b) the area of the sector ADB

c) the area of $ABCD$.

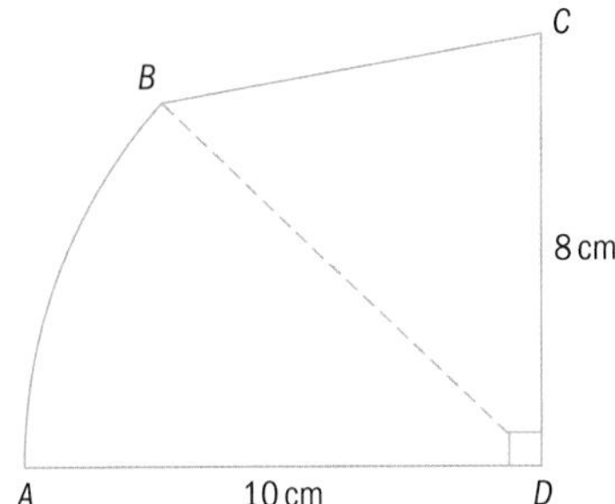

EXAM-STYLE QUESTION

7. A shape is made from joining a triangle ABC to a sector CBD of a circle with radius 7 cm and centre C. The points A, C and D lie on a straight line with $AC = 9$ cm. Angle $BAC = 0.5$ radians. Find

a) the angle ACB giving your answer in radians

b) the area of $ABDC$.

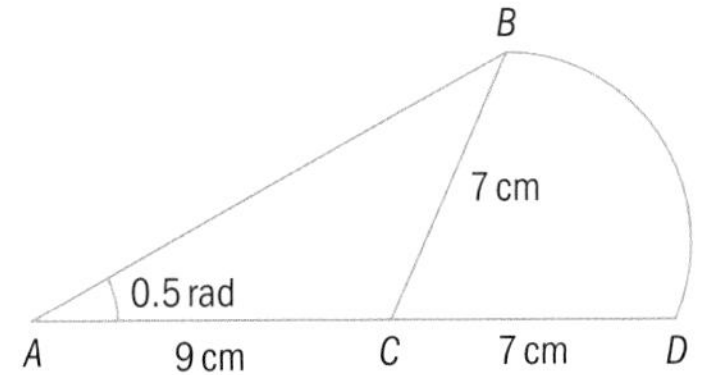

8. The diagram shows the sector OAB of a circle with centre O and radius 20 m.

Angle $AOB = \frac{\pi}{2}$ radians.

The straight line CD divides the sector into two regions which have equal areas and the lengths of OC and OD are equal.

Calculate the exact length of OC.

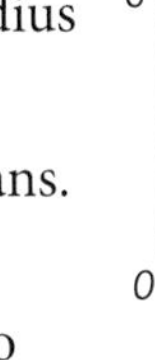

EXAM-STYLE QUESTION

9. The diagram shows a shape made from the arc of a circle with centre O and radius 30 cm and the straight line PQ. The size of angle POQ is $\frac{7\pi}{9}$ radians.

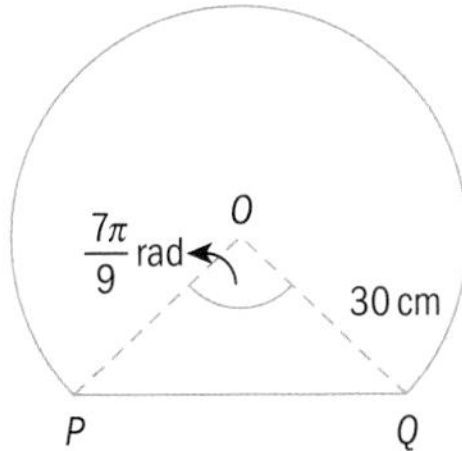

Find the total perimeter of the shape.

10. The diagram shows a sector OPQ of a circle with centre O and radius 12 m.

$PR = QR = OR = 8$ m.

Work out the area of the shaded region PQR.

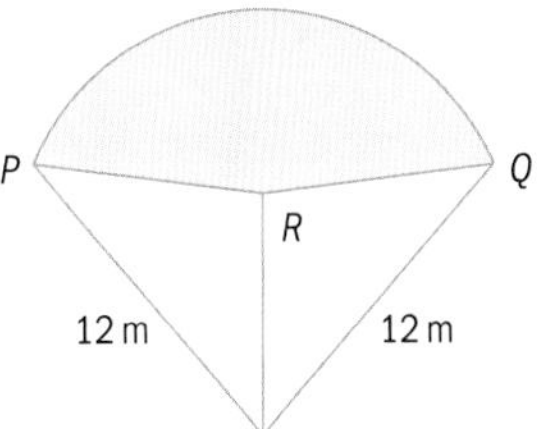

11. AD is an arc of a circle with centre C_1 and radius 6 cm. AB is an arc of a circle with centre C_2. The size of angle AC_2D is $\frac{2\pi}{3}$ radians.

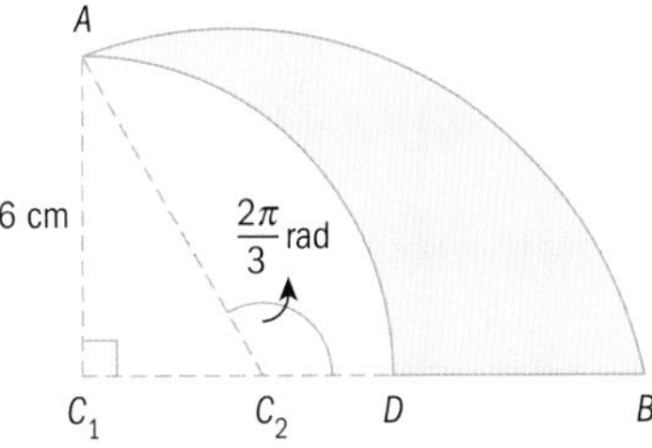

Calculate

a) the radius of the circle with centre C_2

b) the area of the shaded region ABD.

12.

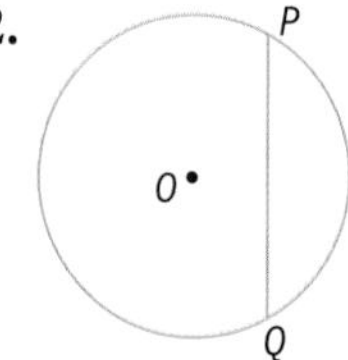

In the diagram, the line PQ divides the circle with centre O and radius r into two segments.

Show that when angle $POQ = 2.6$ radians, the area of the larger segment is approximately twice the area of the smaller segment.

13.

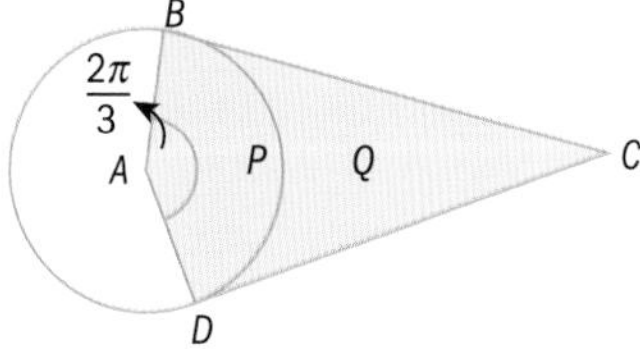

The diagram shows a sector P of a circle with centre A and radius 6 cm. Angle $BAD = \frac{2\pi}{3}$ radians. CB and CD are tangents to the circle at B and D. The shaded region inside the circle is P and the shaded region outside the circle is Q.

i) Find the exact perimeter of Q.

ii) Find the exact area of Q.

14.

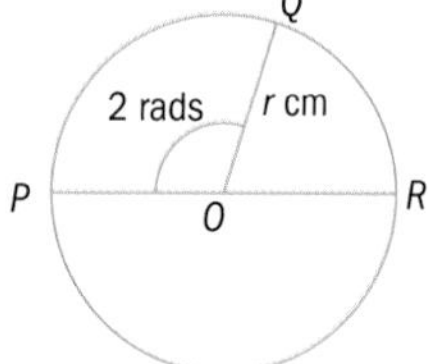

The diagram shows a circle with centre O and radius r cm. The points P, Q, and R lie on the circle. The line PR is a diameter of the circle and angle $POQ = 2$ radians. The area of sector QOR is $30\,\text{cm}^2$ less than the area of sector POQ.

Find the value of r.

15.

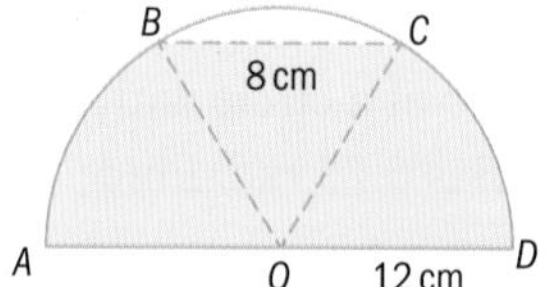

In the diagram, OAB and OCD are sectors of a circle with centre O and radius 12 cm. Angle AOB = angle COD, AOD is a straight line and $BC = 8$ cm.

Calculate

i) angle BOC in radians

ii) the perimeter of the shaded shape

iii) the area of the shaded shape.

16.

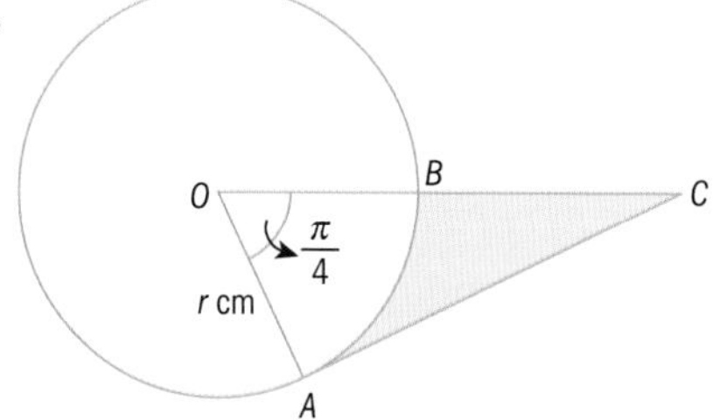

The diagram shows a circle with centre O and radius r cm. Points A and B lie on the circle and angle $AOB = \frac{\pi}{4}$ radians. The tangent to the circle at A intersects the straight line through O and B at the point C.

a) Find expressions in terms of r for

i) the area of sector OAB

ii) the length of AC

iii) the area of triangle OAC.

The area of the shaded region is $100\,\text{cm}^2$.

b) Find the value of r.

Chapter summary

Radians and degrees

- $360° = 2\pi$ radians.
- To convert from degrees to radians, multiply by $\frac{\pi}{180}$.
- To convert from radians to degrees, multiply by $\frac{180}{\pi}$.
-

Degrees	0°	30°	45°	60°	90°	180°	270°	360°
Radians	0	$\frac{\pi}{6}$	$\frac{\pi}{4}$	$\frac{\pi}{3}$	$\frac{\pi}{2}$	π	$\frac{3\pi}{2}$	2π

Arc length and area of a sector

- The angle for a whole circle = 2π radians.
- Arc length = $r\theta$ (θ measured in radians).
- Area of sector = $\frac{1}{2}r^2\theta$ (where θ is measured in radians and r is the radius of the circle).

5 Trigonometry

The Canadarm 2 robotic manipulator is a robotic system which has been operating on the International Space Station since 2001. Its primary uses are moving equipment and supporting astronauts while they are working on the exterior of the station. The machinery is controlled by changing the angles of its joints, which requires repeated use of trigonometric functions.

Objectives

- Sketch and use graphs of the sine, cosine and tangent functions (for angles of any size, and using either degrees or radians).
- Use the exact values of the sine, cosine and tangent of 30°, 45°, 60° and related angles, e.g. $\cos 150° = -\frac{\sqrt{3}}{2}$.
- Use the notations $\sin^{-1} x$, $\cos^{-1} x$ and $\tan^{-1} x$ to denote the principal values of inverse trigonometric relations.
- Use the identities $\frac{\sin\theta}{\cos\theta} \equiv \tan\theta$ and $\sin^2\theta + \cos^2\theta \equiv 1$.
- Find all the solutions of simple trigonometric equations lying in a specified interval.

Before you start

You should know how to:

1. Use the trigonometric ratios and Pythagoras' theorem for a right-angled triangle,

   e.g. $y^2 = 5^2 + 7^2$

   $y = \sqrt{74} = 8.60\text{ cm (3 s.f.)}$

   y, 5 cm, x, 7 cm

   e.g. $\tan x = \frac{5}{7}$

   $x = 35.5°$ (1 d.p.).

2. Convert between degrees and radians,

   e.g. $75° = \frac{\pi}{180} \times 75 = \frac{5\pi}{12} = 1.31$ rad.

Calculators will often need to be converted from degrees to radians and vice versa.

Skills check:

1. Find the side or angle marked x.

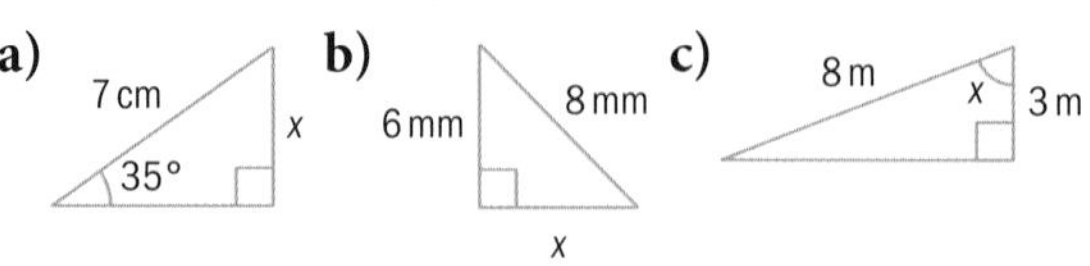

2. **a)** Convert to radians

   i) 90° **ii)** 225° **iii)** 43°.

   b) Convert to degrees

   i) $\frac{3\pi}{4}$ rad **ii)** $\frac{7\pi}{5}$ rad **iii)** 2.5 rad.

3. Solve quadratic equations,

e.g. $x^2 - 4x - 21 = 0$

$(x + 3)(x - 7) = 0; \quad x = -3$ or 7

e.g. $3x^2 + x - 5 = 0$

$$x = \frac{-1 \pm \sqrt{1^2 - 4 \times 3 \times -5}}{6} = \frac{-1 \pm \sqrt{61}}{6}$$

$= -1.47$ or 1.14 (3 s.f.).

3. Solve

a) $2x^2 - 7x - 4 = 0$

b) $x^2 - 5x + 2 = 0$.

5.1 Exact values of trigonometric functions

The unit circle can be used to define the trigonometric ratios.
The angle θ is measured in an anticlockwise direction from the positive x-axis.
$\cos\theta$ can be defined as the x coordinate of P.
$\sin\theta$ can be defined as the y coordinate of P.
$\tan\theta$ can be defined as $\dfrac{\sin\theta}{\cos\theta}$.
The diagram can be used to write down the values of sin, cos and tan for 0°, 90°, 180°, 270° and 360°.

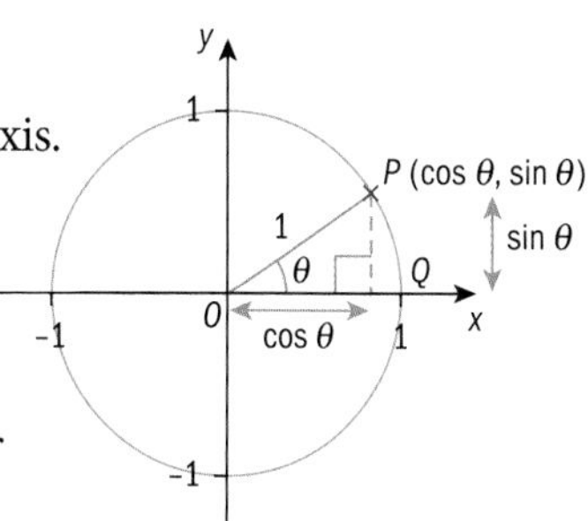

Note:

The word 'cosine' means 'the **sine** of the **co**mplement (of θ)'; 'sineco θ' eventually becoming 'cosine θ'.

The cosine of angle *POQ* is equal to the sine of the complementary angle (*OPQ* in the diagram).
So cosine θ = sine $(90 - \theta)$

or $\cos\theta = \sin(90 - \theta)$

This result can be very useful in helping us to solve problems in trigonometry.

Example 1

Without using a calculator, write down the values of

a) $\cos 0°$ **b)** $\sin 90°$ **c)** $\cos 180°$ **d)** $\cos 270°$ **e)** $\tan 0°$ **f)** $\tan 90°$.

a) $\cos 0° = 1$ — $\cos 0°$ is the x-coordinate of the point P when $\theta = 0°$.

b) $\sin 90° = 1$ — $\sin 90°$ is the y-coordinate of the point P when $\theta = 90°$.

c) $\cos 180° = -1$ — $\cos 180°$ is the x-coordinate of the point P when $\theta = 180°$.

d) $\cos 270° = 0$ — $\cos 270°$ is the x-coordinate of the point P when $\theta = 270°$.

e) $\tan 0° = \dfrac{\sin 0}{\cos 0} = \dfrac{0}{1} = 0$ — Use $\tan\theta = \dfrac{\sin\theta}{\cos\theta}$

f) $\tan 90° = \dfrac{\sin 90°}{\cos 90°} = \dfrac{1}{0}$ — Division by 0 is undefined.

$\tan 90°$ is undefined (or infinitely large)

Some exact values of sine, cosine and tangent can be found easily from the triangles below, and are also shown in the table.

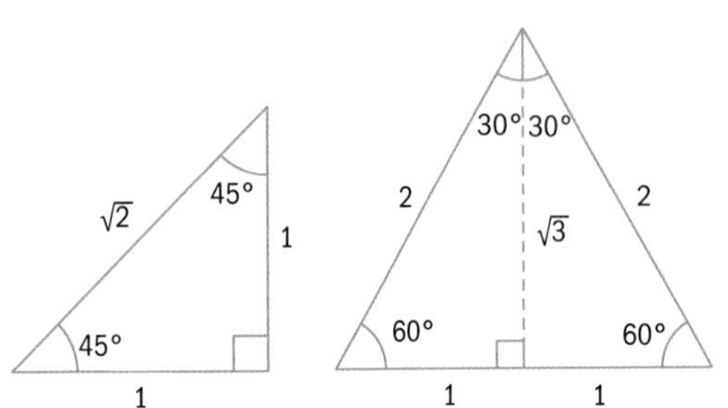

Degrees	Radians	sin	cos	tan
30°	$\frac{\pi}{6}$	$\frac{1}{2}$	$\frac{\sqrt{3}}{2}$	$\frac{1}{\sqrt{3}}$
45°	$\frac{\pi}{4}$	$\frac{1}{\sqrt{2}}$	$\frac{1}{\sqrt{2}}$	1
60°	$\frac{\pi}{3}$	$\frac{\sqrt{3}}{2}$	$\frac{1}{2}$	$\sqrt{3}$

The sine, cosine and tangent of other angles in the range 0° to 360° can be found.

For every angle between 0° and 90° there is a related angle in each of the other three quadrants. These related angles give the same numerical value for each trigonometric function but the signs change.

For example, the angles related to 30° are

$180° - 30° = 150°$
$180° + 30° = 210°$
$360° - 30° = 330°$

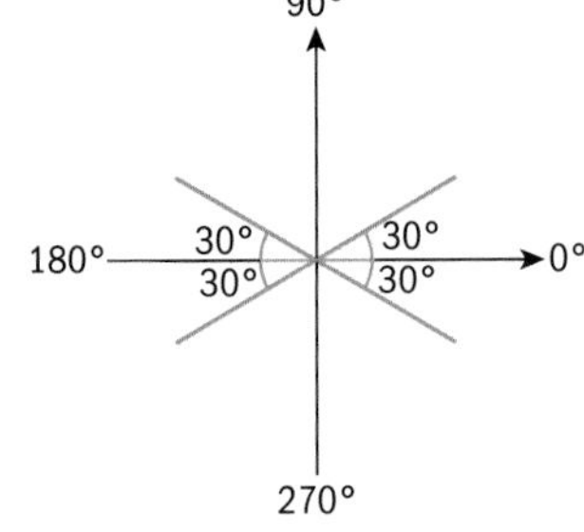

Remember that angles are measured anticlockwise from the positive x-axis.

This diagram shows which functions are positive in each quadrant.

Sine	All
Tangent	Cosine

We can write down the sine, cosine and tangent for related angles, so for example

$\sin 150° = \sin 30° = \frac{1}{2}$ $\quad \cos 150° = -\cos 30° = -\frac{\sqrt{3}}{2}$ $\quad \tan 150° = -\tan 30° = -\frac{1}{\sqrt{3}}$

$\sin 210° = -\sin 30° = -\frac{1}{2}$ $\quad \cos 210° = -\cos 30° = -\frac{\sqrt{3}}{2}$ $\quad \tan 210° = \tan 30° = \frac{1}{\sqrt{3}}$

$\sin 330° = -\sin 30° = -\frac{1}{2}$ $\quad \cos 330° = \cos 30° = \frac{\sqrt{3}}{2}$ $\quad \tan 330° = -\tan 30° = -\frac{1}{\sqrt{3}}$

Use the table of values at the top of the page and the coordinate diagram to help you.

Example 2

Find the exact values of

a) $\sin 120°$ **b)** $\cos 120°$ **c)** $\tan 120°$ **d)** $\sin 225°$ **e)** $\cos 225°$ **f)** $\tan 300°$.

a) $\sin 120° = \sin 60°$
$= \frac{\sqrt{3}}{2}$

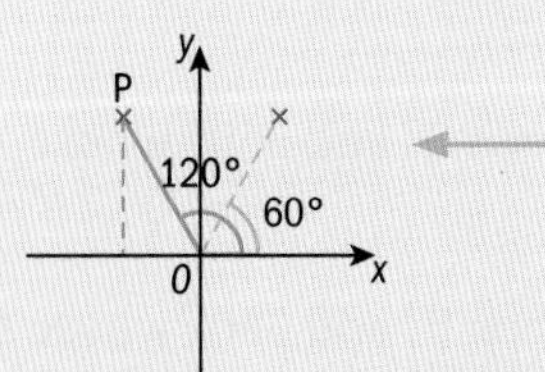

Find the y-coordinate of P when $\theta = 120°$, which is equal to the y-coordinate when $\theta = 60°$.

b) $\cos 120° = -\cos 60°$
$= -\frac{1}{2}$

Find the x-coordinate of P when $\theta = 120°$, which is the negative of the x-coordinate when $\theta = 60°$.

c) $\tan 120° = -\tan 60°$
$= -\sqrt{3}$

When $\theta = 120°$, $\sin\theta > 0$, $\cos\theta < 0$ so $\tan\theta < 0$.

d) $\sin 225° = -\sin 45°$
$= -\frac{1}{\sqrt{2}}$

Use the angle with the (negative) x-axis.

e) $\cos 225° = -\cos 45°$
$= -\frac{1}{\sqrt{2}}$

f) $\tan 300° = -\tan 60°$
$= -\sqrt{3}$

Use the angle with the (positive) x-axis.

Exercise 5.1

1. Find the exact values of

a) $\cos 240°$ **b)** $\tan 135°$ **c)** $\sin 300°$ **d)** $\cos 315°$.

2. Find the exact values of

a) $\sin\frac{3\pi}{4}$ **b)** $\sin\frac{4\pi}{3}$ **c)** $\sin\frac{3\pi}{2}$ **d)** $\tan\frac{5\pi}{3}$.

3. **a)** Without using a calculator, state whether each of these is positive or negative.

i) $\sin 130°$ **ii)** $\cos 130°$ **iii)** $\sin 255°$ **iv)** $\cos 255°$

b) Use your calculator to find the value of

i) $\sin 130°$ **ii)** $\cos 130°$ **iii)** $\sin 255°$ **iv)** $\cos 255°$.

4. Express the following in terms of the related acute angle.

a) $\sin 132°$ **b)** $\cos 310°$ **c)** $\tan 215°$ **d)** $\sin 220°$

e) $\cos 153°$ **f)** $\tan 148°$ **g)** $\cos 195°$ **h)** $\sin 335°$

5. a) Use your calculator to find the angle (to the nearest degree) between $0°$ and $90°$ whose sine is 0.36.

   b) Hence find another angle between $0°$ and $360°$ whose sine is 0.36.

6. Find two angles (to the nearest degree) in the range $0° < x° < 360°$ such that $\cos x° = 0.3$.

7. Find all the angles (to the nearest 0.1°) between $0°$ and $360°$ whose cosine is 0.7660.

8. Find all the angles (to the nearest 0.1°) between $0°$ and $360°$ whose sine is -0.3636.

9. Solve each of these equations, giving all solutions between $0°$ and $360°$ to the nearest degree.

   a) $\sin x° = 0.9$ b) $\cos x° = 0.9$ c) $\sin x° = -0.6$

   d) $\cos x° = 0.33$ e) $\tan x° = 0.25$ f) $\tan x° = -0.44$

10. Solve each of these equations, giving all solutions between 0 and 2π in radians correct to 3 significant figures.

    a) $\sin x = 0.4$ b) $\cos x = 0.4$ c) $\sin x = -0.8$

    d) $\cos x = -0.21$ e) $\tan x = 0.75$ f) $\tan x = -0.36$

5.2 Graphs of trigonometric functions

In Section 5.1 we saw how to define sin, cos and tan using the unit circle. These definitions apply for all values of θ, both positive (anticlockwise) and negative (clockwise).

An angle outside the interval $0° \leq x° < 360°$ has an equivalent angle in the interval $0° \leq x° < 360°$.

For example, an angle of $420°$ is equivalent to an angle of $60°$
and an angle of $-120°$ is equivalent to an angle of $240°$.

The graph of $y = \sin x$ shows that the curve is infinite but repeats every $360°$ (or 2π radians). We say $y = \sin x$ is periodic and has a **period** of $360°$ (or 2π radians).

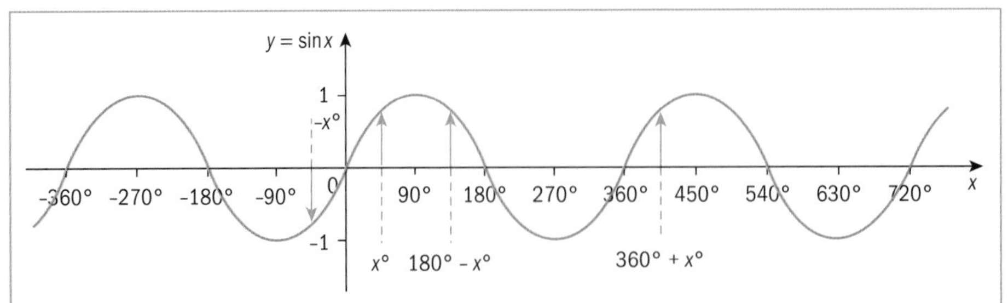

The graph can be drawn using degrees or radians.

From the graph above we can write down some relationships.

For example, $\sin(-x°) = -\sin x°$

$\sin(180 - x)° = \sin x°$

$\sin(360 + x)° = \sin x°$

These equivalences can also be seen from the unit circle shown in Section 5.1.

The graph of $y = \cos x$ shown below behaves in a similar way.

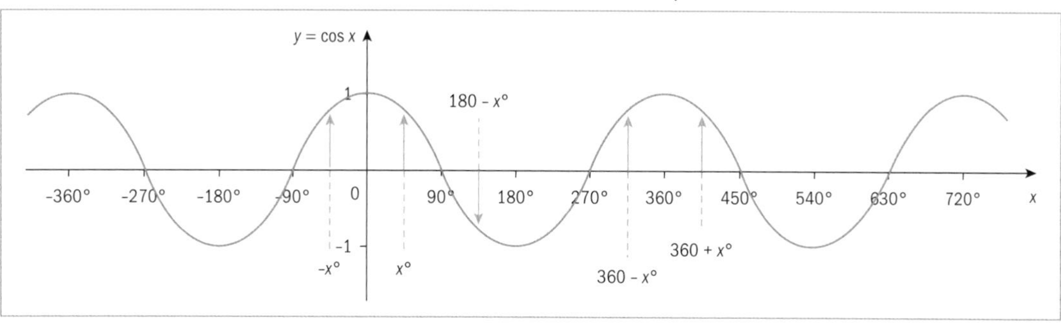

$\cos(-x°) = \cos x°$

$\cos(180 - x)° = -\cos x°$

$\cos(360 - x)° = \cos x°$

$\cos(360 + x)° = \cos x°$

We have seen that $\tan x = \dfrac{\sin x}{\cos x}$.

There is a problem when $x = 90°, 270°, \dots$ because $\cos x° = 0$ at these points and so the value of $\tan x$ is undefined (or infinite) at these points. There are **asymptotes** at these values of x on the graph of $y = \tan x$.

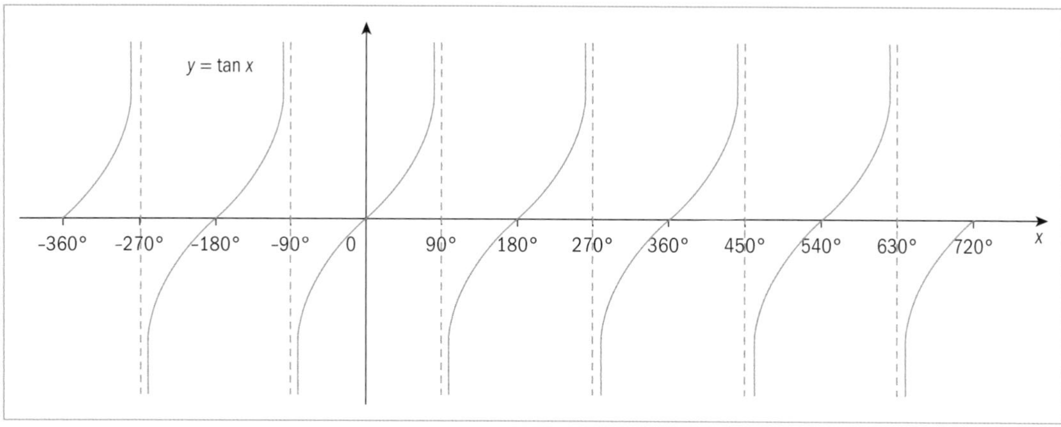

The graph of $y = \tan x$ is periodic and has a period of 180° (or π radians).

Examples of relationships for this function are

$\tan(-x°) = -\tan x°$

$\tan(90 + x)° = -\tan(90 - x)°$

$\tan(180 + x)° = \tan x°$

Use the graph to check that you understand why.

The graphs of sine, cosine and tangent can be used to help you find all the angles, x, within a given interval for which $\sin x = k$ or $\cos x = k$ $(-1 \le k \le 1)$ or $\tan x = k$ (k any real number).

From the sine and cosine graphs we see that there are two angles, x, between 0° and 360° for each value of k. Further angles can be found by taking each of these two values and adding or subtracting multiples of 360°.

For the tangent graph we see that there is one angle, x, between −90° and 90° for each value of k. Further angles can be found by taking this value and adding or subtracting multiples of 180°.

A calculator will give you only one angle (called the principal value), you can sketch the graph to find other angles. Radians may be used instead of degrees.

Example 3

Find all the angles x (to the nearest degree), where $0° < x < 360°$, such that $\sin x = 0.88$.

$x = \sin^{-1}(0.88)$
$= 62°$ ← Use a calculator to find the principal value.

$x = 62°, 180° - 62°$
$x = 62°, 118°$ ← Use a graph sketch to find the second angle in the stated interval.

Example 4

Solve $\tan x = 2.35$ for $0° < x < 720°$.
Give your answers correct to one decimal place.

$x = \tan^{-1}(2.35)$
$= 66.9°$ (1 d.p.) ← This is the principal value given by a calculator.

$x = 66.9°, 180° + 66.9°, 360° + 66.9°, 540° + 66.9°$
$x = 66.9°, 246.9°, 426.9°, 606.9°$ (1 d.p.) ← Add on multiples of 180° to give all values in the interval $0° < x < 720°$.

Exercise 5.2

Use graph sketches in each question to help you find the angles required.

1. Given that $\sin 40° = 0.643$ (3 s.f.)
   a) write down another angle between 0° and 360° whose sine is 0.643
   b) write down two angles between 360° and 720° whose sine is 0.643.
2. Given $\tan \frac{\pi}{4} = 1$, write down all the angles between 0 and 6π whose tangent is 1.
3. Find all the angles between 0° and 360° whose cosine is −0.766.
4. Find all the angles between 0° and 360° whose sine is −0.25.
5. Solve each of these equations, giving all solutions between 0° and 360°.
   a) $\sin x = 0.384$ b) $\tan x = 1.988$ c) $\cos x = 0.379$ d) $\sin x = -0.2$

6. Solve each of these equations, giving all solutions between 0 and 4π.

a) $\sin x = \frac{1}{2}$ b) $\cos\theta = -\frac{1}{\sqrt{2}}$ c) $\sin x = -1$ d) $\tan\theta = -1$

7. a) Express $\sin(180° + x)$ in terms of $\sin x$. b) Express $\cos(180° - x)$ in terms of $\cos x$.

c) Express $\tan(180° + x)$ in terms of $\tan x$. d) Express $\tan(360° - x)$ in terms of $\tan x$.

8. a) Express $\cos(3\pi + x)$ in terms of $\cos x$. b) Express $\sin(x + 4\pi)$ in terms of $\sin x$.

c) Express $\sin(x - \pi)$ in terms of $\sin x$. d) Express $\tan(x - \pi)$ in terms of $\tan x$.

9. Solve the equation $3\sin x = 2\cos x$, giving all solutions between 0° and 360°.

Hint: Use $\tan x = \frac{\sin x}{\cos x}$

5.3 Inverse trigonometric functions

In Chapter 2 we met the idea of an inverse function. We learnt that we sometimes have to restrict the domain of a function so that it is one-to-one and therefore has an inverse which is itself a function.

We also know that we can draw the graph of any inverse function by reflecting the graph of the function in the line $y = x$.

Here are the graphs of $\sin x$, $\cos x$ and $\tan x$, together with the graphs of the inverse functions $\sin^{-1}x$, $\cos^{-1}x$, and $\tan^{-1}x$.

Note: The range of a function may be written in different ways.
For example, the range of $f(x) = \sin x$ can be written in any of the forms
$-1 \le \sin x \le 1$ or $-1 \le f(x) \le 1$
or $-1 \le y \le 1$

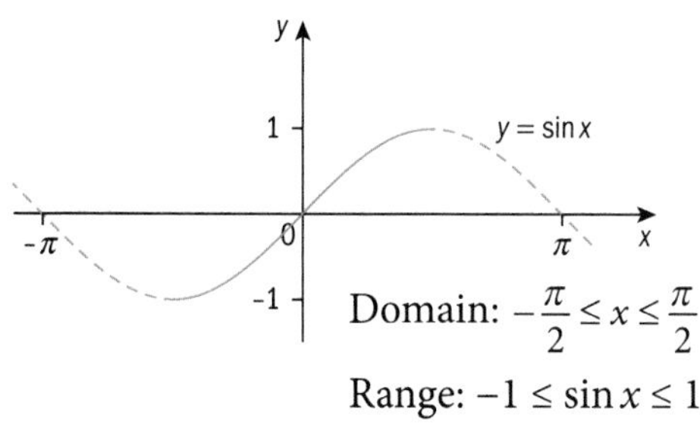

Domain: $-\frac{\pi}{2} \le x \le \frac{\pi}{2}$

Range: $-1 \le \sin x \le 1$

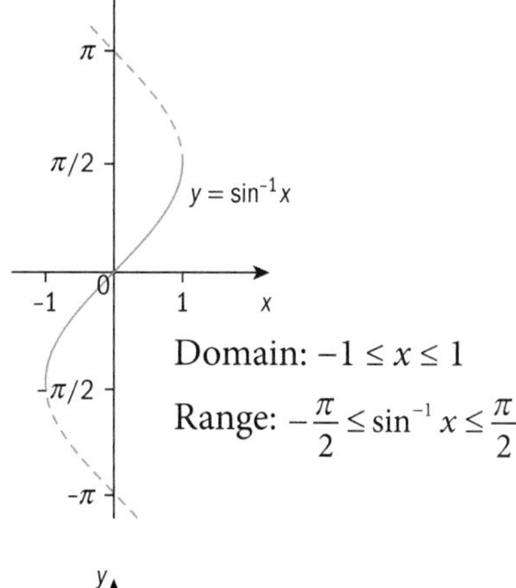

Domain: $-1 \le x \le 1$

Range: $-\frac{\pi}{2} \le \sin^{-1} x \le \frac{\pi}{2}$

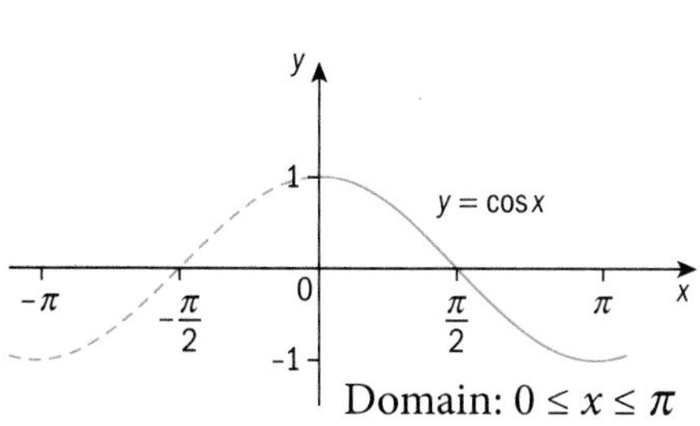

Domain: $0 \le x \le \pi$

Range: $-1 \le \cos x \le 1$

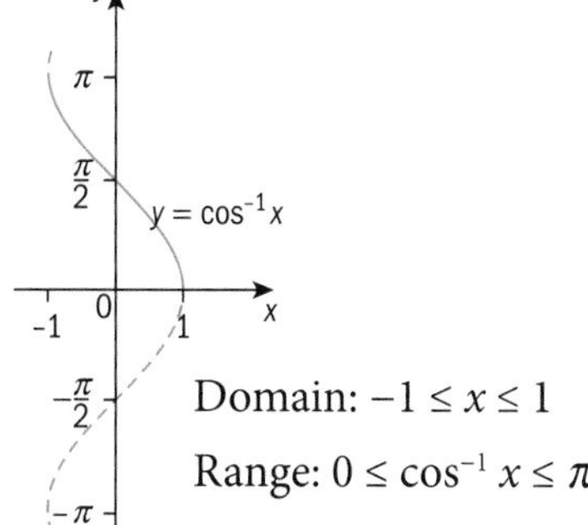

Domain: $-1 \le x \le 1$

Range: $0 \le \cos^{-1} x \le \pi$

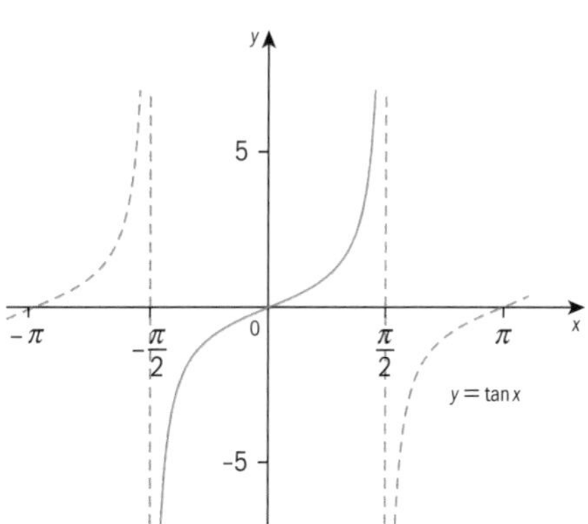

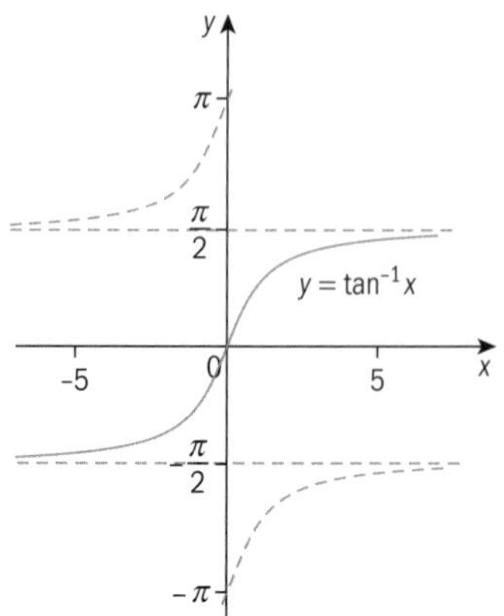

Domain: $-\frac{\pi}{2} < x < \frac{\pi}{2}$

Range: $\tan x \in \mathbb{R}$

Domain: $x \in \mathbb{R}$

Range: $-\frac{\pi}{2} < \tan^{-1} x < \frac{\pi}{2}$

In Section 5.2 we saw how to solve simple trigonometric equations. We used a calculator to find the principal angle for any given value of $\sin x$, $\cos x$ or $\tan x$. This principal angle is the angle which lies in the range of the inverse trigonometric function (and in the restricted domain of the function).

A calculator will give only this principal value. In order to find all the solutions of a trigonometric equation we use a graph sketch or a diagram of the unit circle.

Example 5

a) Write down the value of

i) $\sin\left(\sin^{-1}\left(\frac{\sqrt{3}}{2}\right)\right)$ **ii)** $\sin\left(\sin^{-1}\left(-\frac{\sqrt{3}}{2}\right)\right)$ **iii)** $\tan(\tan^{-1}1)$ **iv)** $\cos\left(\cos^{-1}\left(\frac{1}{\sqrt{2}}\right)\right)$.

b) Is $\sin(\sin^{-1}(x)) = x$ for all values of x?

a) i) $\sin\left(\sin^{-1}\left(\frac{\sqrt{3}}{2}\right)\right) = \sin\left(\frac{\pi}{3}\right) = \frac{\sqrt{3}}{2}$

ii) $\sin\left(\sin^{-1}\left(-\frac{\sqrt{3}}{2}\right)\right) = \sin\left(-\frac{\pi}{3}\right) = -\frac{\sqrt{3}}{2}$

iii) $\tan(\tan^{-1}1) = \tan\left(\frac{\pi}{4}\right) = 1$

iv) $\cos\left(\cos^{-1}\left(\frac{1}{\sqrt{2}}\right)\right) = \cos\left(\frac{\pi}{4}\right) = \frac{1}{\sqrt{2}}$

b) Yes

Exercise 5.3

1. Without using a calculator write down in degrees, where possible, the principal value of

   a) $\sin^{-1} 0$ b) $\sin^{-1}(-1)$ c) $\sin^{-1}\left(-\frac{\sqrt{3}}{2}\right)$ d) $\sin^{-1}\left(\frac{1}{\sqrt{2}}\right)$

   e) $\cos^{-1} 1$ f) $\cos^{-1}\left(\frac{\sqrt{3}}{2}\right)$ g) $\cos^{-1} 2$ h) $\cos^{-1}(-1)$

   i) $\tan^{-1} 1$ j) $\tan^{-1} 0$ k) $\tan^{-1}\left(\frac{1}{\sqrt{3}}\right)$ l) $\tan^{-1}\left(-\frac{1}{\sqrt{3}}\right)$.

2. Use a calculator to find, in degrees correct to 1 decimal place, the principal value of

   a) $\cos^{-1}(0.6)$ b) $\sin^{-1}(0.25)$ c) $\tan^{-1}\left(\frac{19}{20}\right)$

   d) $\tan^{-1}(10)$ e) $\cos^{-1}(-0.83)$ f) $\sin^{-1}\left(-\frac{3}{8}\right)$.

3. a) Use a calculator to find, in radians correct to 3 significant figures, the principal value of

   i) $\tan^{-1}(0.4)$ ii) $\tan^{-1}(-0.4)$ iii) $\sin^{-1}\left(\frac{2}{3}\right)$

   iv) $\sin^{-1}\left(-\frac{2}{3}\right)$ v) $\cos^{-1}(0.62)$ vi) $\cos^{-1}(-0.62)$.

   b) What do you notice about your answers to **(a)(i)** and **(a)(ii)**?

   c) What do you notice about your answers to **(a)(iii)** and **(a)(iv)**?

   d) What do you notice about your answers to **(a)(v)** and **(a)(vi)**?

4. Solve each of the equations where possible.

   a) $\sin^{-1} x = \frac{\pi}{6}$ b) $\sin^{-1} x = \frac{\pi}{3}$ c) $\sin^{-1} x = -\frac{\pi}{6}$ d) $\sin^{-1} x = \pi$

5. a) Write down the value of

   i) $\sin^{-1}\left(\sin\frac{\pi}{3}\right)$ ii) $\sin^{-1}\left(\sin\frac{7\pi}{3}\right)$ iii) $\cos^{-1}\left(\cos\left(\frac{\pi}{4}\right)\right)$ iv) $\cos^{-1}\left(\cos\left(-\frac{\pi}{4}\right)\right)$.

   b) Is $\sin^{-1}(\sin x) = x$ for all values of x?

6. Here is the graph of $f(x) = \cos 2x$.

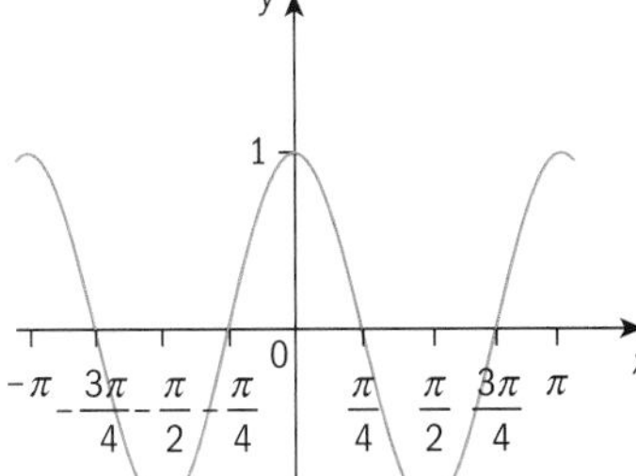

   a) Which of the following domains ensures that the function $f(x) = \cos 2x$ is one-to-one?

   $-\frac{\pi}{4} \le x \le \frac{\pi}{4}$ $0 \le x \le \pi$ $0 \le x \le \frac{\pi}{2}$

   b) What is the range of $f(x)$ for this domain?

   c) Find an expression for a possible inverse function for $f(x) = \cos 2x$.

   d) Write down the domain and range of the inverse function $f^{-1}(x)$.

5.4 Composite graphs

The graph of $y = \sin x$ has period 360°.
The range of values for $y = \sin x$ is $-1 \le y \le 1$.

The graph of $y = a\sin x$ has period 360°.
It is a stretch of $y = \sin x$, factor a, parallel to the y-axis.
The range of values for $y = a\sin x$ is $-a \le y \le a$.

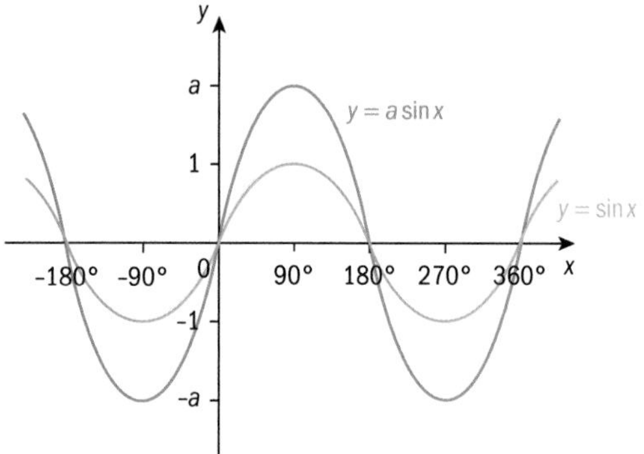

The graph of $y = \sin bx$ has period $\frac{360°}{b}$.
It is a stretch of $y = \sin x$, factor $\frac{1}{b}$, parallel to the x-axis.
The range of values for $y = \sin bx$ is $-1 \le y \le 1$.

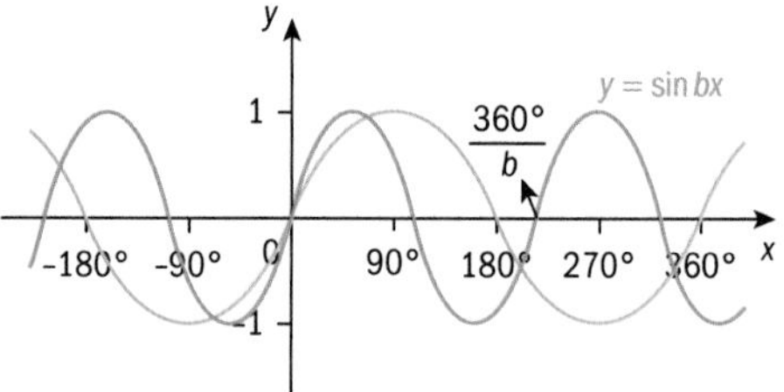

The graph of $y = \sin(x + c°)$ has period 360°.
It is a translation of $y = \sin x$ by $\begin{pmatrix} -c \\ 0 \end{pmatrix}$.
The range of values for $y = \sin(x + c°)$ is $-1 \le y \le 1$.

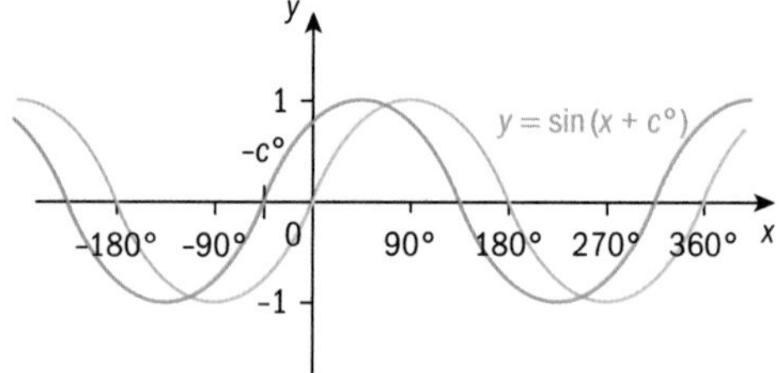

The graph of $y = \sin x + d$ has period 360°.
It is a translation of $y = \sin x$ by $\begin{pmatrix} 0 \\ d \end{pmatrix}$.
The range of values for $y = \sin x + d$ is $-1 + d \le y \le 1 + d$.

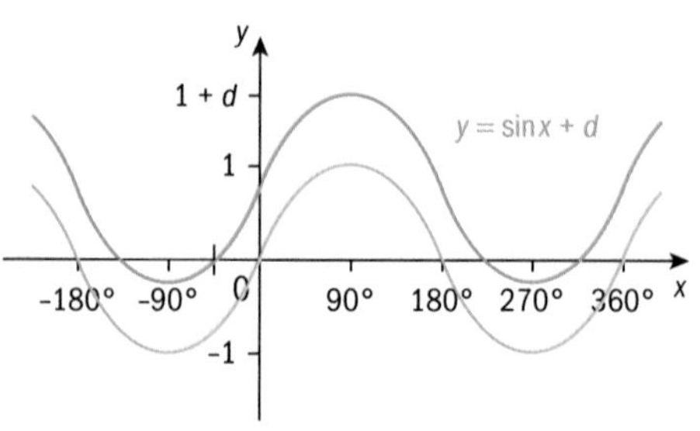

The graphs of $y = \cos x$ and $y = \tan x$ can be transformed in the same way.

Example 6

Sketch the graph of $y = 3\cos x$ for $0° \le x \le 360°$.

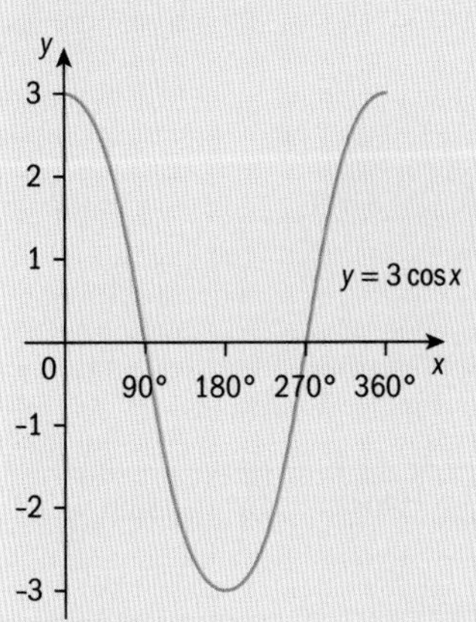

Stretch the graph of $y = \cos x$ parallel to the y-axis, factor 3.

Example 7

Sketch the graph of $y = \sin 2x$ for $0 \le x \le 2\pi$.

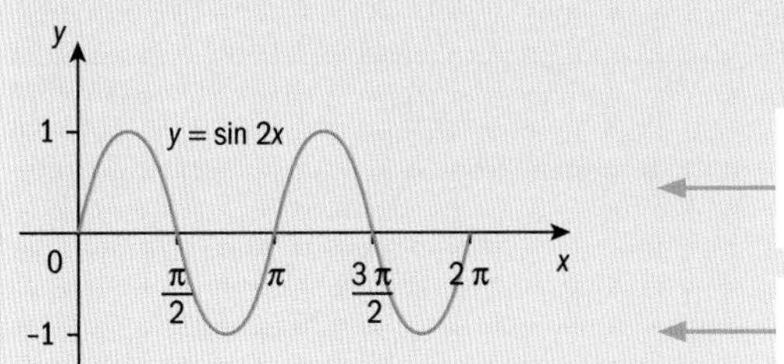

Stretch the graph of $y = \sin x$ parallel to the x-axis, factor $\frac{1}{2}$.

Use radians to label the x-axis.

Exercise 5.4

1. Find the period of each of these functions.

a) $y = \sin 2x$ **b)** $y = \cos 5x$ **c)** $y = \cos x + 1$

d) $y = 5\sin(x - 30)°$ **e)** $y = \tan \frac{1}{2}x$ **f)** $y = 2\tan 3x - 4$

g) $y = \tan(x + 45)°$ **h)** $y = 3\sin(2x - 60)°$

2. Sketch, on separate diagrams, the graphs of

a) $y = \sin 3x$ for $0° \le x \le 360°$

b) $y = -\cos 2x$ for $0° \le x \le 360°$

c) $y = \tan \frac{1}{2}x$ for $-360° \le x \le 360°$

d) $y = 2\sin 4x$ for $0° \le x \le 180°$

e) $y = 3\tan(x + 30°)$ for $0° \le x \le 360°$

f) $y = 3\sin 2x - 1$ for $0° \le x \le 360°$.

3. Sketch, on separate diagrams, the graphs of the following functions for $0 \le \theta \le 2\pi$.

a) $y = 6\cos\theta + 2$ **b)** $y = \tan\left(\theta - \frac{\pi}{4}\right)$

c) $y = \sin\left(\theta + \frac{\pi}{2}\right)$ **d)** $y = \cos\left(2\theta - \frac{\pi}{2}\right)$

4. Find an equation for each graph.

a)

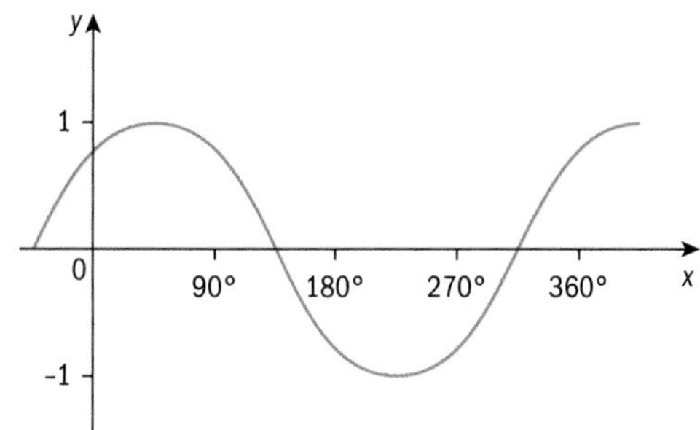

b)

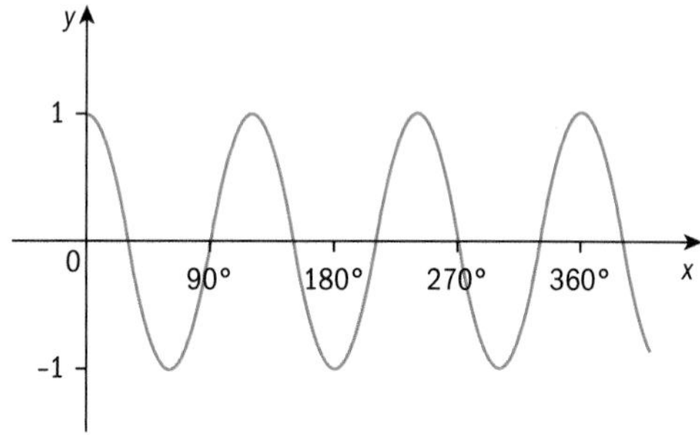

c)

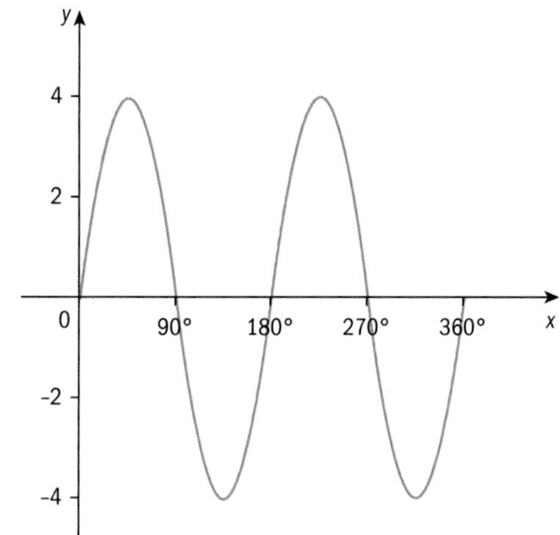

d)

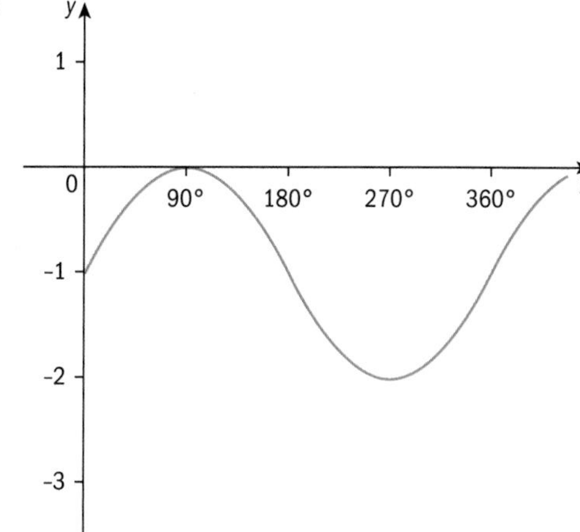

5. a) Sketch on the same diagram the graphs of $y = \sin 2\theta$ and $y = \cos\theta$ for $0° \le \theta \le 180°$.

b) State the number of roots of the equation $\sin 2\theta = \cos\theta$ for which $0° \le \theta \le 180°$.

c) Find the roots of the equation $\sin 2\theta = \cos\theta$ for which $0° \le \theta \le 360°$.

5.5 Trigonometric equations

We can use related angles to help us solve more complex trigonometric equations.

Example 8

Solve the equation $4\sin x + 3 = 5$ for $0° \leq x \leq 360°$.

$4\sin x + 3 = 5$

$4\sin x = 2$ ← First find the value of $\sin x$.

$\sin x = \frac{1}{2}$

$x = 30°, 180° - 30°$

$x = 30°, 150°$

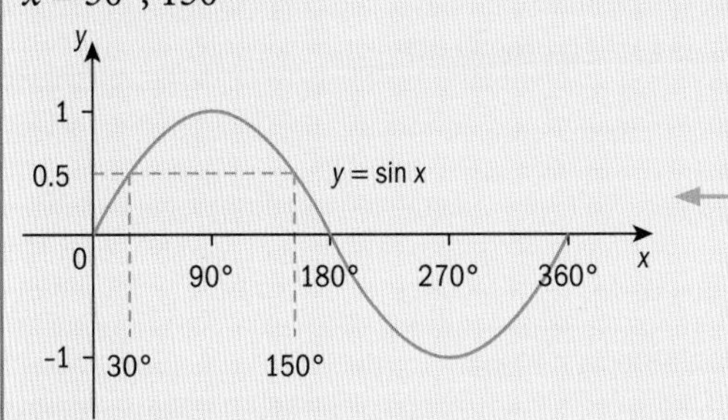

← Use a sketch graph to help you find all the angles in the required interval.

Example 9

Solve the equation $3\cos 2\theta = -1$ for $0° \leq \theta \leq 360°$.

Values for $0° \leq \theta \leq 360°$ are required, so values for $0° \leq 2\theta \leq 720°$ will be needed.

$3\cos 2\theta = -1$

$\cos 2\theta = -\frac{1}{3}$

$2\theta = 109.47...°, \quad 360° - 109.47...°$ ← Use your calculator to find the principal value 109.47...°

$360° + 109.47...°, \quad 720° - 109.47...°$

$2\theta = 109.47...°, \quad 250.52...°$

$469.47...°, \quad 610.52...°$

$\theta = 54.7°, 125.3°, 234.7° \ 305.3°$

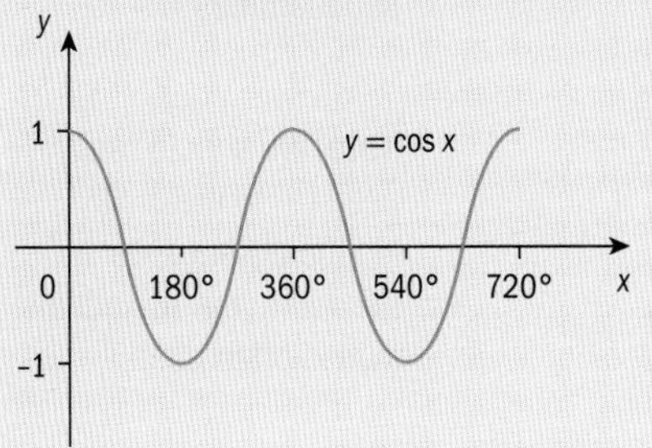

← Use the graph to find the other angles.

Note: You should normally give angles in degrees correct to 1 decimal place, so make sure you work with more than this to avoid rounding errors. Remember to work out all the values from the graph before you divide by 2.

Exercise 5.5

1. Solve these equations for $0° \le x \le 360°$.

   a) $4\sin x = 3$ b) $5\tan x - 1 = 9$ c) $6\cos x + 5 = 8$

   d) $8\sin(x + 20°) = 5$ e) $10 - 3\cos x = 9$ f) $9\sin(x - 15°) = -4$

2. In the following equations, find all the values of θ between $0°$ and $360°$.

   a) $\cos\theta = 0.776$ b) $\sin 2\theta = -0.364$ c) $\tan 3\theta = 1.988$

   d) $\cos\frac{\theta}{2} = -0.379$ e) $\tan\frac{\theta}{2} = -1.030$ f) $\sin\frac{3\theta}{2} = 0.664$.

3. Solve the following equations for $0 \le x \le 2\pi$.

   a) $3\tan x + 2 = 5$ b) $\sin\left(x - \frac{\pi}{6}\right) = \frac{\sqrt{3}}{2}$ c) $\sin 2x = \frac{1}{2}$

   d) $6\cos 3x - 3 = 0$ e) $7 - 2\tan 4x = 13$ f) $10\cos\frac{x}{3} = 1$

4. Solve $6\cos 30x - 3 = 0$ for $0° \le x < 24°$.

5. Solve $3\tan x = \sqrt{27}$ for $-180° \le x < 180°$.

6. Solve $5 + 2\cos\left(3\theta - \frac{\pi}{4}\right) = 6$ for $-\pi \le \theta < \pi$.

5.6 Trigonometric identities

An **identity** is a result that is true for any value of a variable.
For example $x + x \equiv 2x$ is an identity.

In Section 5.1 we met the identity

$$\tan\theta = \frac{\sin\theta}{\cos\theta}$$

This result is true for all values of θ and can be found from the diagram shown.

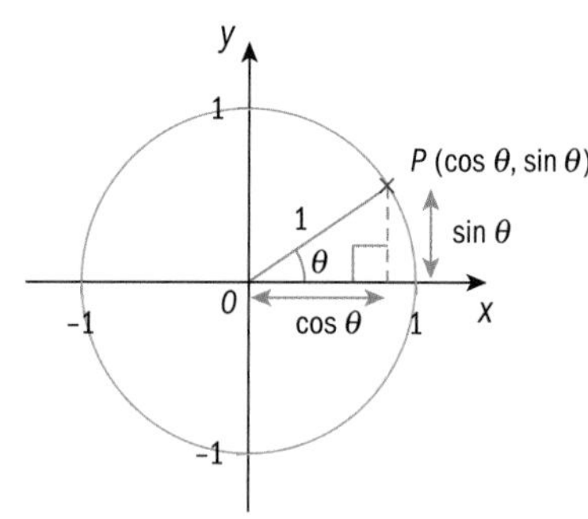

By using Pythagoras' theorem we can also write down a second identity from the diagram.

$$(\sin\theta)^2 + (\cos\theta)^2 = 1$$

This is usually written as

$$\sin^2\theta + \cos^2\theta = 1$$

You should learn these two identities.

Note: We should really use '≡' to mean 'is identically equal to' in this work, but the equals sign is often used.

The two identities can be used to simplify expressions and prove other identities.

Example 10

Simplify **a)** $\dfrac{\sin^3\theta+\sin\theta\cos^2\theta}{\cos\theta}$ **b)** $\dfrac{1+\cos\theta}{\sin\theta}+\dfrac{\sin\theta}{1+\cos\theta}$.

a) $\dfrac{\sin^3\theta+\sin\theta\cos^2\theta}{\cos\theta}=\dfrac{\sin\theta(\sin^2\theta+\cos^2\theta)}{\cos\theta}$ ← Factorise the numerator.

$=\dfrac{\sin\theta}{\cos\theta}$ ← Use $\sin^2\theta+\cos^2\theta=1$.

$=\tan\theta$ ← Use $\dfrac{\sin\theta}{\cos\theta}=\tan\theta$.

b) $\dfrac{1+\cos\theta}{\sin\theta}+\dfrac{\sin\theta}{1+\cos\theta}=\dfrac{(1+\cos\theta)^2+\sin^2\theta}{\sin\theta(1+\cos\theta)}$ ← Add the fractions.

$=\dfrac{1+2\cos\theta+\cos^2\theta+\sin^2\theta}{\sin\theta(1+\cos\theta)}$ ← Multiply out $(1+\cos\theta)^2$.

$=\dfrac{1+2\cos\theta+1}{\sin\theta(1+\cos\theta)}$ ← $\sin^2\theta+\cos^2\theta=1$

$=\dfrac{2(1+\cos\theta)}{\sin\theta(1+\cos\theta)}$ ← Factorise the numerator.

$=\dfrac{2}{\sin\theta}$ ← Divide the numerator and denominator by $(1+\cos\theta)$, for $\cos\theta\neq-1$.

Example 11

Prove the identity $\dfrac{\sin\theta}{\tan\theta(1-\cos\theta)}=\dfrac{\cos\theta}{1-\cos\theta}$, $\sin\theta\neq0$.

$\dfrac{\sin\theta}{\tan\theta(1-\cos\theta)}=\dfrac{\sin\theta}{\dfrac{\sin\theta}{\cos\theta}(1-\cos\theta)}$ ← Start with the left hand side and replace $\tan\theta$ with $\dfrac{\sin\theta}{\cos\theta}$.

$=\dfrac{\cancel{\sin\theta}^{1}\cos\theta}{\cancel{\sin\theta}_{1}(1-\cos\theta)}$ ← $\div\dfrac{\sin\theta}{\cos\theta}$ is the same as $\times\dfrac{\cos\theta}{\sin\theta}$.

$=\dfrac{\cos\theta}{1-\cos\theta}$ ← Divide numerator and denominator by the common factor $\sin\theta$.

Exercise 5.6

1. Express $4\cos^2x-\sin^2x$ in terms of $\cos x$.

Use the identity $\sin^2x+\cos^2x=1$ to replace $\sin^2x$ with $1-\cos^2x$.

2. Express in terms of $\sin\theta$

a) $\sin^2\theta-\cos^2\theta$ **b)** $2\cos^2\theta-4\sin\theta$.

3. Simplify

a) $\dfrac{\sin^3\theta + \sin\theta\cos^2\theta}{\sin\theta}$ b) $1 + \sin\theta + \dfrac{\cos^2\theta}{1+\sin\theta}$.

4. Prove the identity $\dfrac{\sin^4\theta}{\tan\theta} = \sin\theta\cos\theta - \sin\theta\cos^3\theta$.

5. Prove the following identities

a) $\cos^5\theta = \cos\theta - 2\sin^2\theta\cos\theta + \cos\theta\sin^4\theta$

b) $(4\sin\theta + 3\cos\theta)^2 + (3\sin\theta - 4\cos\theta)^2 = 25$

c) $\tan^2 A - \tan^2 B = \dfrac{\sin^2 A - \sin^2 B}{\cos^2 A\cos^2 B}$.

6. Show that $(1 + \sin\theta + \cos\theta)^2 = 2(1 + \sin\theta)(1 + \cos\theta)$.

7. Simplify

a) $\dfrac{\sin^2\theta - \cos^2\theta}{\sin\theta - \cos\theta}$ b) $\dfrac{\sin^4\theta - \cos^4\theta}{\sin^2\theta - \cos^2\theta}$ c) $\dfrac{\sin^4\theta - \cos^4\theta}{\sin\theta - \cos\theta}$.

8. Prove the following identities.

a) $\tan\theta = \sqrt{\dfrac{\sin^2\theta}{1-\sin^2\theta}}$ b) $\dfrac{1-\sin^2\theta}{\tan\theta} = \dfrac{\cos^3\theta}{\sin\theta}$.

c) $\dfrac{\sin^2\theta}{1-\cos\theta} = 1 + \cos\theta$ d) $(2 - \cos^2\theta)^2 - 4\sin^2\theta = \cos^4\theta$

9. a) Express $2\cos x - \sin^2 x$ in the form $(\cos x + p)^2 + q$, stating the values of p and q.

b) Use your answer to part (**a**) to find the maximum and minimum values of $2\cos x - \sin^2 x$.

10. a) Use the identity $\sin^2 x + \cos^2 x = 1$ to prove that $\dfrac{1 + \sin x\cos x}{\cos^2 x} = \tan^2 x + \tan x + 1$.

b) Use the identity you proved in part (**a**) to find the minimum values of $\dfrac{1 + \sin x\cos x}{\cos^2 x}$.

Trigonometric identities can be used to solve some trigonometric equations.

Example 12

Solve $\sqrt{3}\sin x - \cos x = 0$ for $0 \le x \le 2\pi$.

$\sqrt{3}\sin x = \cos x$ — Rearrange the equation.

$\tan x = \dfrac{1}{\sqrt{3}}$ — Divide by $\cos x$ and use the identity $\dfrac{\sin x}{\cos x} = \tan x$.

$x = \dfrac{\pi}{6}, \dfrac{7\pi}{6}$

Example 13

Solve the equation $2\cos^2 x = 1 - \sin x$ for $0° \le x \le 360°$.

$$2(1 - \sin^2 x) = 1 - \sin x$$

Express $\cos^2 x$ in terms of $\sin^2 x$ to obtain a quadratic equation in $\sin x$.

$$2\sin^2 x - \sin x - 1 = 0$$

Rearrange the equation.

$$(2\sin x + 1)(\sin x - 1) = 0$$

Factorise the quadratic expression.

$$\sin x = -\frac{1}{2}, 1$$

Solve each equation.

When $\sin x = -\frac{1}{2}$, $x = 210°, 330°$

When $\sin x = 1$, $x = 90°$

$$x = 90°, 210°, 330°$$

Example 14

Solve the equation $\sin^2(2x - 60°) = 0.6$ for $0° \le x \le 180°$.

$\sin(2x - 60°) = \pm 0.7745...$ — Don't forget the negative value.

$\sin(2x - 60°) = 0.7745...$ — Solve for each value separately.

$2x - 60° = 50.76...°, 180° - 50.76...°$ — Find all the values for $2x - 60$.

$2x = 110.76...°, 189.23...°$ — Find all the values for $2x$ by adding 60° to each angle.

$x = 55.38...°, 94.61...°$ — Finally, divide each angle by 2.

$\sin(2x - 60°) = -0.7745...$ — Solve for the negative value.

$2x - 60° = -50.76...°, 180° + 50.56...°$ — Find all the values for $-60° \le 2x - 60° \le 300°$.

$2x = 9.23...°, 290.76...°$ — Find all the values for $2x$.

$x = 4.61...°, 145.38...°$ — Find all the values for x.

$x = 4.6°, 55.4°, 94.6°, 145.4°$ — List all solutions correct to 1 decimal place.

Note: In order to find all the values for $0° \le x \le 180°$ we need to find all possible values of $2x - 60°$ between $-60°$ and $300°$.

Exercise 5.7

Solve the equations in questions **1** to **8** for $0° \leq \theta \leq 360°$.

1. $\cos^2\theta - \sin^2\theta = 0.$

2. $2\cos^2\theta + \sin\theta = 1.$

3. $3\sin^2\theta = \cos^2\theta.$

4. $\cos^2\theta - 4\sin^2\theta = 1.$

5. $1 + \sin\theta\cos^2\theta = \sin\theta$

6. $6\sin(\theta + 70)° - 5\cos(\theta + 70)° = 0$

7. $3\cos 2\theta = 4\sin 2\theta.$

8. $2\sin\theta = 3\tan\theta.$

9. Solve the equation $\sin\theta + 5\cos\theta = 3\sin\theta$ for $0 \leq \theta \leq 2\pi$.

10. Solve the equation $2\cos\theta - 5\sin\theta = 4\cos\theta + 3\sin\theta$ for $0 \leq \theta \leq 2\pi$.

11. a) Simplify $\dfrac{6 - 6\cos^2\theta}{2\sin\theta}$.

b) Hence solve the equation $6 - 6\cos^2\theta = 3\sin\theta$ for $0 \leq \theta \leq 2\pi$.

12. Solve the equation $\tan^3\theta - \tan^2\theta - 2\tan\theta = 0$ for $0° \leq \theta \leq 180°$.

Summary exercise 5

1. An angle θ is known to be between 270° and 360° and $\sin^2\theta = \frac{9}{25}$.

   Find **a)** $\sin\theta$ **b)** $\cos\theta$ **c)** $\tan\theta$.

2. Solve the equation $\sin 2\theta = 0.667$ for $0° \le \theta \le 180°$.

EXAM-STYLE QUESTION

3. Solve the equation $2\sin\theta = \cos\theta$ for $0 \le \theta \le \frac{\pi}{2}$.

4. Solve the equation $\cos 2x = \cos 144°$ for $0° \le x \le 360°$.

EXAM-STYLE QUESTIONS

5. Solve the equation $8\cos^2 x = 5 + 2\sin x$ for $0° \le x \le 360°$.

6. Solve the equation $\sin(x - 30°) = 0.7$ for $0° \le x \le 360°$.

7. Prove the identity
   $\tan\theta\sin\theta = \frac{1}{\cos\theta} - \cos\theta$.

8. Show that $(\cos x - \sin x)^2 + (\cos x + \sin x)^2 = 2$.

9. **a)** Sketch the graph of $y = \sin 4x$ for $0° \le x \le 180°$.

   b) Find all the angles x in the interval $0° \le x \le 180°$ for which

   i) $\sin 4x = \frac{\sqrt{3}}{2}$ **ii)** $\sin 4x = -\frac{1}{2}$.

10. Sketch the graph of $f(x) = \sin\left(2x + \frac{\pi}{4}\right)$ for $-\pi \le x \le \pi$.

11. **a)** Given that $2\sin^2\theta - \cos\theta = 1$, show that $2\cos^2\theta + \cos\theta - 1 = 0$.

    b) Hence solve the equation $2\sin^2(\theta - 20°) - \cos(\theta - 20°) - 1 = 0$ for $0° \le \theta \le 180°$.

12. **a)** Solve the equation $\tan\frac{1}{2}x = 3$ for $0 \le x \le 4\pi$.

    b) Solve the equation $\sin\theta(3\cos\theta - \sin\theta) = 0$ for $0 \le \theta \le 2\pi$.

13. **a)** Express $3\sin^2 x - 4\cos^2 x$ in the form $a + b\cos^2 x$, stating the values of a and b.

    b) Hence state the maximum and minimum values of $3\sin^2 x - 4\cos^2 x$.

    c) Solve the equation $3\sin^2 x - 4\cos^2 x + 1 = 0$ for $0° \le x \le 180°$.

14. **a)** Sketch the graph of $y = \sin^2 x$ for $0° \le x \le 360°$.

    b) Prove that if $\sin^2 x > \frac{3}{4}$ then $\cos^2 x < \frac{1}{4}$.

    c) Solve $\sin^2 x > \frac{3}{4}$ for $0° \le x \le 360°$.

15. Solve the following equation for $0° \le \theta \le 360°$.

    a) $3\sin^2\theta + 5\cos\theta - 1 = 0$

    b) $2\sin\theta\tan\theta = 3$

    c) $7\cos^2\theta + \sin^2\theta = 5\cos\theta$

    d) $3\tan\theta = 5\sin\theta$

    e) $\cos^2\theta - \sin^2\theta = 0$

    f) $1 + \sin\theta\cos^2\theta = \sin\theta$

16. **a)** Sketch, on the same diagram, the graphs of $y = 2\cos x$ and $y = \cos^2 x$ for $-360° \le x \le 360°$.

    b) State the number of solutions of the equation $2\cos x = \cos^2 x$ for $-360° \le x \le 360°$.

17. a) Sketch, on the same diagram, the graphs of $y = \frac{1}{2}\sin\theta$ and $\sin\frac{1}{2}\theta$ for $0 \le \theta \le 2\pi$.

b) State the number of solutions of the equation $\sin\theta = 2\sin\frac{1}{2}\theta$ for $0 \le \theta \le 2\pi$.

18. a) Sketch, on the same diagram, the graphs of $y = \tan\theta$ and $y = \tan(-\theta)$ for $-270° \le \theta \le 270°$.

b) State the solutions of the equation $\tan\theta = \tan(-\theta)$ for $-270° \le \theta \le 270°$.

19. a) Sketch the curves $y = 2\cos x$ and $y = 3\sin x$ on the same axes for $0° \le x \le 360°$.

b) Find the range of values of x in the interval $0° \le x \le 360°$ for which $3\sin x \ge 2\cos x$.

20. a) i) Given that $\frac{\sin\theta - \cos\theta}{\cos\theta} = 2$, show that $\tan\theta = 3$.

ii) Solve the equation $\frac{\sin\theta - \cos\theta}{\cos\theta} = 2$ for $0° \le \theta \le 360°$.

iii) Solve the equation $\frac{\sin(\theta - 20°) - \cos(\theta - 20°)}{\cos(\theta - 20°)} = 2$ for $0° \le \theta \le 360°$.

b) i) Given that $2\sin^2 2\theta - \cos 2\theta = 1$, show that $2\cos^2 2\theta + \cos^2\theta - 1 = 0$.

ii) Solve the equation $2\sin^2 2\theta - \cos 2\theta = 1$ for $0° \le \theta \le 720°$.

21. i) Prove the identity $\left(\frac{1}{\cos\theta} - \tan\theta\right)^2 \equiv \frac{1 - \sin\theta}{1 + \sin\theta}$.

ii) Hence, solve the equation $\left(\frac{1}{\cos\theta} - \tan\theta\right)^2 = 2$ for $0° \le \theta \le 360°$.

22. i) Show that the equation $2\tan x = \cos x$ can be written as a quadratic equation in $\sin x$.

ii) Solve the equation $2\tan x = \cos x$ for $0° \le x \le 180°$.

iii) Hence, state the set of values of x for which $2\tan(x - 30°) = \cos(x - 30°)$ for $0° \le x \le 180°$.

23. i) Prove the identity $\frac{\sin\theta\cos\theta}{(\cos\theta - \sin\theta)(\cos\theta + \sin\theta)} \equiv \frac{\tan\theta}{(1 - \tan^2\theta)}$.

ii) Solve the equation $5\sin\theta\cos\theta = (\cos\theta - \sin\theta)(\cos\theta + \sin\theta)$ for $0° \le \theta \le 180°$.

24. i) Show that the equation $2\sin^2 x - \cos x = 1$ can be written as a quadratic equation in $\cos x$.

ii) Solve the equation $2\sin^2 x - \cos x = 1$ for $0° \le x \le 180°$.

iii) Hence solve the equation $2\sin^2\left(\frac{x}{2}\right) - \cos\left(\frac{x}{2}\right) = 1$ for $0° \le x \le 180°$.

25. The function $f: x \mapsto 1 + \cos 2x$ is defined for the domain $0 \le x \le \frac{\pi}{2}$.

i) Solve the equation $2f(x) = 3$.

ii) Sketch the graph of $y = f(x)$.

iii) State the range of the function f.

iv) State the domain and range of the function f^{-1}.

The function $g: x \mapsto 3 - x$ is defined for the domain $x \in \mathbb{R}$.

v) Sketch the graph of $y = gf(x)$.

Chapter summary

Exact values of trigonometric functions

Degrees	Radians	sin	cos	tan
30°	$\frac{\pi}{6}$	$\frac{1}{2}$	$\frac{\sqrt{3}}{2}$	$\frac{1}{\sqrt{3}}$
45°	$\frac{\pi}{4}$	$\frac{1}{\sqrt{2}}$	$\frac{1}{\sqrt{2}}$	1
60°	$\frac{\pi}{3}$	$\frac{\sqrt{3}}{2}$	$\frac{1}{2}$	$\sqrt{3}$

Angles in the range 0° to 360°

- 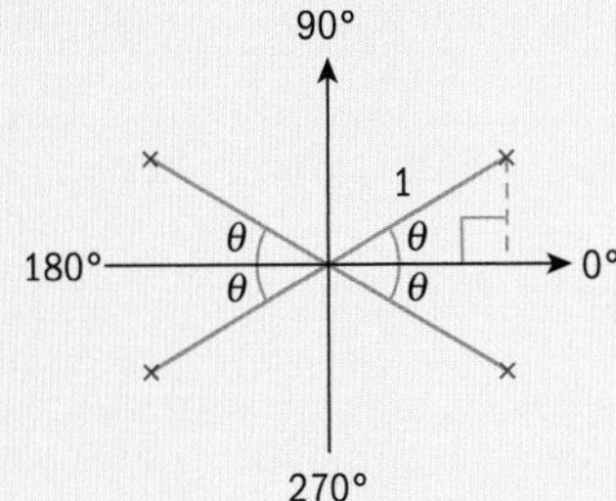

$\sin(180° - \theta) = \sin\theta$
$\cos(180° - \theta) = -\cos\theta$
$\tan(180° - \theta) = -\tan\theta$

$\sin(180° + \theta) = -\sin\theta$
$\cos(180° + \theta) = -\cos\theta$
$\tan(180° + \theta) = \tan\theta$

$\sin(360° - \theta) = -\sin\theta$
$\cos(360° - \theta) = \cos\theta$
$\tan(360° - \theta) = -\tan\theta$

- This diagram shows which functions are positive in each quadrant.

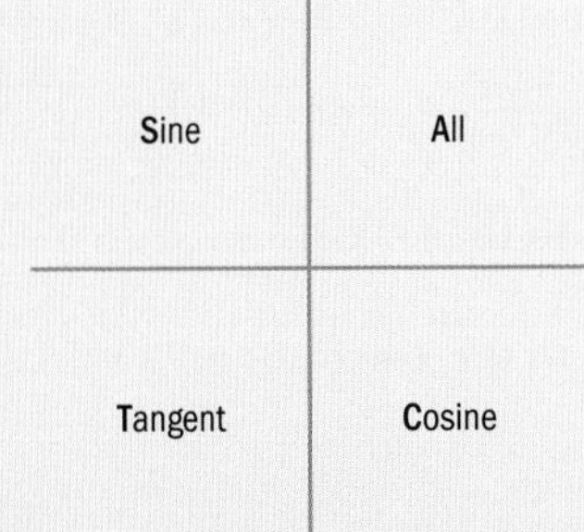

Graphs of trigonometric functions

- 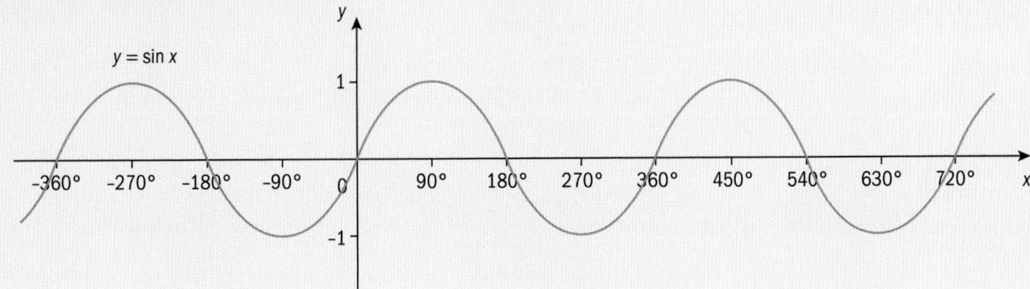

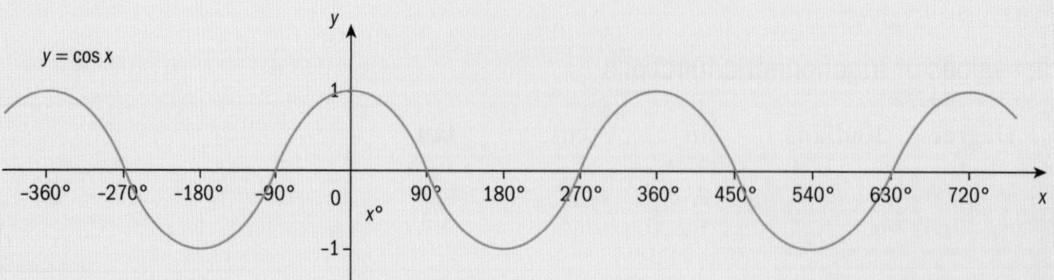

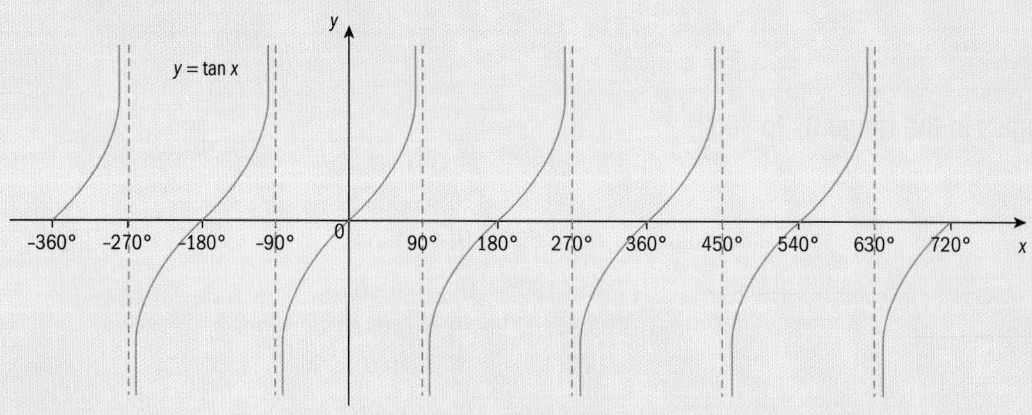

Composite graphs

- The graph of $y = \sin(x + c°)$ is a translation of $y = \sin x$ by $\begin{pmatrix} -c \\ 0 \end{pmatrix}$.
- The graph of $y = \sin x + d$ is a translation of $y = \sin x$ by $\begin{pmatrix} 0 \\ d \end{pmatrix}$.
- The graph of $y = a\sin x$ is a stretch of $y = \sin x$, factor a, parallel to the y-axis.
- The graph of $y = \sin bx$ is a stretch of $y = \sin x$, factor $\frac{1}{b}$, parallel to the x-axis.

Trigonometric identities

- $\tan\theta = \frac{\sin\theta}{\cos\theta}$
- $\sin^2\theta + \cos^2\theta = 1$

6 Binomial expansion

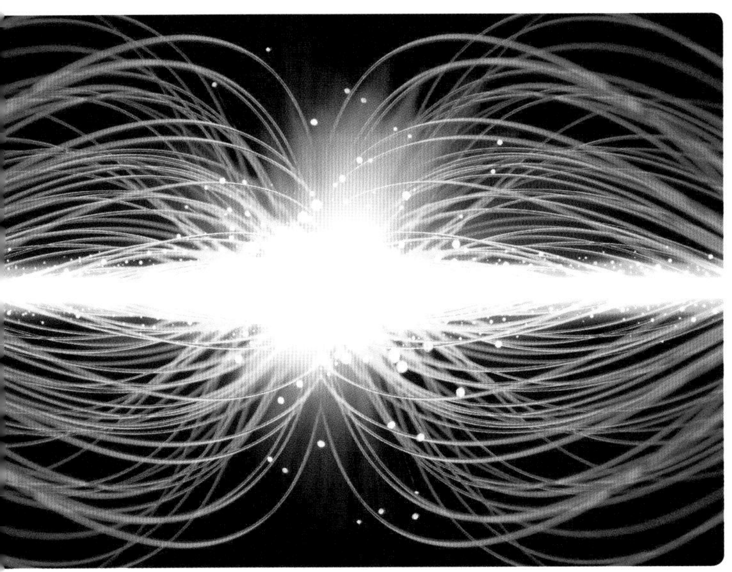

The binomial theorem is used by scientists to make approximations in calculations. For example, physicists use it to analyse the collision of particles in the process known as nuclear fission.

At the Large Hadron Collider at CERN in Switzerland, scientists also make use of these kinds of equations in computer models to calculate the collision of high-energy particles.

Objectives:

- Understand the notation $\binom{n}{r}$ and $n!$
- Use the expansion of $(a + b)^n$, where n is a positive integer.

Before you start

You should know how to:

1. Expand brackets,

e.g. $(2x - 5)^2 = (2x - 5)(2x - 5)$

$= 4x^2 - 10x - 10x + 25 = 4x^2 - 20x + 25$

e.g. $(2x - 5)^3 = (2x - 5)(2x - 5)^2$

$= (2x - 5)(4x^2 - 20x + 25)$

$= 8x^3 - 40x^2 + 50x - 20x^2 + 100x - 125$

$= 8x^3 - 60x^2 + 150x - 125$

2. Simplify indices,

e.g. $(3x^2)^5$

$= 3^5(x^2)^5 = 243x^{10}$

e.g. $(-2y^7)^6$

$= (-2)^6(y^7)^6 = 64y^{42}$

Skills check:

1. Expand

a) $(x + 7)^2$

b) $(4x - 1)^2$

c) $(2x + 3)^3$

d) $(5 - y)^2$

e) $(3x - 10)^3$

f) $(1 - 2x)^3$.

2. Simplify

a) $(2x^{10})^4$ **b)** $(-3x^4)^3$ **c)** $(-5x^3)^4$

d) $-8(2x^5)^7$ **e)** $3(-x^9)^5$ **f)** $-5(3x^6)^3$.

6.1 Pascal's triangle

We know from expanding brackets that:

$(1 + x)^1 = 1 + 1x$

$(1 + x)^2 = 1 + 2x + 1x^2$

$(1 + x)^3 = 1 + 3x + 3x^2 + 1x^3$

$(1 + x)^4 = 1 + 4x + 6x^2 + 4x^3 + 1x^4$

$(1 + x)^5 = 1 + 5x + 10x^2 + 10x^3 + 5x^4 + 1x^5$

If we look at the coefficients of the terms in each expansion, we get

1	1				
1	2	1			
1	3	3	1		
1	4	6	4	1	
1	5	10	10	5	1

This table of coefficients can be extended and written in a triangle.

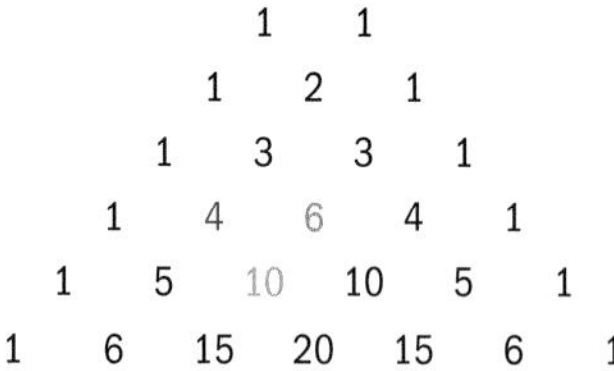

In the triangle, we can see that each number comes from adding the two numbers above it.
e.g. 4 + 6 = 10

This triangle is known as **Pascal's triangle**.

If we wanted to find the expansion of $(1 + x)^5$ we could use Pascal's triangle to do this.
We can see that the next row of coefficients is 1, 6, 15, 20, 15, 6, 1.
Hence, $(1 + x)^6 = 1 + 6x + 15x^2 + 20x^3 + 15x^4 + 6x^5 + 1x^6$.
We say that the expansion in ascending powers of x of $(1 + x)^6$ is $1 + 6x + 15x^2 + 20x^3 + 15x^4 + 6x^5 + 1x^6$.

We can use Pascal's triangle to give the coefficients in the expansion of $(a + b)^n$.
For example, $(a + b)^4$ uses the fourth row of coefficients: 1, 4, 6, 4, 1.
Hence, $(a + b)^4 = 1a^4 + 4a^3b^1 + 6a^2b^2 + 4a^1b^3 + 1b^4$.

Example 1

Expand $(3 + 2x)^5$.

We can use the coefficients **1, 5, 10, 10, 5, 1** ← This is the 5th row of Pascal's triangle.

$$(3 + 2x)^5 = \mathbf{1}(3)^5 + \mathbf{5}(3)^4(2x)^1 + \mathbf{10}(3)^3(2x)^2 + \mathbf{10}(3)^2(2x)^3 + \mathbf{5}(3)^1(2x)^4 + \mathbf{1}(2x)^5$$

$$= 1(243) + 5(81)(2x) + 10(27)(4x^2) + 10(9)(8x^3) + 5(3)(16x^4) + 1(32x^5)$$

$$= 243 + 810x + 1080x^2 + 720x^3 + 240x^4 + 32x^5$$

Example 2

Expand $(5x-2)^4$.

We can use the coefficients **1, 4, 6, 4, 1.**

This is the 4th row of Pascal's triangle.

$$(5x-2)^4 = \mathbf{1}(5x)^4 + \mathbf{4}(5x)^3(-2)^1 + \mathbf{6}(5x)^2(-2)^2 + \mathbf{4}(5x)^1(-2)^3 + \mathbf{1}(-2)^4$$

Make sure the minus sign is **inside** the bracket.

$$= 1(625x^4) + 4(125x^3)(-2) + 6(25x^2)(4) + 4(5x)(-8) + 1(16)$$

$$= 625x^4 - 1000x^3 + 600x^2 - 160x + 16$$

$(a-b)^n$ so the signs will alternate + – + –.

Exercise 6.1

1. Use Pascal's triangle to expand
   a) $(2+3x)^4$ b) $(2x+1)^6$ c) $(1-4x)^5$ d) $(x-3)^7$.
2. Write down the first three terms in the expansion $(2-10x)^5$ in ascending powers of x.
3. Write down the last three terms in the expansion $(4+x)^6$ in ascending powers of x.
4. The 2nd term in ascending powers of x in the expansion of $(1-5x)^n$ is $-25x$. Find the value of n.
5. Use Pascal's triangle to find the value of $(1.03)^4$.

Hint: $1.03 = 1 + \frac{3}{100}$

6.2 Binomial notation

You can use factorial notation ($n!$) and your calculator to help you expand binomial expressions.

$$n! = n(n-1)(n-2)\ldots(n-r)\ldots(3)(2)(1)$$

Or you can use the  button on your calculator.

e.g. $6! = 6 \times 5 \times 4 \times 3 \times 2 \times 1 = 720$

You can also use combinations to help you expand binomial expressions.

Note: $0! = 1$

The number of ways of choosing r items from a set of n items is written as ${}^nC_r$ or $\binom{n}{r}$ or $\frac{n!}{(n-r)!r!}$

$${}^nC_r = \binom{n}{r} = \frac{n!}{(n-r)!r!}$$

We write 5!, rather than 5 × 4 × 3 × 2 × 1, as we know this will cancel with the 5! in the denominator.

e.g. $\binom{8}{5} = {}^8C_5 = \frac{8!}{(8-5)!5!} = \frac{8!}{(3)!5!} = \frac{8\times7\times6\times\cancel{5!}}{3\times2\times\cancel{5!}} = \frac{8\times7\times\cancel{6}}{\cancel{3\times2}} = 56$

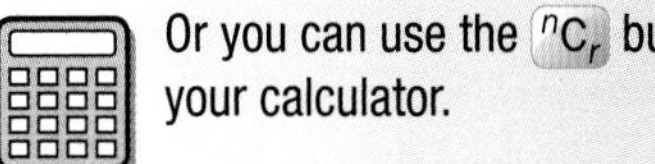

Exercise 6.2

1. **Calculate**

   a) 5! b) 10! c) $\frac{20!}{17!}$ d) $\frac{8!}{2!6!}$.

2. Calculate

   a) $\binom{9}{3}$ b) $\binom{12}{10}$ c) $\binom{7}{1}$ d) $\binom{5}{5}$.

6.3 Binomial expansion

$(1 + x)^1$	$1 + 1x$
$(1 + x)^2$	$1 + 2x + 1x^2$
$(1 + x)^3$	$1 + 3x + 3x^2 + 1x^3$
$(1 + x)^4$	$1 + 4x + 6x^2 + 4x^3 + 1x^4$

$$(1 + x)^n = 1 + nx + \frac{n(n-1)}{2!}x^2 + \frac{n(n-1)(n-2)}{3!}x^3 + \ldots + x^n$$

This is known as the **binomial expansion** of $(1 + x)^n$ where n is a positive integer.

This can be written as

$$(1 + x)^n = 1 + \binom{n}{1}x + \binom{n}{2}x^2 + \binom{n}{3}x^3 + \ldots \binom{n}{r}x^r + \ldots + x^n$$

Binomial means 'two terms'. It is called the binomial expansion because there are two terms in the brackets that are expanded.

$$\binom{n}{r} = \frac{n!}{(n-r)!r!} = {}^nC_r$$

$\binom{n}{r} = \frac{n!}{(n-r)!r!}$ = the coefficient of the term in x^r = the $(r + 1)$th term.

$$(x + y)^n = \binom{n}{0}x^ny^0 + \binom{n}{1}x^{n-1}y^1 + \binom{n}{2}x^{n-2}y^2 + \ldots + \binom{n}{n-1}\ x^1y^{n-1} + \binom{n}{n}x^0y^n$$

This is known as the **binomial theorem** where n is a positive integer.

Example 3

Write down the 5th term, in ascending powers of x, in the expansion of $(1 + x)^7$.

5th term: $\binom{7}{4} 1^{7-4}x^4$ — 5th term is the $(r + 1)$th term so $r = 4$.

$= \frac{7!}{(7-4)!4!}x^4 = 35x^4$ — Use your calculator to get 35.

Example 4

Find the first three terms, in descending powers of x, in the expansion of $(4x - 3)^5$.

1st term $= 1(4x)^5 = 1024x^5$

2nd term $= \binom{5}{1}(4x)^4(-3)^1$ — The two indices add up to 5 for each term.

$= 5(256x^4)(-3) = -3840x^4$

3rd term $= \binom{5}{2}(4x)^3(-3)^2$ — The two indices add up to 5 for each term.

$= 10(64x^3)(9) = 5760x^3$

$(4x - 3)^5 = 1024x^5 - 3840x^4 + 5760x^3 + \ldots$.

Example 5

Find the coefficient of x^2 in the expansion of $(2 + 3x)^7$.

3rd term $= \binom{7}{2}(2)^5(3x)^2$ — 1st term has no x, 2nd term has x, 3rd term has x^2.

$= 21(32)(9x^2) = 6048x^2$ — Remember to square 3 as well as x.

Thus the coefficient is 6048. — Do not put x^2 with the answer.

Exercise 6.3

1. Find the first three terms in the expansion of $(1 - x)^{10}$ in ascending powers of x.

2. Find the first four terms in the expansion of $(3 + x)^8$ in ascending powers of x.

3. Find the coefficient of the term in x^3 in the expansion of $(1 - 6x)^7$.

4. Using the binomial expansion, expand each of these expressions.

   a) $(3x + 4y)^4$ **b)** $(x - 2y)^3$ **c)** $\left(1 - \frac{x}{2}\right)^5$

   d) $(2 + 5x)^4$ **e)** $\left(\frac{x}{3} + 1\right)^3$ **f)** $\left(3 - \frac{3}{x}\right)^5$

5. Find the coefficient of the 6th term in the expansion of $(2 - x)^{10}$ in ascending powers of x.

6. Find the coefficient of the term in x^4 in the expansion of $(3 + 2x)^8$.

7. Find the first three terms in the expansion of $(1 + ax)^7$ in ascending powers of x.

8. Find the term independent of x in the expansion of $\left(x^2 + \frac{1}{x^2}\right)^6$.

9. Find the 5th term in the expansion of $\left(4 - \frac{3x}{2}\right)^9$ in ascending powers of x.

10. Expand, in ascending powers of x, as far as the term in x^2.

    a) $(5 - x)^6$ **b)** $(3 + 2x)^9$ **c)** $\left(2 + \frac{x}{5}\right)^7$ **d)** $\left(1 - \frac{2x}{3}\right)^{12}$

11. Find the coefficient of x^5 in the expansion of $(2x - 1)^{15}$.

12. **a)** Find the first four terms in the expansion of $\left(1 + \frac{x}{2}\right)^{10}$ in ascending powers of x.

    b) Hence determine the value of $(1.05)^{10}$ correct to 2 decimal places.

13. Find, in ascending powers of x, the first three terms of each of the following.

    a) $(1 - 3x)^8$ **b)** $(3 + 2x)^7$ **c)** $\left(2 + \frac{x}{5}\right)^5$ **d)** $\left(1 - \frac{x}{3}\right)^{12}$

14. Find the first four terms in the expansion of $\left(1 - \frac{x}{a}\right)^6$ in ascending powers of x.

15. Use the binomial expansion to find the exact value of $(1.001)^4$.

16. Find the term in y^3 in the expansion of $(2 - y)^6$.

17. Find the coefficient of y^5 in the expansion of $(3 + 2y)^8$.

18. Find the term independent of x in the expansion of $\left(x^2 - \frac{2}{x}\right)^6$.

19. a) Write down the expansion of $(1 + x)^5$.

b) By letting $x = y + y^2$, find the coefficient of y^8 in the expansion of $(1 + y + y^2)^5$.

20. a) Find the first four terms in the expansion of $\left(1 - \frac{x}{2}\right)^8$ in ascending powers of x.

b) Use these terms to find the value of $(0.99)^8$ giving your answer correct to 4 decimal places.

6.4 More complex expansions

Example 6

a) Find the first three terms in the expansion in descending powers of x of $\left(2x - \frac{4}{x}\right)^6$.

b) Hence determine the coefficient of x^3 in the expansion of $(1 + 2x)\left(2x - \frac{4}{x}\right)^6$.

a) 1st term $= 1(2x)^6 = 64x^6$ ← Remember to make 2 to the power of 6 as well.

2nd term $= \binom{6}{1}(2x)^5\left(-\frac{4}{x}\right)^1$

$= 6(32x^5)\left(-\frac{4}{x}\right)$

$= -768x^4$ ← The answer is negative as $(-)^1$ is negative.

3rd term $= \binom{6}{2}(2x)^4\left(-\frac{4}{x}\right)^2$

$= 15(16x^4)\left(\frac{16}{x^2}\right)$

$= 3840x^2$ ← The answer is positive as $(-)^2$ is positive.

$\left(2x - \frac{4}{x}\right)^6 = 64x^6 - 768x^4 + 3840x^2$.

b) Term with coefficient of x^3 in the expansion of $(1 + 2x)\left(2x - \frac{4}{x}\right)^6$

$= (2x)(3840x^2)$ ← There is no x^3 term in **(a)** so we cannot have $1 \times x^3$ term too.

coefficient $= 7680$

Example 7

Find the first five terms, in descending powers of x, in the expansion of $(x-2)\left(x+\frac{3}{x}\right)^9$.

$$\left(x+\frac{3}{x}\right)^9 = (x)^9 + \binom{9}{1}(x)^8\left(\frac{3}{x}\right)^1 + \binom{9}{2}(x)^7\left(\frac{3}{x}\right)^2 + \ldots$$

First expand $\left(x+\frac{3}{x}\right)^9$.

$$= x^9 + 27x^7 + 324x^5 + \ldots$$

Note: We know that the highest power of x will be x^{10} from $x \times x^9$.
The final expansion will have x^{10} to x^6.
x^6 comes from $x \times x^5$.
Thus we only need to expand the 2nd bracket to the term in x^5.

$$\begin{aligned}(x-2)\left(x+\frac{3}{x}\right)^9 &= x(x^9 + 27x^7 + 324x^5 + \ldots) - 2(x^9 + 27x^7 + 324x^5 + \ldots)\\ &= x^{10} + 27x^8 + 324x^6 + \ldots - 2x^9 - 54x^7 - 648x^5 + \ldots\\ &= x^{10} - 2x^9 + 27x^8 - 54x^7 + 324x^6 + \ldots\end{aligned}$$

Example 8

Given that the coefficient of x^3 in the expansion of $(1 + ax + 2x^2)(2 - x)^7$ is 560, determine the value of a.

$$(2-x)^7 = 2^7 + \binom{7}{1}(2)^6(-x) + \binom{7}{2}(2)^5(-x)^2 + \binom{7}{3}(2)^4(-x)^3 + \ldots$$

$$= 128 - 448x + 672x^2 - 560x^3 + \ldots$$

Terms in x^3 in the expansion of $(1 + ax + 2x^2)(2 - x)^7$

$$= 1(-560x^3) + ax(672x^2) + 2x^2(-448x)$$

Coefficient of $x^3 = -560 + 672a - 896 = 672a - 1456 = 560$

$$672a = 2016$$

$$a = 3$$

Exercise 6.4

1. Find, in ascending powers of x, the first three terms of each of the following.
   a) $(2 + x)(1 + x)^8$
   b) $(1 - 2x)(3 + 2x)^9$

2. Find, in ascending powers of x, the first three terms in the expansion of $(1-2x)\left(1+\frac{x}{2}\right)^{10}$.

3. Find the 3rd term in descending powers of x, in the expansion of $(x + 3)(2x - 1)^5$.

4. Find the term independent of x in the expansion of $\left(3 + \frac{x}{2}\right)\left(2 + \frac{3}{x}\right)^6$.

5. The coefficient of x^2 in the expansion of $\left(1 - \frac{x}{2}\right)^5 - (a + x)^4$ is $2a$ where $a > 0$. Find a.

6. a) Write down the first three terms in the expansion of $(3 + x)^7$ in ascending powers of x.
   b) Use this result to write down the first three terms in the expansion of $[(3 + (y + y^2)]^7$ in ascending powers of y.

7. Find the coefficient of x^4y^2 in the binomial expansion of $(2x - y)^6$.

8. Find the coefficient of x^3 in the expansion of $(1 + x - 2x^2)\left(x - \frac{2}{x}\right)^7$.

9. Find the coefficient of x^4 in the expansion of $(1 - x^2)\left(2x + \frac{1}{x}\right)^6$.

10. Expand $(2 + x + x^2)^4$ in ascending powers of x up to and including the term in x^2.

Summary exercise 6

1. Write down the first three terms in the expansion of $\left(1 - \frac{x}{2}\right)^6$ in ascending powers of x.

2. Find the 3rd term in the binomial expansion of $(2 - 3x)^{10}$ in ascending powers of x.

3. Find the term independent of x in the expansion of $\left(2x^2 - \frac{1}{x}\right)^9$.

EXAM-STYLE QUESTIONS

4. Use the binomial expansion to show that the value of $(0.98)^5$ is 0.903921 correct to 6 decimal places.

5. $(2 - ax)^6 = 64 + px + 135x^2 + \ldots$ with $p > 0$ and $a > 0$. Find the value of p and the value of a.

6. The 3rd term in ascending powers of x in the expansion of $\left(1 + \frac{1}{x}\right)^n$, $n > 2$, is $\frac{6}{x^2}$. Find the value of n.

7. The binomial expansion of $\left(3 - \frac{x}{4}\right)^8$ is $6561 + px + qx^2 + \ldots$ Find the value of p and the value of q.

8. Find the 6th term in the expansion of $\left(\frac{x}{2} + 1\right)^{20}$ in ascending powers of x.

9. Find $(1 + x)^4 - (1 - x)^4$. Write your answer in ascending powers of x.

10. a) Find the first five terms in the expansion of $(2 + x)^6$ in ascending powers of x.

    b) Using your expansion, show that $(2.03)^6 = 69.9804$ correct to 4 decimal places.

11. Find the ratio of the coefficient of the x^3 term to the coefficient of the x^4 term in the expansion of $\left(2x + \frac{1}{2}\right)^6$.

12. Find the 5th term in the expansion of $\left(2 - \frac{1}{3x}\right)^{11}$ in ascending powers of x.

13. $(1 + kx)^8 = 1 + 12x + ax^2 + bx^3 + \ldots$

    a) Find the value of k, the value of a and the value of b.

    b) Use your values of k, a and b to find the coefficient of the term in x^3 in the expansion of $(1 - x)(1 + kx)^8$.

14. a) Expand $(1 + 2x)^3$.

    b) Show that $(1 + 2x)^3 + (1 - 2x)^3 = 2(1 + 12x^2)$.

    c) Hence solve $(1 + 2x)^3 - (1 - 2x)^3 = 8$.

15. a) Write down the first three terms, in ascending powers of x, in the expansion of $(1 + ax)^n$ where $a \neq 0$ and $n > 2$.

    b) The coefficient of the x^2 term is half the coefficient of the term in x. Show that $n = \frac{1+a}{a}$.

    c) Find the coefficient of the x^2 term when $a = \frac{1}{5}$.

16. The first three terms, in ascending powers of x, in the expansion of $\left(1 + \frac{a}{x}\right)^n$ are $1 + \frac{24}{x} + \frac{252}{x^2}$.
Find the value of a and the value of n.

17. a) Write down the binomial expansion of $(1 + x)^5$.

    b) Use your answer to **(a)** to express $(1 - \sqrt{2})^5$ in the form $p + q\sqrt{2}$ where p and q are integers.

18. Find the coefficient of x^2 in the expansion of $(2 - 3x + x^2)\left(1 - \frac{x}{2}\right)^{10}$.

19. $(2 - ax)^8 = 2^b - 5120x + cx^2 + \ldots$
Find the value of a, the value of b and the value of c.

20. $(1 + x)^4(1 + ax)^7 = 1 + 74x^2 + bx^2 + \ldots$
Find the value of a and the value of b.

Chapter summary

Pascal's triangle

- We can use **Pascal's triangle** to find the coefficients of the terms in the expansion of expressions such as $(1 + x)^6$ i.e 1, 6, 15, 20, 15, 6, 1
  Hence, $(1 + x)^6 = 1 + 6x + 15x^2 + 20x^3 + 15x^4 + 6x^5 + 1x^6$
- Pascal's triangle

```
          1   1
        1   2   1
      1   3   3   1
    1   4   6   4   1
  1   5  10  10   5   1
1   6  15  20  15   6   1
```

Factorial notation

- $n! = n(n-1)(n-2)\ldots(n-r)\ldots(3)(2)(1)$
- $0! = 1$

Binomial expansion

- The **binomial expansion** of $(1 + x)^n$ is $1 + nx + \frac{n(n+1)}{2!}x^2 + \frac{n(n+1)(n+2)}{3!}x^3 + \ldots + x^n$ where n is a positive integer.
- The coefficient of the term in x^r = the $(r + 1)$th term $= \binom{n}{r} = \frac{n!}{(n-r)!r!} = {}^n\mathbf{C}_r$.
- $(x+y)^n = \binom{n}{0}x^ny^0 + \binom{n}{1}x^{n-1}y^1 + \binom{n}{2}x^{n-2}y^2 + \ldots + \binom{n}{n-1}x^1y^{n-1} + \binom{n}{n}x^0y^n$

  This is known as the **binomial theorem** where n is a positive integer.

7 Series

There are many sequences found in nature. The Italian mathematician Leonardo of Pisa (c.1170–c.1250), better known as Fibonacci, first discovered the sequence of numbers which would become known as the Fibonacci series. This sequence can be used to describe the spirals which appear in sea shells, the arrangement of seeds in a flowering plant, and many other sequences found throughout nature.

Objectives

- Recognise arithmetic and geometric progressions.
- Use the formula for the nth term and for the sum of the first n terms to solve problems involving arithmetic and geometric progressions.
- Use the condition for the convergence of a geometric progression and the formula for the sum to infinity of a convergent geometric progression.

Before you start

You should know how to:

1. Find the coefficient of an equation,
   e.g. for the expression $-60x^2 + 150x - 125$ the **coefficient** of x^2 is -60, the coefficient of the term in x is $+150$ and the constant term is -125.

2. Find the nth term of a linear sequence,
   e.g. the nth term of $-5, 8, 21, 34, 47$ is $13n - 18$ (the terms are increasing by 13 so $13n$, term before -5 is -18).
   e.g. nth term of $7, 4, 1, -2, -5$ is $-3n + 10$ (the terms are decreasing by 3 so $-3n$, term before 7 is 10).

Skills check:

1. $5x^3 + x^2 - 12x - 1$
   Find
   a) the term independent of x
   b) the coefficients of the term in x^2 and in x.

2. Find the nth term of these linear sequences
   a) $2, 7, 12, 17, 22, \ldots$
   b) $-3, -7, -11, -15, -19, \ldots$
   c) $-4, 2, 8, 14, 20, \ldots$
   d) $-1, -10, -19, -28, -37, \ldots$

7.1 Sequences

When a set of numbers follows a pattern and there is a clear rule for finding the next number in the pattern, then we have a **sequence**.

The following are examples of sequences:

1, 3, 5, 7, 9, …
1, 4, 9, 16, 25, …
2, −4, 8, −16, 32, …

Each number in the sequence is called a **term** of the sequence.

u_1 is the 1st term of a sequence.
u_2 is the 2nd term of a sequence.
u_n is the nth term of a sequence.

u_n can be used to describe any term in the sequence.
u_{n+1} is the term after u_n.

Example 1

Write down the first three terms of each sequence.

a) $u_n = n(n + 1)$

b) $u_{n+1} = 2 - \frac{5}{u_n}$ where $u_1 = 1$

a) $u_1 = 1(1 + 1) = 2$
$u_2 = 2(2 + 1) = 6$
$u_3 = 3(3 + 1) = 12$

We substitute for $n = 1$, $n = 2$ and $n = 3$ in $n(n + 1)$ to find the first three terms.

b) $u_1 = 1$
$u_2 = 2 - \frac{5}{1} = -3$
$u_3 = 2 - \frac{5}{-3} = 3\frac{2}{3}$

$u_2 = 2 - \frac{5}{u_1}$ and $u_3 = 2 - \frac{5}{u_2}$

Example 2

Write down the nth term of each sequence.

a) 3, 7, 11, 15, 19, …

b) 2, 5, 10, 17, 26, …

a) nth term $= 4n - 1$

$4n$ because the difference between each term is 4; and −1 because the 1st term is $4(1) - 1 = 3$.

b) nth term $= n^2 + 1$

Each term is 1 more than the square numbers 1, 4, 9, 16, 25, …

Exercise 7.1

1. Write down the first three terms of each sequence.

   a) $u_n = 3n + 5$
   b) $u_{n+1} = 4u_n - 7$ and $u_1 = 3$
   c) $u_n = (n - 3)(n + 1)$
   d) $u_1 = -1, u_{n+1} = \dfrac{3}{u_n + 2}$
   e) $u_n = 8 - \dfrac{2n}{3+n}$
   f) $u_1 = 1, u_{n+1} = (3u_n + 2)^2$
   g) $u_n = (-2)^n$
   h) $u_{n+1} = \dfrac{u_n}{u_n + 2}$ and $u_1 = 2$
   i) $u_{n+1} = 5(u_n + 3)$, and $u_1 = 2$
   j) $u_n = (-1)^n 7n^2$

2. An expression for the nth triangular number is $\dfrac{n(n+1)}{2}$.
   Write down the 20th triangular number.

3. Write down the nth term of each sequence.

   a) 7, 13, 19, 25, 31, …
   b) 7, 4, 1, −2, −5, …
   c) $\frac{1}{2}, \frac{1}{3}, \frac{1}{4}, \frac{1}{5}, \frac{1}{6}, \ldots$
   d) −15, −4, 7, 18, 29, …

4. The nth term of the sequence 2, 0, −2, −4, −6, … is 4 times the nth term of the sequence −22, −20, −18, −16, −14, … Work out the value of n.

7.2 Finite and infinite series

A **series** is formed when the terms of a sequence are added.
e.g. 12 + 9 + 6 + 3 + 0 + −3 + −6 + …

A series is **finite** if it stops after a finite number of terms.

e.g. 5 + 9 + 13 + 17 + 21 + 25 is a finite series. It has six terms.

A series is **infinite** if it continues indefinitely.

e.g. 5 + 9 + 13 + 17 + 21 + …

The **general term** of the series 5 + 9 + 13 + 17 + 21 + … is $4r + 1$.
We write the general term in terms of r. We work it out as we would work out the nth term.

The finite series 5 + 9 + 13 + 17 + 21 + 25 can be written as $\sum_{r=1}^{6}(4r+1)$ and we read this as:

The sum of the terms in the sequence $(4r + 1)$ with r going from 1 to 6, i.e. the first six terms in the sequence defined by $(4r + 1)$.

As 5 + 9 + 13 + 17 + 21 + 25 = 90 we can say $\sum_{r=1}^{6}(4r+1) = 90$

The infinite series 5 + 9 + 13 + 17 + 21 + … can be written as $\sum_{r=1}^{\infty}(4r+1)$.

The Greek letter $\sum$ is called **sigma**.

Example 3

Write down the first three terms, the last term and the nth term of the series $\sum_{r=3}^{10}(2^r+1)$.

1st term $= 2^3 + 1 = 9$ — For the 1st term $r = 3$.

2nd term $= 2^4 + 1 = 17$

3rd term $= 2^5 + 1 = 33$

Last term $= 2^{10} + 1 = 1025$ — For the last term $r = 10$.

nth term $= 2^n + 1$ — For the nth term $r = n$.

Example 4

Write the following series using the sigma notation.

a) $7 + 9 + 11 + 13 + 15 + \dots$ **b)** $1 - x + x^2 - x^3 + x^4 - x^5$

a) $= \sum_{r=1}^{\infty}(2r+5)$ — This is the same as $\sum_{r=2}^{\infty}(2r+3)$ or $\sum_{r=3}^{\infty}(2r+1)$ etc.

b) $= \sum_{r=0}^{5}(-1)^r x^r$ — It is useful to remember that when the signs change with each term then we can use $(-1)^r$ to show this.

Exercise 7.2

1. Write down the first three terms, the last term (if the series is finite) and the nth term of the following series.

a) $\sum_{r=1}^{9}(3r-2)$ **b)** $\sum_{r=1}^{7}2^r$ **c)** $\sum_{r=1}^{n}r(r+1)$ **d)** $\sum_{r=4}^{10}(r^2+2r+1)$

2. Work out the 15th term in the series $\sum_{r=1}^{\infty}(2r+1)$.

3. Write the following series using the sigma notation.

a) $2 + 5 + 8 + 11 + \dots$ **b)** $1 + 4 + 9 + 16 + \dots + 121$ **c)** $7 + 3 - 1 - 5 + \dots$

d) $x^2 - x^4 + x^6 - x^8 + x^{10} - x^{12}$ **e)** $1 + 2 + 3 + 4 + 5 + 6 + 7 + 8$ **f)** $12 + 15 + 18 + 21 + \dots$

4. Given that $\sum_{r=1}^{4}6r = 2\sum_{r=1}^{n}5r$, work out the value of n.

7.3 Arithmetic progressions

An **arithmetic progression** is a sequence that has a **common difference** between each term.

e.g. 1, 4, 7, 10, ... This has a common difference of 3.
e.g. 11, 7, 3, −1, ... This has a common difference of −4.

a = 1st term $(= u_1)$
d = common difference $(= u_2 - u_1 = u_3 - u_2 = u_4 - u_3$ etc.)
l = last term in a finite series

An arithmetic progression with n terms can be written as
$$a, (a + d), (a + 2d), (a + 3d), \ldots, \ldots, \ldots, [a + (n - 1)d]$$

The ***n*th term** of an arithmetic progression can be written as
$$\boldsymbol{u_n = a + (n - 1)d}$$

Note: We can use this to write $l = a + (n - 1)d$

We can use this formula to write down an expression for any term.

e.g. 8th term = $a + 7d$ e.g. 20th term = $a + 19d$

Example 5

The 1st term of an arithmetic progression is −6 and the common difference is 5.
Find the 4th term and the nth term.

$a = -6 \quad d = 5$

4th term $= a + 3d$
$= -6 + (3)5 = 9$ ← Substitute $a = -6$ and $d = 5$

nth term $= a + (n - 1)d$
$= -6 + (n - 1)5 = 5n - 11$ ← Substitute $a = -6$ and $d = 5$

Example 6

The nth term of an arithmetic progression is $7n - 2$. Find the 1st term and the common difference, d.

The 1st term has $n = 1$ $u_1 = 7(1) - 2 = 5$ ← Substitute $n = 1$

The 2nd term has $n = 2$ $u_2 = 7(2) - 2 = 12$ ← Substitute $n = 2$

d = 2nd term − 1st term $d = 12 - 5 = 7$

The **sum of an arithmetic progression** can be written as

$S_n = \frac{n}{2}(a + l)$ or $S_n = \frac{n}{2}[2a + (n - 1)d]$

Proof:

$S_n = a + (a + d) + (a + 2d) + \ldots + (l - 2d) + (l - d) + l$

$S_n = l + (l - d) + (l - 2d) + \ldots + (a + 2d) + (a + d) + a$ ← Rearrange the 1st line in the opposite order.

thus,

$2S_n = (a + l) + (a + l) + (a + l) + \ldots + (a + l) + (a + l) + (a + l)$

$= n(a + l)$ ← We have n lots of $(a + l)$.

$S_n = \frac{n}{2}(a + l)$ ← Divide both sides of the equation by 2.

but, $l = a + (n - l)\,d$

thus, $S_n = \frac{n}{2}[a + a + (n - l)\,d]$ ← Substitute l in S_n.

$= \frac{n}{2}[2a + (n - 1)\,d]$

Example 7

The 5th term of an arithmetic progression is 21 and the sum of the first six terms is 90. Find the 18th term.

5th term $= a + 4d = 21$

$S_6 = \frac{6}{2}[2a + (6 - 1)\,d] = 90$

Thus, $6a + 15d = 90$

Solving $a + 4d = 21$ and $6a + 15d = 90$ simultaneously gives $a = 5$ and $d = 4$.

18th term $= a + 17d$

$= 5 + 17(4) = 73$

Example 8

The sum of the first n terms of an arithmetic progression is $n^2 + 5n$.

a) Find the nth term. **b)** Write down the first four terms.

a) $S_1 = 1^2 + 5(1) = 6 =$ first term ← Substitute $n = 1$ into $n^2 + 5n$.

$S_2 = 2^2 + 5(2) = 14 =$ first term + second term ← Substitute $n = 2$ into $n^2 + 5n$.

Second term $= S_2 - S_1 = 14 - 6 = 8$

$a = 6, d = 2,$

nth term $= 6 + (n - 1)2 = 2n + 4$

Or

The nth term $= S_n - S_{n-1} = n^2 + 5n - [(n - 1)^2 + 5(n - 1)]$ ← For S_{n-1} put $n - 1$ instead of n.

$= n^2 + 5n - [n^2 - 2n + 1 + 5n - 5]$

$= 2n + 4$

b) First four terms $= 2(1) + 4, 2(2) + 4, 2(3) + 4, 2(4) + 4$

$= 6, 8, 10, 12$

Note: It is useful to remember that the nth term of any progression is $S_n - S_{n-1}$.

Exercise 7.3

1. The 9th term of an arithmetic progression is 8 and the 4th term is 18. Find the 1st term and the common difference, d.

2. The 3rd term of an arithmetic progression is 1 and the 6th term is 10. Find the 4th term.

3. In an arithmetic progression the 8th term is 39 and the 4th term is 19. Find the 1st term and the sum of the first 12 terms.

4. The sum of the terms of an arithmetic progression is 38.5. The 1st term is 1 and the last term is 6. Work out how many terms there are in the arithmetic progression.

5. The nth term of an arithmetic progression is $\frac{1}{2}(7 - n)$. Write down the first three terms and the 20th term.

6. The sum of the first n terms of a series is given by $S_n = 2n(n + 3)$. Show that the terms of the series form an arithmetic progression.

7. In an arithmetic progression the 8th term is twice the 4th term and the 20th term is 40. Find the common difference, d, and the sum of the first 10 terms.

8. In an arithmetic progression the sum of the 1st term and the 5th term is 18. The 5th term is 6 more than the 3rd term. Find the sum of the first 12 terms.

9. Find the sum of the first 30 terms of the arithmetic progression 2, −5, −12, −19, …

10. The sum of the first n terms of an arithmetic progression is $n^2 - 3n$. Write down the 10th term.

11. Show that the sum of the integers from 1 to n is $\frac{1}{2}n(n + 1)$.

12. Find the sum of all the odd numbers between 20 and 100.

13. The 20th term of an arithmetic progression is −41 and the sum of the first 20 terms is −440. Work out the sum of the first 10 terms.

14. Mo has seven pieces of wood. Each piece of wood has a different length. The lengths are in arithmetic progression. The length of the largest piece of wood is 5 times the length of the smallest piece of wood. The total length of all seven pieces of wood is 630 cm. Work out the length of the largest piece.

15. Megan borrows some money from her aunt. She gives her aunt \$10 the first month, \$12 the second month, \$14 the third month. The payments continue to rise by \$2 each month. The final payment she makes to her aunt is \$48. Work out the total number of payments Megan makes to her aunt.

16. The first three terms of an arithmetic progression are $(m - 3)$, $(m + 1)$ and $(5m + 5)$.
    - **a)** Work out the value of m.
    - **b)** Work out the sum of the first 10 terms.

17. **a)** The 15th term of an arithmetic progression is 24. The sum of the first 15 terms is 570. Work out the first three terms.

    b) The sum of the first N terms of this arithmetic progression is 0. Work out the value of N.

18. The 1st term of an arithmetic progression is $3x + 2$. The last term is $13x + 8$ and the common difference is 2.

    Find, in terms of x,

    a) the number of terms

    b) the sum of all the terms.

19. The first four terms of an arithmetic progression are 2, 6, 10, 14. Find the least number of terms needed so that the sum of the terms is greater than 2000.

20. Find the sum of the integers between 1 and 500 that are divisible by 6.

7.4 Geometric progressions

A **geometric progression** is a sequence that has a **common ratio** between each term.

e.g. 3, 6, 12, 24, … This has a common ratio of 2.
e.g. 2, −6, 18, −54, … This has a common ratio of −3.

a = 1st term $(= u_1)$
r = common ratio $(= u_2 \div u_1 = u_3 \div u_2 = u_4 \div u_3 \text{ etc.})$
l = last term in a finite series

A geometric progression with n terms can be written as

$a, ar, ar^2, ar^3, \ldots, \ldots, \ldots, ar^{n-1}$

The **nth term** of a geometric progression can be written as $\boldsymbol{ar^{n-1}}$.

Note: We can use this to write $l = ar^{n-1}$.

We can use this formula to write down an expression for any term.
e.g. 8th term = ar^7 e.g. 20th term = ar^{19}

Example 9

The 3rd term of a geometric progression is 9 and the common ratio is −3.
Find the 6th term and the nth term.

3rd term $= ar^2 = 9$ $\quad a \times (-3)^2 = 9$ $\quad a = \frac{9}{9} = 1$ ← Substitute $r = -3$.

6th term $= ar^5 = 1(-3)^5 = -243$

nth term $= ar^{n-1} = (1)(-3)^{n-1} = (-3)^{n-1}$

Example 10

The nth term of a geometric progression is $10\left(\frac{1}{2}\right)^{n-1}$. Find the 1st term and the common ratio.

1st term $= u_1 = 10\left(\frac{1}{2}\right)^0 = 10$ ← Substitute $n = 1$.

$a = 10$

2nd term $= u_2 = 10\left(\frac{1}{2}\right)^1 = 5$ ← Substitute $n = 2$.

$r =$ 2nd term ÷ 1st term $= 5 \div 10$

$r = \frac{1}{2}$

The **sum of a geometric progression** can be written as $S_n = \frac{a(1-r^n)}{1-r}$, $r \neq 1$.

We use this formula when $-1 < r < 1$.

When we do not have $-1 < r < 1$, we use the formula $S_n = \frac{a(r^n - 1)}{r-1}$.

Proof:

$S_n = a + ar + ar^2 + \ldots + ar^{n-3} + ar^{n-2} + ar^{n-1}$

$rS_n = \quad ar + ar^2 + \ldots + ar^{n-3} + ar^{n-2} + ar^{n-1} + ar^n$ ← Multiply each term by r.

$S_n - rS_n = a - ar^n$ ← Subtract the 2nd line from the lst line.

$S_n(1 - r) = a(1 - r^n)$ ← Factorise each side.

$S_n = \frac{a(1-r^n)}{1-r}$ ← Use when $|r| < 1$.

If we multiply by $\frac{-1}{-1}$ we get $S_n = \frac{a(r^n-1)}{r-1}$ ← Use when $|r| > 1$.

Note: This is not defined when $r = 1$.

Example 11

The 3rd term of a geometric progression is 16 and the 6th term is -128.

Find the 1st term and the sum of the first seven terms.

$ar^2 = 16 \qquad ar^5 = -128$

$\frac{ar^5}{ar^2} = -\frac{128}{16}$ ← With geometric progressions we **divide** the equations to solve simultaneously.

$r^3 = -8 \qquad r = -2$

$ar^2 = 16 \qquad a(-2)^2 = 16 \qquad a = 16 \div 4 = 4$

$S_7 = \frac{a(r^6-1)}{r-1} = \frac{4[(-2)^7-1]}{-2-1}$ ← Substitute $a = 4$, $r = -2$ and $n = 6$.

$= \frac{4[(-128-1]}{-3} = 172$

Example 12

Mabintou deposits \$100 in a bank account at the start of each year.
She earns 4% compound interest.

a) Work out how much money she has in her bank account at the end of the 3rd year.

b) Work out how much money she has in her account at the start of the 30th year after she has made her annual deposit.

a) At the end of the third year the amount in her account is
$\$100(1.04) + \$100(1.04)^2 + \$100(1.04)^3$
$= \$104 + \$108.16 + \$112.4864 = \$324.65$

b) At the start of the 30th year, in \$, she has
$100 + 100(1.04) + 100(1.04)^2 + \ldots + 100(1.04)^{29}$

$a = 100 \qquad r = 1.04 \qquad n = 30$

$S_{30} = \dfrac{100(1.04^{30} - 1)}{1.04 - 1} = \$5608.49$ (to the nearest cent)

Example 13

An arithmetic progression has first term a and common difference d.
The 1st, 3rd and 13th terms of the arithmetic progression form the first three terms of a geometric progression with common ratio r.

a) Express d in terms of a, and find the value of r.

b) The 4th term in the geometric progression also appears in the arithmetic progression. Determine which term it is.

a) a, $a + 2d$, $a + 12d$ are the first three terms in the geometric progression.

So $(a + 2d)^2 = a(a + 12d)$

$a^2 + 4ad + 4d^2 = a^2 + 12ad$

$4d^2 = 8ad$

$d = 2a$

So $a + 2d = a + 2 \times 2a = 5a$

The 2nd term of the geometric progression is $5a$, so $r = 5$.

b) The 4th term in the geometric progression is $ar^3 = 125a$.

If this is the nth term of the arithmetic progression with first term a and common difference $2a$:

$125a = a + (n - 1)(2a) \Rightarrow n = 63$

So the 4th term in the geometric progression is the 63rd term in the arithmetic progression.

These are the 1st, 3rd and 13th terms of an arithmetic progression.

If a, b and c are consecutive terms in the geometric progression, then $\dfrac{b}{a} = \dfrac{c}{b}$, so $b^2 = ac$.

Expand the bracket and simplify the arithmetic progression.

Substitute $d = 2a$.

2nd term ÷ 1st term = $5a \div a$

Substitute $r = 5$.

Exercise 7.4

1. Find which of the following series are geometric progressions.
   If they are, write down the common ratio.

   a) $5 + 15 + 45 + 135 + \dots$
   b) $-60 + 30 - 15 + 7.6 - 3.8 + \dots$
   c) $\frac{1}{2} + \frac{1}{3} + \frac{1}{4} + \frac{1}{5} + \dots$
   d) $\frac{1}{3} + \frac{1}{12} + \frac{1}{48} + \frac{1}{192}$
   e) $1 - 1 + 1 - 1 + \dots$
   f) $0.8 - 0.16 + 0.032 - 0.0064 + \dots$

2. The 1st and 2nd terms of a geometric progression are 4 and 12, respectively.
   Find the sum of the first ten terms.

3. The 3rd and 4th terms of a geometric progression are 5 and -20, respectively.
   Find the 1st term, the common ratio, r, and the 9th term.

4. The 4th and 6th terms of a geometric progression are 200 and 800, respectively. Find two possible values for the common ratio, r, and two possible values for the 1st term.

5. Find the 13th term of the geometric progression 1.1, 1.21, 1.331, …

6. Find the sum to 11 terms of the geometric progression $\frac{1}{2} - \frac{1}{4} + \frac{1}{8} - \frac{1}{16} + \dots$

7. The 1st term of a geometric progression is 1 and the 5th term is 0.0016.
   Find the value of r and the sum of the first eight terms.

8. A geometric progression has a common ratio of $\frac{2}{5}$. The sum of the first four terms is $2\frac{13}{25}$.

   Find the 1st term of the progression.

9. The 4th term of a geometric progression is 9 times the 6th term. The 5th term is 3 and $r > 0$.
   Find the sum of the first six terms.

10. The nth term of a geometric progression is $\left(-\frac{1}{3}\right)^n$. Find the 1st term and the common ratio, r.

11. The sum of the first nine terms of a geometric sequence is 684 and the common ratio is -2.
    Find the 1st and the 8th terms.

12. A geometric sequence is $1 - 2x + 4x^2 - 8x^3 + \dots$ Find an expression for the nth term.

13. $x + 5$, x and $x - 4$ are three consecutive terms of a geometric progression.
    Find the value of x.

14. The sum to n terms of a geometric progression is $3^n - 1$.
    Find the first four terms and the common ratio, r.

15. A clock has a pendulum. The time it takes for the pendulum's first swing is 5 seconds.
    The time for each successive swing is $\frac{3}{5}$ of the time of the previous swing.
    Find how long it takes the pendulum to do six swings.
    Give your answer to the nearest second.

16. Kwame asks his father for some money. He asks for 1¢ on the first day, 2¢ on the second day, 4¢ on the third day, 8¢ on the fourth day, etc. He wants his father to continue to double the money each day. Calculate how much money he would get from his father after 30 days, if his father agreed to pay him. Give your answer in dollars ($).

17. The first four terms of a geometric progression are 1, 3, 9, 27.
Find the smallest number of terms that will give a total greater than 265 000.

18. The sum of the 1st and 2nd term of a geometric progression is 1.
The sum of the 4th and 5th terms is 27. Work out the common ratio, r, and the 1st term.

19. Each year the value of a company increases by 2%. At the start of 2012, the value of the company was $24 000. Estimate the value of the company at the end of 2025. Write your answer to the nearest $1000.

20. Three consecutive terms of a geometric progression are $x - 1$, $x^2 - 1$ and $x + 1$.
Find two possible values of x.

7.5 Infinite geometric progressions

Consider the geometric progression $a + ar + ar^2 + \ldots + ar^{n-1}$

We know that $S_n = \frac{a(1-r^n)}{1-r}$ when $|r| < 1$.

As n gets larger and larger, r^n gets smaller and smaller.
This means that as $n \to \infty$, $r^n \to 0$.

Thus, for an infinite series, $S_n \to \frac{a(1-0)}{1-r} = \frac{a}{1-r}$.

S_∞ is the sum to infinity of a geometric progression.

Note: This is only the case when $|r| < 1$ as if $|r| > 1$, r^n will be very large and S_n will tend to ∞.

$$S_\infty = \frac{a}{1-r} \text{ provided } |r| < 1$$

Example 14

The nth term of a geometric progression is $81\left(-\frac{1}{3}\right)^n$. Find the sum to infinity.

1st term: $u_1 = 81\left(-\frac{1}{3}\right)^1 = -27$ ← Substitute $n = 1$.

$a = -27$

2nd term: $u_2 = 81\left(-\frac{1}{3}\right)^2 = 9$ ← Substitute $n = 2$.

$r = \frac{u_2}{u_1} = \frac{9}{-27} = -\frac{1}{3}$

$S_\infty = \frac{a}{1-r} = -\frac{27}{1-\left(-\frac{1}{3}\right)} = -20\frac{1}{4}$ ← Substitute $a = -27$ and $r = -\frac{1}{3}$.

Example 15

a) Write the recurring decimal $0.4\dot{5}$ as the sum of a geometric progression.

b) Use your answer to part (**a**) to show that $0.4\dot{5}$ can be written as $\frac{41}{90}$.

a) $0.4\dot{5} = 0.4 + 0.0\dot{5} = 0.4 + \frac{5}{100} + \frac{5}{1000} + \frac{5}{10000} + \ldots$

b) Consider $\frac{5}{100} + \frac{5}{1000} + \frac{5}{10000} + \ldots$ $\quad a = \frac{5}{100}\ r = \frac{1}{10}$

$$S_\infty = \frac{a}{1-r} = \frac{\frac{5}{100}}{1-\frac{1}{10}} = \frac{5}{100} \times \frac{10}{9} = \frac{5}{90}$$

Thus $0.4\dot{5} = 0.4 + 0.0\dot{5} = \frac{2}{5} + \frac{5}{90} = \frac{41}{90}$

Exercise 7.5

1. The first four terms of a geometric progression are $3, -1, \frac{1}{3}, -\frac{1}{9}$.
   Find the sum to infinity.

2. A geometric progression has a common ratio of $\frac{3}{4}$ and a sum to infinity of 92. Find the 1st term.

3. A geometric progression has a sum to infinity of -30. The 1st term of the geometric progression is -9. Find the common ratio, r.

4. **Calculate** the sum to infinity of the series $125 + 75 + 45 + 27 + \ldots$

5. By expressing the recurring decimal $0.\dot{3}\dot{6}$ as the sum of a geometric progression, write $0.\dot{3}\dot{6}$ as a fraction in its simplest form.

6. The first four terms of a geometric progression are $1, \frac{x}{x+1}, \frac{x^2}{(x+1)^2}, \frac{x^3}{(x+1)^3}$.
   Find the sum to infinity of this geometric progression.

7. The nth term of a geometric progression is $(-1)^n\,(2)^n$.

   a) Find the first three terms.

   b) Explain why this geometric progression does not have a sum to infinity.

8. A geometric progression has the sum to infinity equal to twice the 1st term.
   Find the common ratio, r.

9. The 1st and 3rd term of a geometric progression are 18 and 2, respectively.
   Find two possible values of the common ratio, r, and the sum to infinity for each.

10. The sum of the first two terms of a geometric progression is 10. The sum to infinity of the geometric progression is 18. Work out the two possible values of the common ratio, r.

11. All the terms of a geometric progression are positive. The sum of the first two terms is 100 and the sum to infinity is 180.
Find the first term and the common ratio.

Summary exercise 7

EXAM-STYLE QUESTIONS

1. The 10th and the 20th terms of an arithmetic progression are 18 and 88, respectively.
Find the 1st term and the sum of the first 16 terms.

2. A geometric progression has a 1st term of 80. The sum of the first two terms is 40. Find the common ratio, r, and the sum to infinity of the geometric progression.

3. Work out $\sum_{r=3}^{6}(2r-1)^2$.

4. The 1st term and the last term of an arithmetic progression are 15 and -20, respectively. The sum of all the terms is -40. Work out how many terms there are in the arithmetic progression.

5. The 1st term of a geometric progression is 5. The sum to infinity of the geometric progression is 20. Find the common ratio, r.

6. The sum of the first six terms of an arithmetic progression is 51 and the sum of the first 12 terms of the same progression is 282. Find the 1st term and the common difference, d.

7. Find the sum of the first nine terms of a geometric sequence that has a 3rd term of 45 and a 6th term of 1215.

8. The sum of the first three terms of an arithmetic progression is 15. The product of the 1st and 3rd terms is -24.
Find two possible values for the 1st term.

9. The 4th term of a geometric series is 6.
The 7th term is 162.

a) Find the 1st term and the common ratio, r.

b) Show that the sum of the first n terms can be written as $S_n = \frac{3^n - 1}{9}$.

10. The first three terms of an arithmetic progression are $\frac{1}{2}$, x and 25. The first three terms of a geometric progression are $x + \frac{1}{4}$, $32\frac{1}{2}$ and y, where x and y are positive numbers. Find the value of x and the value of y.

11. Find the first three terms of the sequence defined by $u_{n+1} = 1 - \frac{3}{1-2u_n}$ where $u_1 = 1$.

12. Nita invests \$50 at the beginning of each year for 10 years at 4% compound interest. Calculate the total value of the investment at the end of the 10 years.

13. A geometric progression has its 1st term and 3rd term as $-\frac{3}{7}$ and $-\frac{12}{175}$, respectively. The common ratio, r, is positive. Find r and the sum to infinity of this geometric progression.

14. Work out the number of positive terms in the arithmetic progression $200 + 194 + 188 + 182 + \ldots$

15. The nth term of a geometric progression is $(-1)^{n+1}(3)^n(-x)^{-n}$. Find the common ratio, r.

16. The 1st term of a finite arithmetic progression is 18 and the common difference, d, is $-2\frac{1}{2}$. The sum of the terms is -12. Find the number of terms.

17. Find the sum of all the positive integers from 1 to 300 that are not divisible by 3.

18. The 16th term of an arithmetic progression is 53 and the 20th term is 65. Find the sum of all the odd numbers in the sequence that are less than 100.

19. The first three terms of a series are $x - 3$, $x + 1$ and $3x - 5$.

a) If the series is an arithmetic progression, find the first three terms.

b) **i)** If the series is a geometric progression, find two possible values of x.

ii) Explain why there is no sum to infinity for either of these values of x.

20. The 7th term of an arithmetic progression is 20. The 1st, 3rd and 11th terms of the arithmetic progression are the first three terms of a geometric progression. Find the common ratio, r, of this geometric progression.

21. A geometric progression has first term a and common ratio r. The 1st, 2nd and 4th terms of the geometric progression form the first three terms of an arithmetic progression.

a) Show that $(r - 1)(r^2 + r - 1) = 0$.

b) Given that the sum to infinity of the geometric progression is finite, find the exact value of r.

c) Find the sum to infinity of the geometric progression in terms of a.

22. In a geometric progression, the sum to infinity is five times the first term.

a) Find the common ratio.

b) The 2nd term is 12. Find the 1st term.

c) Find the percentage difference between the sum of the first ten terms and the sum to infinity, correct to 3 significant figures.

23. The first term of a geometric progression is 10 and the common ratio is -2.

The first term of an arithmetic progression is 10 and the common difference is -2.

Find the difference between the sums of the first ten terms of the two sequences.

24. The 12th term of an arithmetic progression is equal to three times the 2nd term. The 17th term of the progression is 60.
Find the sum of the first 50 terms.

25. The first three terms of a geometric progression are $3p$, $5p + 5$, $8p + 20$ respectively.

a) Show that $p^2 - 10p + 25 = 0$.

b) Hence show that $p = 5$ and find the common ratio of this progression.

c) Find the sum of the first 10 terms of the progression.

Chapter summary

Terms of a sequence

- u_n can be used to describe any term in a sequence. u_1 is the first term.
- The term after u_n is u_{n+1}.
- The term before u_n is u_{n-1}.

Finite and infinite series

- A series is **finite** if it has a finite number of terms, e.g. $5 + 9 + 13 + 17 + 21$.
- A series is **infinite** if it continues indefinitely, e.g. $5 + 9 + 13 + 17 + 21 + \ldots$
- The **general term** of the series $5 + 9 + 13 + 17 + 21 + \ldots$ is $4r + 1$.
- The finite series $5 + 9 + 13 + 17 + 21 + 25$ can be written as $\sum_{r-1}^{6}(4r+1)$.
- The infinite series $5 + 9 + 13 + 17 + 21 + \ldots$ can be written as $\sum_{r-1}^{\infty}(4r+1)$.

Arithmetic progressions

- a = 1st term $(= u_1)$
- d = common difference $(= u_2 - u_1 = u_3 - u_2 = u_4 - u_3$ etc.)
- l = last term in a finite series
- An arithmetic progression with n terms can be written as $a, (a + d), (a + 2d), (a + 3d), \ldots, \ldots, \ldots, [a + (n - 1)d]$.
- The ***n*th term** of an arithmetic progression can be written as $\mathbf{[a + (n - 1)d]}$.
- The ***n*th term** is $l = a + (n - 1)d$.
- We can use this formula to write down an expression for any term. e.g. 8th term $= a + 7d$ e.g. 20th term $= a + 19d$.

Geometric progressions

- A **geometric progression** is a series that has a **common ratio** between each term.
- a = 1st term $(= u_1)$
- r = common ratio $(= u_2 \div u_1 = u_3 \div u_2 = u_4 \div u_3$ etc.)
- l = last term in a finite series
- A geometric progression with n terms can be written as $a, ar, ar^2, ar^3, \ldots, \ldots, \ldots, ar^{n-1}$.
- The ***n*th term** of a geometric progression can be written as ar^{n-1}.
- The ***n*th term** is $l = ar^{n-1}$.
- We can use this formula to write down an expression for any term. e.g. 8th term $= ar^7$ e.g. 20th term $= ar^{19}$
- The **sum of a geometric progression** can be written as $S_n = \dfrac{a(1 - r^n)}{1 - r}$.
- The sum to infinity of a geometric progression is $S_\infty = \dfrac{a}{1 - r}$ provided $|r| < 1$.

Maths in real-life

Infinity

People have been interested in the existence of infinity and its relation to reality as far back as Aristotle.

Aristotle distinguished between two types of infinity. Potential infinities are never-ending lists, for example the set of real numbers. Actual infinities are when something that can be measured becomes infinite, for example a black hole which is formed of an infinitely dense mass at a single point.

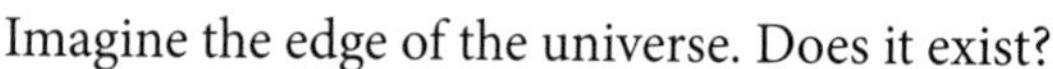

Imagine the edge of the universe. Does it exist?
It is tempting to conclude that the universe does not have an edge and is therefore infinite. However, we can travel around the world without ever finding an edge and yet the earth has a finite area. Whether the universe is finite or infinite is a question that still hasn't been answered.

Zeno of Elea (5th century BC) was a Greek philosopher whose paradoxes, known through the writings of Aristotle, may have significantly influenced the Greeks' perception of magnitude and number.
The four paradoxes of motion are concerned with whether or not space and time are fundamentally made up of minute indivisible parts.

In the paradox of Achilles and the Tortoise, Achilles is in a race with the tortoise. Achilles allows the tortoise a head start. Achilles is faster than the tortoise but when he reaches the tortoise's starting point he finds that the tortoise has moved on – and when Achilles gets to that point, the tortoise has moved on further. This argument can be repeated infinitely many times – so it appears that Achilles cannot catch or pass the tortoise. However this is obviously incorrect, as we know Achilles will pass the tortoise.

The fallacy in the argument is that the distance covered (and the time taken) are infinite sums of decreasing terms which have a finite sum – and time and distance move on beyond those points in time or space.

Georg Cantor (1845–1918) was a German mathematician who was responsible for the establishment of set theory and for profound developments in the notion of the infinite. In 1873, he showed that the set of rational numbers is countable. He also showed that the set of real numbers is not countable, and he showed that the power set of a set (the set of all subsets of the set) cannot be put into a one–one correspondence with the elements of the set. This leads us to the strange conclusion that there are different degrees (sizes) of infinite numbers.

Another German mathematician, David Hilbert (1862–1943), further illuminated the strange nature of infinite, but countable sets by stating his 'Infinite hotel paradox'. The paradox is this: if a hotel with infinitely many rooms is full and another guest arrives, that guest can be accommodated by each existing guest moving from their current room to the room with the next highest number, leaving room number 1 free for the new arrival. If an infinite number of extra guests arrive, they could be accommodated by each existing guest moving to the room number which is twice their existing room number, leaving an infinite number of odd numbered rooms available for the new arrivals. It therefore seems that whatever we add to infinity, we are still left with infinity.

However, what happens when a guest leaves? Are we still left with infinitely many guests? What happens when an infinite number of guests leave?

Infinity is a very strange concept and cosmologists are struggling to resolve dilemmas similar to Zeno's paradoxes because the mathematics in their current theories are leading them to counter-intuitive conclusions, such as the existence of a multiverse – an infinite number of universes! Because this is an area of science that we have no way of testing empirically, physicists are having to wrestle with big philosophical questions about infinity and the nature of the universe (or multiverse).

8 Differentiation

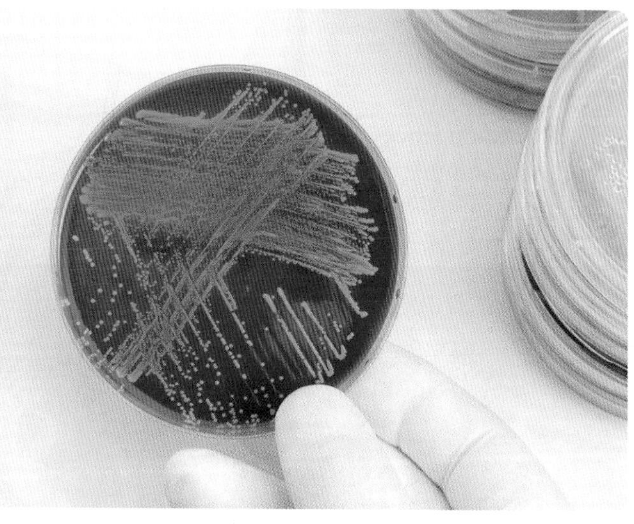

As calculus is a study of the rates of change between variables it is used in a huge number of the real-world applications of mathematics. Scientists use differentiation to analyse the growth of bacteria, population change, the number of hours of daylight throughout the year, radioactive decay and inflation rates in finance to name a few.

Objectives

- Understand the idea of gradient of a curve at a point as the limit of the gradients of a suitable sequence of chords.
- Use the notations $f'(x)$, $f''(x)$, $\frac{dy}{dx}$ and $\frac{d^2y}{dx^2}$ for first and second derivatives.
- Use the derivative of x^n (for any rational n) together with constant multiples, sums, differences of functions and of composite functions using the chain rule.
- Apply differentiation to gradients, tangents and normals.

Before you start

You should know how to:

1. Write rational expressions in the form ax^n,

e.g. $\frac{3}{x^2} = 3x^{-2}$ **e.g.** $\sqrt{x} = x^{\frac{1}{2}}$

e.g. $\frac{4}{\sqrt{x}} = 4x^{-\frac{1}{2}}$ **e.g.** $\frac{1}{\sqrt[3]{x^5}} = x^{-\frac{5}{3}}$

2. Deal with indices, for example making the powers of x one less,

e.g. x^3; $x^{3-1} = x^2$ **e.g.** x^{-2}; $x^{-2-1} = x^{-3}$

e.g. $x^{\frac{1}{2}}$; $x^{\frac{1}{2}-1} = x^{-\frac{1}{2}}$ **e.g.** $\sqrt{x^5}$; $x^{\frac{5}{2}-1} = x^{\frac{3}{2}}$

e.g. $\frac{1}{\sqrt{x}}$; $x^{-\frac{1}{2}-1} = x^{-\frac{3}{2}}$

Skills check:

1. Write the following in the form ax^n.

a) $\frac{2}{x^3}$ **b)** $4\sqrt{x}$ **c)** $\frac{2}{\sqrt{x}}$

d) $\frac{3}{\sqrt[3]{x}}$ **e)** $\sqrt{x^5}$ **f)** $\frac{1}{\sqrt{x^3}}$

2. Make the powers of x one less.

a) x^8 **b)** x^{-3} **c)** $x^{\frac{3}{2}}$

d) $\frac{1}{x}$ **e)** x^{-4} **f)** $\sqrt{x^3}$

g) $\sqrt[3]{x}$ **h)** $\frac{1}{x^2}$ **i)** $\frac{1}{\sqrt{x^7}}$

8.1 The gradient of the tangent

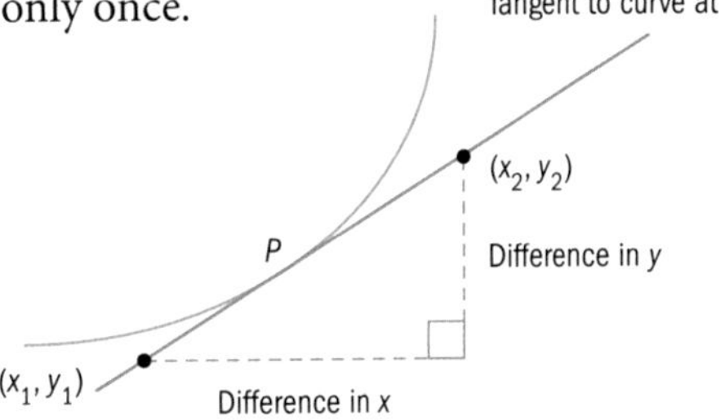

A tangent to a curve is a straight line that touches the curve only once. The tangent to a curve at any point P is the straight line that touches the curve at P.

The gradient **of the curve** at any point P is defined as the gradient of the tangent at P.

The gradient of the tangent at $P = \dfrac{y_2 - y_1}{x_2 - x_1}$

and represents the rate of change of y with respect to x.

Example 1

Estimate the gradient of the tangent to the curve $y = (x - 2)^2$ at the point (4, 4).

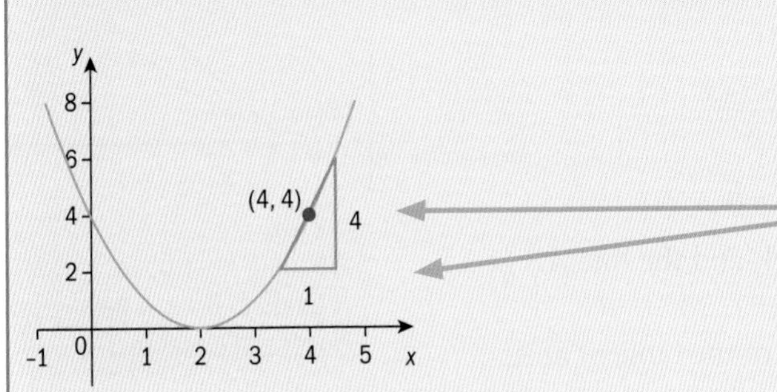

We can only *estimate* the gradient as it is difficult to draw the tangent accurately.

Hint: Draw the base of the triangle with an integer value.

Gradient of the tangent at (4, 4) $= \dfrac{4}{1} = 4$

Exercise 8.1

1. $y = x^2 + x - 2$
   a) Using a scale of 1 cm to 1 unit, plot the graph of $y = x^2 + x - 2$ for $-3 \le x \le 3$.
   b) Estimate the gradient of the tangent to the curve at the point (2, 4).
   c) Estimate the gradient of the tangent to the curve at the point (−1, −2).

2. The grid shows the graph of $y = x^3 - 3x + 1$.
   a) Estimate the gradient of the tangent to the curve at the point where $x = -1.5$.
   b) Write down the coordinates of the points on the curve where the gradient of the tangent equals 0.

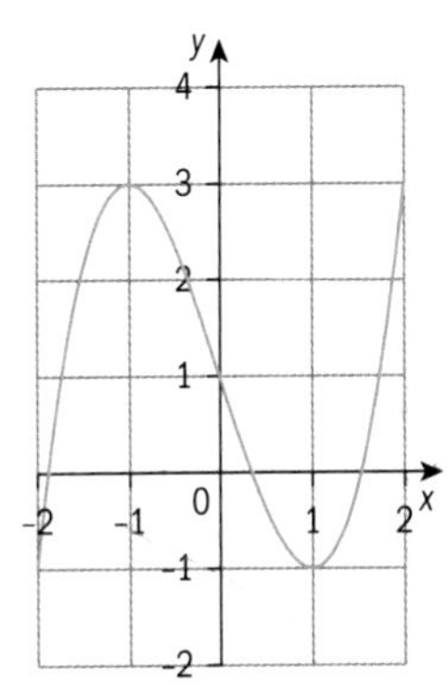

8.2 Gradient of a tangent as a limit

Consider the problem of finding the gradient of the tangent at point A on the curve $y = \mathrm{f}(x)$.
If B is another point on the curve, fairly close to A, then the gradient of the chord AB gives an approximate value for the gradient of the tangent at A.
As we move B closer and closer to A (i.e. B_1, B_2, B_3, B_4), the approximate value for the gradient of the tangent at A gets more accurate.

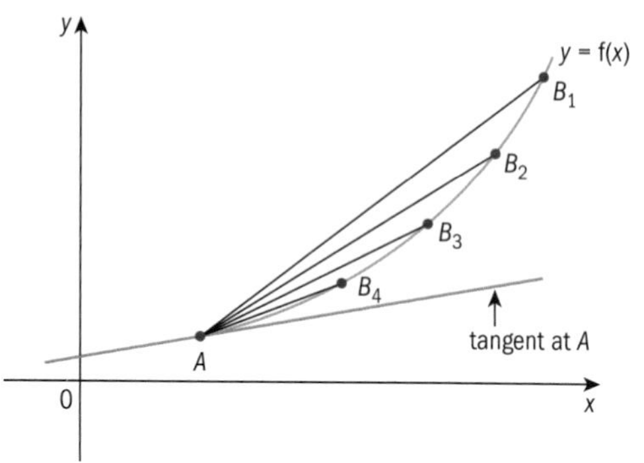

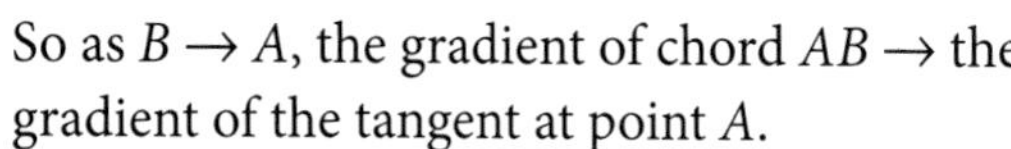

So as $B \rightarrow A$, the gradient of chord $AB \rightarrow$ the gradient of the tangent at point A.

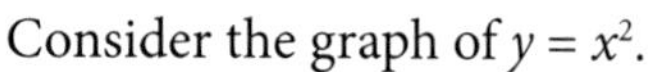

Consider the graph of $y = x^2$.
The point A has coordinates (x, x^2).
The point B has coordinates $((x + h), (x + h)^2)$.

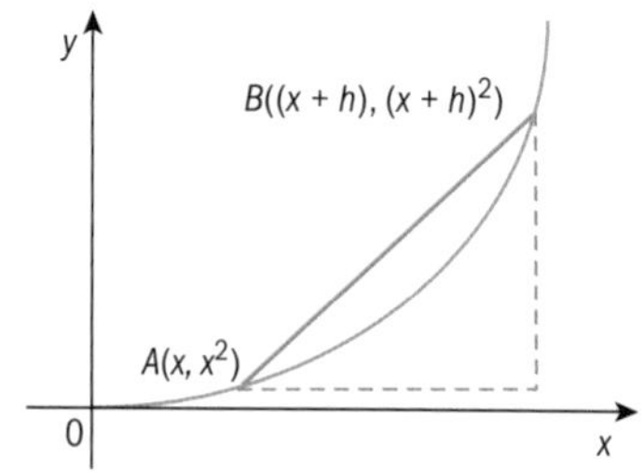

The gradient of $AB = \dfrac{(x+h)^2 - x^2}{(x+h) - x} = \dfrac{x^2 + 2xh + h^2 - x^2}{x + h - x}$

$$= \frac{2xh + h^2}{h} = \frac{h(2x+h)}{h} = 2x + h.$$

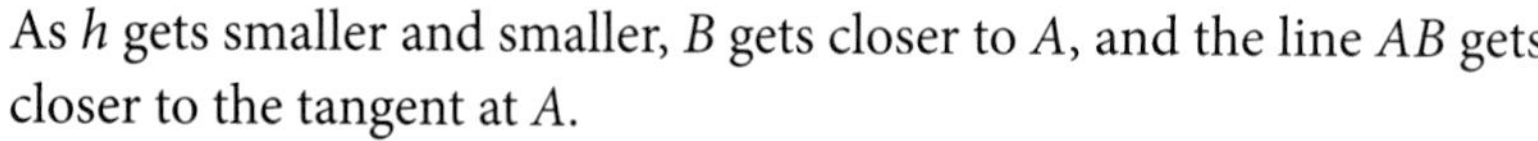

As h gets smaller and smaller, B gets closer to A, and the line AB gets closer to the tangent at A.
We say that as $h \rightarrow 0$, we can find the gradient of the tangent at A.

As $h \rightarrow 0$, the gradient of $AB \rightarrow 2x$.

Thus, the gradient of the tangent at any point on the curve $y = x^2$ is $2x$.

We read '$h \rightarrow 0$' as **h tends to 0**, i.e. h gets closer and closer to 0.

The definition of $\mathrm{f}'(x)$ is given as $\lim\limits_{h \rightarrow 0} \dfrac{\mathrm{f}(x+h) - \mathrm{f}(x)}{h}$.

Note: The gradient of the tangent at any point is the rate of change of y with respect to x.

Example 2

Find an expression for $\mathrm{f}'(x)$ when $\mathrm{f}(x) = x^3$.

$$\mathrm{f}'(x) = \lim_{h \rightarrow 0} \frac{\mathrm{f}(x+h) - \mathrm{f}(x)}{h}$$

$$= \lim_{h \rightarrow 0} \frac{(x+h)^3 - x^3}{h}$$

$\mathrm{f}(x) = x^3$ so $\mathrm{f}(x + h) = (x + h)^3$

$$= \lim_{h \rightarrow 0} \frac{x^3 + 3x^2h + 3xh^2 + h^3 - x^3}{h}$$

$(x + h)^3 = x^3 + 3x^2h + 3xh^2 + h^3$

$$= \lim_{h \rightarrow 0} \frac{h(3x^2 + 3xh + h^2)}{h}$$

Simplify and factorise.

$$= \lim_{h \rightarrow 0} 3x^2 + 3xh + h^2$$

$$= 3x^2$$

As $h \rightarrow 0$, we get $3x^2 + 0 + 0$.

When $\mathrm{f}(x) = x^3$, $\mathrm{f}'(x) = 3x^2$.

Exercise 8.2

By considering the gradient of the chord joining $(x, f(x))$ and $((x + h), f(x + h))$ for the following functions $f(x)$, deduce what $f'(x)$ will be.

1. $f(x) = 2x^2$

Hint: $f(x + h) = 2(x + h)^2$

2. $f(x) = 8x$

3. $f(x) = x^4$

Hint: $(x + h)^4 = x^4 + 4x^3h + 6x^2h^2 + 4xh^3 + h^4$

4. $f(x) = x^2 + 3x$

5. $f(x) = \frac{1}{x}$

6. $f(x) = x^n$

8.3 Differentiation of polynomials

Notation:

For functions written in the form $y = f(x)$, the gradient function (the derivative) is $f'(x)$.

For functions written in the form y is a function in x, the derived function is $\frac{dy}{dx}$.

$\frac{dy}{dx}$ is the derivative of y with respect to x.

$\frac{d}{dx}$ means differentiate with respect to x.

We can differentiate terms in x using these rules:

If $y = x^n$ then $\frac{dy}{dx} = nx^{n-1}$

If $y = ax^n$ then $\frac{dy}{dx} = anx^{n-1}$

Example 3

Find the derived function in each case.

a) $f(x) = x^8$

b) $f(x) = \frac{6}{x^4}$

c) $f(x) = 3x$

d) $f(x) = 5$

▶ Continued on the next page

a) $f(x) = x^8$

$f'(x) = 8x^{8-1} = 8x^7$

Use nx^{n-1}.

b) $f(x) = \frac{6}{x^4} = 6x^{-4}$

First write the power as a negative power.

$f'(x) = -24x^{-5}$

$= -\frac{24}{x^5}$

$-4 \times 6 = -24$ and -5 is one less than -4.

c) $f(x) = 3x$

$f'(x) = 3$

$f(x) = 3x^1$, $f'(x) = 3x^{1-1} = 3x^0 = 3 \times 1 = 3$

d) $f(x) = 5 = 5x^0$

$f'(x) = 0$

Note: When we differentiate a constant, c, we always get 0. This makes sense as the graph of $f(x) = c$ is a horizontal straight line.

Example 4

Find $\frac{dy}{dx}$ when $y = 3\sqrt{x}$.

$y = 3\sqrt{x} = 3x^{\frac{1}{2}}$

First write $\sqrt{x}$ as $x^{\frac{1}{2}}$.

$\frac{dy}{dx} = \frac{3}{2}x^{-\frac{1}{2}}$

$= \frac{3}{2\sqrt{x}}$

$x^{-\frac{1}{2}} = \frac{1}{\sqrt{x}}$

Exercise 8.3

1. Find the derived function in each case.

a) $f(x) = x^3$ b) $f(x) = x^9$ c) $f(x) = 4x^2$ d) $f(x) = -15x$

e) $f(x) = 7x^{-4}$ f) $f(x) = 3x^{-1}$ g) $f(x) = -\frac{1}{2}x$ h) $f(x) = \frac{4}{5}x^{10}$

i) $f(x) = \sqrt{x}$ j) $f(x) = -4\sqrt{x}$ k) $f(x) = \frac{2}{x}$ l) $f(x) = \frac{7}{x^2}$

2. Find $\frac{dy}{dx}$ in each case.

a) $y = 2x^{\frac{1}{2}}$ b) $y = -\frac{5}{x}$ c) $y = \frac{\sqrt{x}}{3}$ d) $y = 12x^7$

e) $y = \frac{8}{x^3}$ f) $y = \frac{3}{\sqrt{x}}$ g) $y = \frac{1}{\sqrt[3]{x}}$ h) $y = \frac{2}{9}x$

i) $y = x \times x^5$ j) $y = 12x^8 \div 3x$ k) $y = 4x \times 5x^2$ l) $y = -4x$

Hint: Write the expression in the form ax^n before differentiating.

3. Find

a) $\dfrac{\mathrm{d}}{\mathrm{d}x}(x^{-5})$ b) $\dfrac{\mathrm{d}}{\mathrm{d}x}\left(x^{\frac{1}{3}}\right)$ c) $\dfrac{\mathrm{d}}{\mathrm{d}x}\left(4x^{-\frac{1}{4}}\right)$

d) $\dfrac{\mathrm{d}}{\mathrm{d}x}(-2x)$ e) $\dfrac{\mathrm{d}}{\mathrm{d}x}\left(\dfrac{1}{x^2}\right)$ f) $\dfrac{\mathrm{d}}{\mathrm{d}x}(5\sqrt{x})$

g) $\dfrac{\mathrm{d}}{\mathrm{d}x}\left(-\dfrac{1}{2x^3}\right)$ h) $\dfrac{\mathrm{d}}{\mathrm{d}x}\left(\sqrt[3]{8x^2}\right)$ i) $\dfrac{\mathrm{d}}{\mathrm{d}x}(x \div x^6)$

j) $\dfrac{\mathrm{d}}{\mathrm{d}x}\left(\dfrac{x^9}{x^3}\right)$ k) $\dfrac{\mathrm{d}}{\mathrm{d}x}(x^7 \times x^4)$ l) $\dfrac{\mathrm{d}}{\mathrm{d}x}\left(\dfrac{8x^{12}}{2x^4}\right)$.

4. Find the derivative in each case.

a) $y = 4x^2$ b) $y = -\dfrac{2}{3}x$ c) $y = \dfrac{5}{7}x^{21}$ d) $y = 16\sqrt{x}$

e) $y = \dfrac{9}{\sqrt[3]{x}}$ f) $y = -\dfrac{4}{\sqrt{x}}$ g) $y = -5x$ h) $y = 25x^{\frac{1}{5}}$

i) $y = (5x^9 \times 2x)$ j) $y = \left(x^{\frac{3}{2}} \div x^{\frac{1}{2}}\right)$ k) $y = -x^2 \times -3x^3$ l) $y = -x$

Hint: In parts **i)**, **j)** and **k)**, simplify your expression before differentiating.

8.4 Differentiation of more complex functions

We can differentiate more complex functions by first ensuring we write them as the sum of powers of x.
We then differentiate each term using the same rules as before.

Example 5

Find $\dfrac{\mathrm{d}y}{\mathrm{d}x}$ when

a) $y = x^5 + 4x^4$ b) $y = (2x - 5)^2$ c) $y = \dfrac{3x^6 + 8}{x^2}$.

a) $y = x^5 + 4x^4$

$\dfrac{\mathrm{d}y}{\mathrm{d}x} = 5x^4 + 16x^3$ ← Differentiate term by term.

b) $y = (2x - 5)^2 = 4x^2 - 20x + 25$ ← First expand the brackets.

$\dfrac{\mathrm{d}y}{\mathrm{d}x} = 8x - 20$ ← When you differentiate a constant it is 0.

c) $y = \dfrac{3x^6 + 8}{x^2} = \dfrac{3x^6}{x^2} + \dfrac{8}{x^2} = 3x^4 + 8x^{-2}$  First divide each term by x^2.

$\dfrac{\mathrm{d}y}{\mathrm{d}x} = 12x^3 - 16x^{-3}$

$= 12x^3 - \dfrac{16}{x^3}$

Example 6

Find $\frac{d}{dx}\left(24\sqrt[3]{x} - \frac{1}{2\sqrt{x}}\right)$.

$$\frac{d}{dx}\left(24\sqrt[3]{x} - \frac{1}{2\sqrt{x}}\right) = \frac{d}{dx}\left(24x^{\frac{1}{3}} - \frac{1}{2}x^{-\frac{1}{2}}\right)$$

First write all terms in the form ax^n.

$$= 8x^{-\frac{2}{3}} + \frac{1}{4}x^{-\frac{3}{2}}$$

$$= \frac{8}{x^{\frac{2}{3}}} + \frac{1}{4x^{\frac{3}{2}}}$$

Write the expression for the derivative in terms of algebraic fractions.

$$= \frac{8}{\sqrt[3]{x^2}} + \frac{1}{4\sqrt{x^3}}$$

Exercise 8.4

1. Find the derived function in each case.

a) $f(x) = x^7 + x^3$ **b)** $f(x) = 6x^8 - 4x$ **c)** $f(x) = 3x^2 + 5x - 9$

d) $f(x) = (3x + 7)(2x - 9)$ **e)** $f(x) = x^{-4}(x^2 + x^6)$ **f)** $f(x) = 5x^2(3x^2 - 7x - 2)$

g) $f(x) = \frac{x^5 + x^8}{x^2}$ **h)** $f(x) = \frac{9x^7 - 2x^5}{x^3}$ **i)** $f(x) = \frac{x^7 + 2x}{x^5}$

j) $f(x) = (5 - x)^2$

2. Find $\frac{dy}{dx}$ in each case.

a) $y = x^4 + \frac{1}{x} - \frac{3}{\sqrt{x}}$ **b)** $y = \frac{10x^6 - 3x^5}{2x^2}$ **c)** $y = \frac{7x + 3 + \sqrt{x}}{x}$

d) $y = \left(\sqrt{x} + \frac{1}{\sqrt{x}}\right)^2$ **e)** $y = 2x^3(1 - 4x - 8x^2)$ **f)** $y = (\sqrt{x} + 3)(2\sqrt{x} + 5)$

g) $y = \frac{5\sqrt{x} - 3x}{\sqrt{x}}$ **h)** $y = x^{-\frac{1}{2}} + 4x^{-\frac{3}{2}} - x^{\frac{1}{2}}$ **i)** $y = \frac{(2x - 3)^2}{x^2}$

j) $y = \frac{4}{x} + \frac{3}{x^2}$

3. Find

a) $\frac{d}{dx}(3x^4 - 2x^3 + 5x^2 - x - 1)$ **b)** $\frac{d}{dx}\left(x^{\frac{2}{3}} + \sqrt{x}\right)$ **c)** $\frac{d}{dx}4\sqrt{x}(3\sqrt{x} + x)$

d) $\frac{d}{dx}\left(\frac{7x - 4 - 2x^5}{x^3}\right)$ **e)** $\frac{d}{dx}(x^2 + 7)^2$ **f)** $\frac{d}{dx}(ax^2 + bx + c)$

g) $\frac{d}{dx}\sqrt{x}(x^2 + 8)$ **h)** $\frac{d}{dx}\left(\frac{12x^2 - x^3}{3x^5}\right)$.

4. Find $f'(x)$ in each case.

a) $f(x) = \frac{6}{\sqrt[3]{x}} + 4\sqrt{x}$

b) $f(x) = \frac{1}{3}x^{\frac{3}{2}} - \frac{1}{5}x^{\frac{5}{2}}$

c) $f(x) = x(x-2)^2$

d) $f(x) = \frac{x+3}{\sqrt{x}}$

e) $f(x) = \frac{12x^3 - 5x}{3x^4}$

f) $f(x) = (\sqrt{x} - 4)^2$

g) $f(x) = 5x - \frac{1}{2\sqrt{x}}$

h) $f(x) = \frac{(\sqrt{x} - x)^2}{\sqrt{x}}$

8.5 The chain rule (differentiating a function of a function)

In Chapter 2 we dealt with composite functions.
e.g. If $f(x) = x^7$ and $g(x) = (x^2 - 3)$ then $fg(x) = f(x^2 - 3) = (x^2 - 3)^7$.

Consider $y = (x^2 - 3)^7$. If we let $u = x^2 - 3$, then $y = u^7$.

$\frac{dy}{du} = 7u^6 = 7(x^2 - 3)^6$ and $\frac{du}{dx} = 2x$

The **chain rule** states that:
If y is a function of u, and u is a function of x,
$$\frac{dy}{dx} = \frac{dy}{du} \times \frac{du}{dx}$$

Thus if $y = (x^2 - 3)^7$ we can write $\frac{dy}{dx} = \frac{dy}{du} \times \frac{du}{dx} = 7(x^2 - 3)^6 \times 2x = 14x(x^2 - 3)^6$.

Example 7

Find $\frac{dy}{du}$ when $y = \sqrt{4 - 3x}$.

Let $u = 4 - 3x$ — Let u = the expression in the square root.

$\frac{du}{dx} = -3$

$y = \sqrt{u} = u^{\frac{1}{2}}$

$\frac{dy}{du} = \frac{1}{2}u^{-\frac{1}{2}}$

$= \frac{1}{2}(4 - 3x)^{-\frac{1}{2}}$ — Now substitute $u = 4 - 3x$.

$\frac{dy}{dx} = \frac{dy}{du} \times \frac{du}{dx} = \frac{1}{2}(4 - 3x)^{-\frac{1}{2}} \times -3$ — Apply the chain rule.

$= -\frac{3}{2}(4 - 3x)^{-\frac{1}{2}}$ — Simplify your answer.

$= -\frac{3}{2\sqrt{4 - 3x}}$

Example 8

Find $\frac{dy}{dx}$ when $y = \frac{2}{(3x^2 + 1)^4}$.

Let $u = 3x^2 + 1$

$\frac{du}{dx} = 6x$

$y = 2u^{-4}$

$\frac{dy}{du} = -8u^{-5}$

$= -8(3x^2 + 1)^{-5}$

$\frac{dy}{dx} = \frac{dy}{du} \times \frac{du}{dx} = -8(3x^2 + 1)^{-5} \times 6x$

$= -48x(3x^2 + 1)^{-5}$

$= -\frac{48x}{(3x^2 + 1)^5}$

Let u = the expression in the bracket.

Now substitute $u = 3x^2 + 1$.

Apply the chain rule.

Simplify your answer.

Exercise 8.5

1. Find $\frac{dy}{dx}$ in each case.

a) $y = \sqrt{4x - 1}$ **b)** $y = (x^2 + 4)^5$ **c)** $y = \frac{1}{(1 - x)^3}$

d) $y = \frac{5}{2x + 7}$ **e)** $y = \frac{3}{\sqrt{5 + 2x}}$ **f)** $y = \frac{2}{3 - 4x}$

g) $y = \frac{1}{\sqrt[3]{4x - 3}}$ **h)** $y = \sqrt{5 + 3x^3}$ **i)** $y = \frac{6}{2x^2 + 3x}$

j) $y = \frac{1}{2(1 - 3x^2)^4}$

2. Differentiate each expression with respect to x.

a) $\frac{1}{9x + 8}$ **b)** $\sqrt{8 - 7x}$ **c)** $\sqrt{2x^4 + 7}$

d) $\frac{3}{\sqrt{3 - x^2}}$ **e)** $\frac{1}{(4x - 5)^4}$ **f)** $\frac{2}{\sqrt[4]{5 + 3x}}$

g) $(3x^2 - 2)^4$ **h)** $\sqrt{5x - 2x^2}$ **i)** $\frac{1}{x(5 - 3x^2)}$

j) $\frac{1}{1 + \sqrt{x}}$

8.6 Finding the gradient of the tangent using differentiation

We can find the gradient of the tangent to the curve at a specific point on the curve.

For example, consider the curve $y = x^3$ at the point where $x = 2$.

Point A has coordinates $(2, 2^3)$.

Point B has coordinates $((2 + h), (2 + h)^3)$.

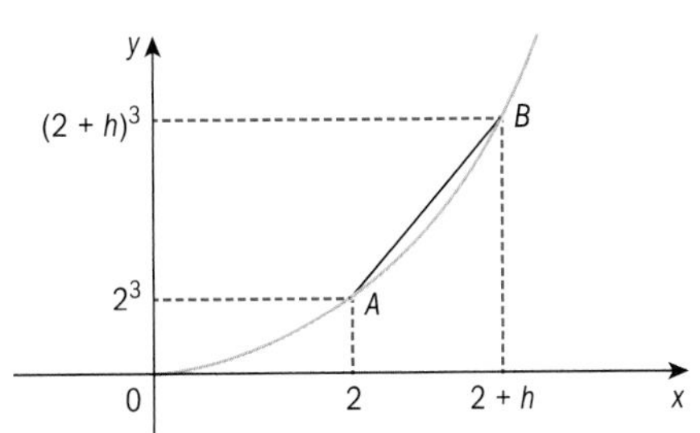

The gradient of $AB = \dfrac{(2+h)^3 - 2^3}{(2+h)-2} = \dfrac{8+12h+6h^2+h^3-8}{2+h-2}$

$= \dfrac{h^3+6h^2+12h}{h} = h^2 + 6h + 12$

As h gets smaller and smaller, the line AB gets closer to the tangent at A.

As $h \to 0$, the gradient of the tangent at $A \to 0 + 0 + 12 = 12$

Another way of doing this is: $y = x^3$ $\qquad \dfrac{dy}{dx} = 3x^2$

When $x = 2$, $\dfrac{dy}{dx} = 3(2)^2 = 12$

So the gradient of the tangent to the curve $y = x^3$ at the point where $x = 2$ is 12.

Example 9

Find the gradient of the tangent to the curve $y = 4x^3 - 2x^2 + 1$ at the point $(1, 3)$.

$\dfrac{dy}{dx} = 12x^2 - 4x$ — $\dfrac{dy}{dx}$ gives us the gradient at any point on the curve.

At $x = 1$, $\dfrac{dy}{dx} = 12(1)^2 - 4(1) = 8$ — We only need to use the x-value.

Gradient of the tangent = 8

Example 10

Find the coordinates of the point on the curve with equation $y = 5 - 3x - 4x^2$ where the gradient is 1.

$\dfrac{dy}{dx} = -3 - 8x = 1$ — Put $\dfrac{dy}{dx} = 1$ as the gradient = 1.

$-8x = 4$

$x = -0.5$

When $x = -0.5$,

$y = 5 - 3(-0.5) - 4(-0.5)^2 = 5.5$ — Substitute in the original equation to find y.

Coordinates are $(-0.5, 5.5)$.

Example 11

Find the gradient of the tangent to the curve $y = 4x^2 - 5x + 2$ at each of the points where the curve meets the line $y = 7x - 6$.

The curve meets the line when

$4x^2 - 5x + 2 = 7x - 6$ ← The y-values are equal at the points of intersection.

$4x^2 - 12x + 8 = 0$

$x^2 - 3x + 2 = 0$ ← It is easier to factorise if we divide the equation by the common factor of 4.

$(x - 1)(x - 2) = 0$

$x = 1$ or $x = 2$

$\frac{dy}{dx} = 8x - 5$ ← Differentiate to find an expression for the gradient.

When $x = 1$, the gradient of the tangent $= 8(1) - 5 = 3$.

When $x = 2$, the gradient of the tangent $= 8(2) - 5 = 11$.

Exercise 8.6

1. Find the gradient of the tangent to the curve $y = 5x^3 + 3x^2 - x + 4$ at the point (1, 11).
2. Find the gradient of the tangent to the curve $y = x^3 - 8x^2 - 7$ where $x = -2$.
3. Find the coordinates of the points on the curve with equation $y = 2x^3 + 3x^2 - 12x$ where the gradient is 24.
4. Find the gradient of the tangent to the curve $y = 2x^2 + 3x + 1$ at each of the points where the curve meets the line $y = 4(x + 1)$.
5. Find the gradient of the tangent to the curve $y = (2x + 3)^2$ at the point where $x = 0$.
6. Find the gradient of the tangent to the curve $y = (x + 2)(x - 3)$ at each of the points where the curve crosses the x-axis.
7. Find the coordinates of the points on the curve $y = 2x^3 - 15x + 7$ where the gradient is 9.
8. Find the values of x where the tangents to the curve $y = x^3 - x^2 - 42x - 7$ are parallel to the line $y = -2x$.
9. Find the coordinates of the points on the curve $y = x^4 + 2x^3$ where the gradient is parallel to the x-axis.
10. The gradient of the tangent to the curve $y = 2x^3 + ax^2 - x + 3$ at the point $x = 1$ is 3. Find the value of a.

11. Find the y-coordinate and the gradient of $y = (x - 3)^2$ when $x = -2$.

12. Find the gradient of the tangent to the curve $y = \frac{(4x-1)^2}{x^2}$ at the point $(-1, 25)$.

13. Find the coordinates of the points on the curve $y = 6 + 9x - 3x^2 - x^3$ where the gradient is 9.

14. Find the gradient of the tangent to the curve $y = (\sqrt{x} + 3)(3\sqrt{x} - 5)$ at the point where $x = 1$.

15. Find the coordinates of the point on the curve $y = \frac{2x-5+\sqrt{x}}{x}$ where the gradient is zero.

8.7 The second derivative

We know that if we differentiate y with respect to x, we get $\frac{dy}{dx}$.

If we differentiate $\frac{dy}{dx}$, we get $\frac{d^2y}{dx^2}$.

$\frac{d^2y}{dx^2}$ is called the second derivative **of y with respect to x.**

Similarly, using the alternative notation, if we differentiate $f(x)$ with respect to x, we get $f'(x)$.

If we differentiate $f'(x)$ with respect to x, we get **the second derivative** $f''(x)$.

Example 12

Find $\frac{d^2y}{dx^2}$ when $y = (4x^2 + 1)^2$.

$y = (4x^2 + 1)(4x^2 + 1) = 16x^4 + 8x^2 + 1$ ← First expand the brackets.

$\frac{dy}{dx} = 64x^3 + 16x$ ← Differentiate each of the three terms.

$\frac{d^2y}{dx^2} = 192x^2 + 16$ ← Then differentiate again.

Example 13

Find $\frac{d^2y}{dx^2}$ when $\frac{dy}{dx} = \frac{3}{\sqrt{x}}$.

$\frac{dy}{dx} = \frac{3}{\sqrt{x}} = 3x^{-\frac{1}{2}}$ — Remember $\frac{3}{\sqrt{x}} = \frac{3}{x^{\frac{1}{2}}} = 3x^{-\frac{1}{2}}$

$\frac{d^2y}{dx^2} = -\frac{3}{2}x^{-\frac{3}{2}}$ — We only need to differentiate once when given $\frac{dy}{dx}$.

$= -\frac{3}{2\sqrt{x^3}}$ — $x^{\frac{3}{2}} = \sqrt{x^3}$ or $(\sqrt{x})^3$ or $x\sqrt{x}$

Example 14

Given that $f(x) = 8x^3 - 3x + \frac{4}{x}$, find the value of $f''(-2)$.

$f'(x) = 24x^2 - 3 - \frac{4}{x^2}$ — Differentiate $f(x)$ to get $f'(x)$.

$f''(x) = 48x + \frac{8}{x^3}$ — Differentiate $f'(x)$ to get $f''(x)$.

$f''(-2) = 48(-2) + \frac{8}{(-2)^3}$ — Substitute $x = 2$ in to $f''(x)$.

$= -96 - 1 = -97$

Exercise 8.7

1. Find $\frac{d^2y}{dx^2}$ in each case.

a) $y = 2x^7 + 4x^3$ **b)** $y = 3x^9 - 7x$ **c)** $y = (2x + 1)(x - 8)$

d) $y = x^{-3}(x + 5x^7)$ **e)** $y = 6x^3(2x^2 - x + 1)$ **f)** $y = (9 - x)^5$

g) $y = \frac{x^5 + x^4}{x^2}$ **h)** $y = \frac{24x^7 - 9x^4}{3x^3}$ **i)** $y = 2x^3 + \frac{3}{x}$

2. Find $f''(x)$ in each case.

a) $f(x) = \frac{8x^2 - x}{2x^3}$ **b)** $f(x) = \frac{3x - 5 + \sqrt{x}}{x}$ **c)** $f(x) = \frac{7\sqrt{x} - 8x}{\sqrt{x}}$

d) $f(x) = (\sqrt{x} - 1)(4\sqrt{x} + 3)$ **e)** $f(x) = 8x^{-\frac{1}{4}}$ **f)** $f(x) = \sqrt{4x - 3}$

g) $f(x) = \frac{3}{x^4} + \frac{6}{x^2}$ **h)** $f(x) = \frac{x^{12}}{x^4}$ **i)** $f(x) = \frac{12}{\sqrt[3]{x}}$

3. Given that $y = 8x^3 - 3x + \frac{4}{x}$, find the value of $\frac{d^2y}{dx^2}$ when $x = -2$.

4. Given that $f(x) = \frac{4}{\sqrt{3x-2}}$, find the value of $f''(2)$.

5. Find $\frac{d^2y}{dx^2}$ when $\frac{dy}{dx} = 1 - 7x^2$.

6. Given that $y = 6x^3 - 2x^2$, show that $\frac{d^2y}{dx^2} - 4\frac{dy}{dx} + 20 = 4(4 + 13x - 18x^2)$.

7. Given that $f'(x) = \frac{5}{(5-2x)^8}$, find $f''(x)$.

8. Given that $f(x) = 2x^4 - 3x^3 - x^2$, find the value of

   a) $f'(3)$ b) $f''(-2)$ c) $\frac{1}{f(1)}$.

9. Given that $y = ax^4 - 3x^2$ and $\frac{d^2y}{dx^2} = 42$ when $x = 2$, determine the value of a.

10. Given that $\frac{dy}{dx} = \frac{6x-1}{2x^4}$, find the value of $\frac{d^2y}{dx^2}$ when $x = -1$.

8.8 Equation of the tangent and the normal

We can use differentiation to find the gradient of the tangent to a curve at a particular point. We will then be able to find the **equation of the tangent** and the **equation of the normal** at this point by using our knowledge of coordinate geometry.

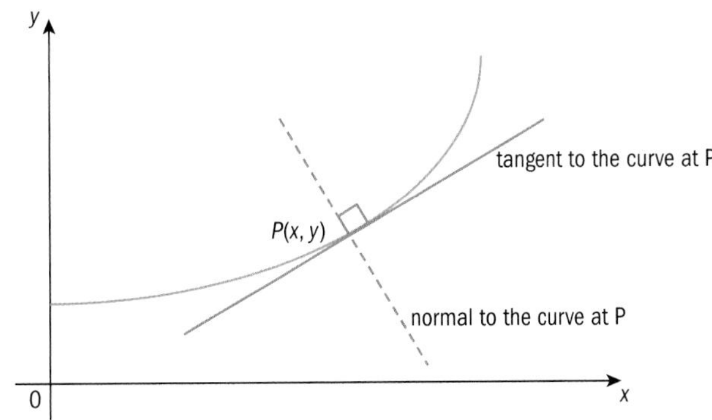

Example 15

Find the equation of the tangent to the curve $y = 3x^3 - 6x^2 + x$ at the point $(1, -2)$.

$y = 3x^3 - 6x^2 + x$

$\frac{dy}{dx} = 9x^2 - 12x + 1$ ← Differentiate to find the gradient of the tangent.

When $x = 1$, $\frac{dy}{dx} = 9(1)^2 - 12(1) + 1 = -2$. ← Substitute to find the gradient at $(1, -2)$.

Tangent goes through $(1, -2)$ with $m = -2$.

$y + 2 = -2(x - 1)$ ← We use $(1, -2)$ and $m = -2$ to find the equation.

$y + 2 = -2x + 2$

Equation of tangent is $y + 2x = 0$.

Example 16

Find the equation of the normal to the curve $y = 4\sqrt{x} + 7$ at the point where $x = 16$.

$y = 4\sqrt{x} + 7 = 4x^{\frac{1}{2}} + 7$

Write the expression using indices.

$\frac{dy}{dx} = 2x^{-\frac{1}{2}} = \frac{2}{\sqrt{x}}$

When $x = 16$, $y = 4\sqrt{16} + 7 = 23$

Substitute $x = 16$ into $y = 4\sqrt{x} + 7$.

and $\frac{dy}{dx} = \frac{2}{\sqrt{16}} = \frac{1}{2}$ so gradient of normal = −2.

Gradient of normal $= -\frac{1}{\text{gradient of tangent}}$

Normal goes through (16, 23) with $m = -2$.

$y - 23 = -2(x - 16)$

Equation of normal is $y = -2x + 55$.

Example 17

The equation of a curve is $y = 4x - x^2$. The normal to the curve at the point $P(1, 3)$. meets the curve again at Q. Find the coordinates of Q.

$y = 4x - x^2$

$\frac{dy}{dx} = 4 - 2x$

When $x = 1$, $\frac{dy}{dx} = 4 - 2 = 2$,

so gradient of the normal $= -\frac{1}{2}$.

Equation of the normal is $y - 3 = -\frac{1}{2}(x - 1)$.

So $y = -\frac{1}{2}x + \frac{7}{2}$

To find where the curve meets the normal

we solve $y = 4x - x^2$ and $y = -\frac{1}{2}x + \frac{7}{2}$ simultaneously.

$$4x - x^2 = -\frac{1}{2}x + \frac{7}{2}$$

$$8x - 2x^2 = -x + 7$$

Multiply both sides by 2.

$$2x^2 - 9x + 7 = 0$$

$$(2x - 7)(x - 1) = 0$$

$x = \frac{7}{2}, y = \frac{7}{4}$ or $x = 1, y = 3$

We already know it goes through (1, 3).

The normal meets the curve again at $\left(\frac{7}{2}, \frac{7}{4}\right)$.

Exercise 8.8

1. Find the equation of the tangent to the curve

a) $y = x^3 - 6x + 3$ at the point $(2, -1)$

b) $y = 2x^4 + 9x^3 + x$ when $x = -2$

c) $y = (x - 7)(x + 4)$ at the point when $x = 1$

d) $y = \dfrac{x^4 - 3x^3}{x}$ at the point $(1, -2)$

e) $y = \sqrt{x}(x^2 - 1)$ at the point $(1, 16)$

f) $y = \sqrt[3]{x}$ at the point when $x = 8$.

2. Find the equation of the normal to the curve

a) $y = 7x^2 - 8x + 9$ when $x = 2$

b) $y = 5x^3 + x^2 - 2$ at the point $(-1, -6)$

c) $y = (3x + 10)^2$ at the point $(-3, 1)$

d) $y = 2x^2(6 - x)$ when $x = 5$

e) $y = \dfrac{6x^2 - x^3}{2x^4}$ when $x = -2$

f) $y = \dfrac{3}{\sqrt{2x + 1}}$ at the point $(4, 1)$.

3. The curve $y = 2x^3 + 6x - 5$ crosses the y-axis at the point P. Find the equation of the tangent to the curve at the point P.

4. Find the equation of the tangent and the normal to the curve $y = 2x^3 - x$ at the point where $x = 2$.

5. Find the equations of the tangents to the curve $y = x^2 - x - 12$ at each of the points where the curve crosses the x-axis.

6. **a)** Find the coordinates of the points on the curve $y = x^2 - x + 2$ at which the gradients are 3 and -3.

b) Find the equations of the tangents at these points.

c) Show that these tangents intersect at the point $\left(\frac{1}{2}, -\frac{1}{2}\right)$.

7. **a)** Find the equation of the tangent to the curve $y = 5 - 2x - x^2$ at the point $(1, 2)$.

b) This tangent meets the x-axis at P and the y-axis at Q. Find the area of triangle OPQ.

8. The curve $y = (5 - x)(2 + x)$ crosses the x-axis at points A and B.
The tangents at the points A and B meet at point C.
Find the coordinates of C.

9. The normal to the curve $y = \sqrt{x} + \sqrt[3]{x}$ at the point where $x = 1$ meets the axes at $(p, 0)$ and $(0, q)$. Find p and q.

10. The normals at the points $(0, 0)$ and $(1, 2)$ to the curve $y = x + x^3$ meet at point P. Find the coordinates of P.

Summary exercise 8

1. Find $\frac{dy}{dx}$ when
   a) $y = x(x + 3)^2$
   b) $y = 6\sqrt{x} + \frac{3}{\sqrt[3]{x}}$
   c) $y = \frac{2x^6 - 8x^3}{2x^2}$
   d) $y = \sqrt{3 - 2x^2}$.

2. Find $f'(x)$ when
   a) $f(x) = \frac{9\sqrt[3]{x^2} + \sqrt{x}}{x}$
   b) $f(x) = \frac{1}{2}x^3\left(x - 4x^2 + \frac{6}{x}\right)$
   c) $f(x) = \frac{3}{(5x - 2)^2}$
   d) $f(x) = \left(\sqrt{x} - \frac{1}{\sqrt{x}}\right)^2$.

3. Find the gradient of the tangent to the curve $y = 5 - 3x + x^3$ at the point (1, 3).

4. Find the equation of the normal to the curve $y = \sqrt{x}(\sqrt{x} - 2)$ at the point where $x = 9$.

EXAM-STYLE QUESTIONS

5. Given that $y = 3x^3 - ax^2$ and $\frac{d^2y}{dx^2} = 54$ when $x = 2$, work out the value of a.

6. Given that $y = 4x^2 - x^3$, show that $\frac{d^2y}{dx^2} + \frac{dy}{dx} - 3y = 8 + 2x - 15x^2 + 3x^3$.

7. The curve $y = \sqrt{9 - 4x}$ cuts the y-axis at the point P. Find the equation of the tangent to the curve at the point P.

8. Find the points on the curve $y = x^3 + 3x^2 - 4$ where the tangent is parallel to the line $y = 9x - 2$.

9. Find the equation of the normal to the curve $y = 3x^2$ at the point $(t, 3t^2)$.

10. a) Find the equation of the tangent to the curve $y = x(2 - x)$ at the origin.
    b) Find a point on the curve where the tangent is perpendicular to the tangent at the origin.

11. Differentiate $\frac{7x\sqrt{x} - 8x}{\sqrt{x}}$ with respect to x.

12. If $\frac{dy}{dx} = \frac{x}{(5 - 3x^2)^6}$, show that $\frac{d^2y}{dx^2} = \frac{5 + 33x^2}{(5 - 3x^2)^7}$.

13. Find the equations of the tangents to the curve $y = x^3 - 6x^2 + 12x + 2$ which are parallel to the line $y = 5 + 3x$.

14. a) Find the equations of the tangent and the normal to the curve $y = 4\sqrt{x}$ at the point $P(9, 12)$.

    The tangent and normal to the curve at P cut the x-axis at Q and R.

    b) Find the area of triangle PQR.

15. Given that $f(x) = 3x^4 + 5x^3 - 6x^2$, find the value of
    a) $f'(-2)$
    b) $f''\left(\frac{1}{2}\right)$
    c) $\frac{24}{f(-1)}$.

16. Given that $y = ax^2 + 3x - a^2$ has a gradient of 5 when $x = -2$, find the value of y when $x = -3$.

Chapter summary

Differentiating $y = x^n$

- If $y = x^n$ then $\frac{dy}{dx} = nx^{n-1}$

Differentiating $y = ax^n$

- If $y = ax^n$ then $\frac{dy}{dx} = anx^{n-1}$

We differentiate when

- we are given $f(x)$ and want to find $f'(x)$
- we are given y and are want to find $\frac{dy}{dx}$
- we want to find $\frac{d}{dx}(f(x))$
- we want to find the gradient of the tangent (or the normal) of a curve given the equation of the curve
- we want to find the coordinates of a point on a curve and are given the value of the gradient at that point
- we want to find the rate of change of y with respect to x given y.

The chain rule (differentiating a function of a function)

- The chain rule states that:
  If y is a function of u, and u is a function of x, $\frac{dy}{dx} = \frac{dy}{du} \times \frac{du}{dx}$

The second derivative

- If we differentiate $\frac{dy}{dx}$ we get $\frac{d^2y}{dx^2}$.
- If we differentiate $f'(x)$ we get $f''(x)$.

9 Further differentiation

Calculus is used by meteorologists to create computer programs which model atmospheric flow. The way atmospheric pressure changes over time can be studied through the use of differentiation in these models. This allows scientists to roughly ascertain the likelihood of rain or a storm.

Objectives

- Apply differentiation to increasing and decreasing functions and rates of change.
- Locate stationary points and determine their nature, and use this information to sketch graphs.

Before you start

You should know how to:

1. Find the gradient of a straight line,

e.g. $y = 3x - 2$

When the line has an equation written in the form $y = mx + c$, coefficient of x is 3 so gradient = 3.

e.g. $x - 3y = 6$

We first rearrange the equation into the form $y = mx + c$.

$x - 6 = 3y$

$y = \frac{1}{3}x - 2$

gradient $= \frac{1}{3}$

Skills check:

1. Find the gradient of these lines.

a) $y = 5 - x$

b) $2x + y = 1$

c) $4x - 8y = 3$

d) $3y + 2x = 4$

e) $\frac{2y}{3} = 6x + 5$

f) $y = \frac{5-2x}{3}$

9.1 Increasing and decreasing functions

When the gradient of a function is positive we say the function is an **increasing function**.
A function $f(x)$ is increasing for $a < x < b$ if $f'(x) > 0$ for $a < x < b$.

When the gradient of a function is negative we say the function is a **decreasing function**.
A function $f(x)$ is decreasing for $a < x < b$ if $f'(x) < 0$ for $a < x < b$.

Example 1

Find the values of x for which $f(x) = 5 - 4x - x^2$ is a decreasing function.

$f'(x) = -4 - 2x$ — First differentiate to give $f'(x)$.

Decreasing function so $-4 - 2x < 0$.

$-4 < 2x$ — Then solve for x in an inequality.

$x > -2$

$f(x)$ is a decreasing function when $x > -2$.

Example 2

$y = x^3 - x^2$. Find the range of values of x for which y is increasing.

$\frac{dy}{dx} = 3x^2 - 2x$ — First differentiate to find $\frac{dy}{dx}$.

$= x(3x - 2) > 0$ — y is **increasing** so put $\frac{dy}{dx} > 0$.

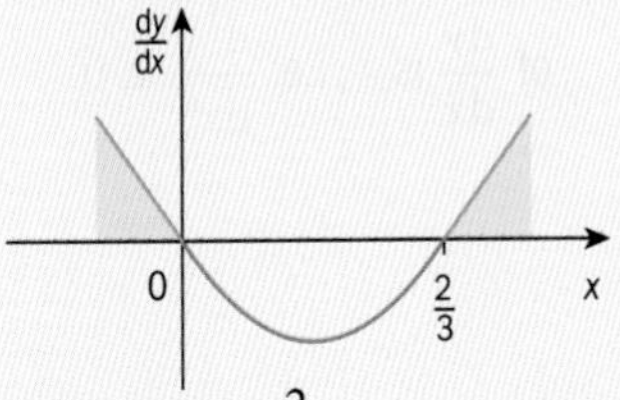

Draw a diagram. Note that this is the graph of $\frac{dy}{dx}$.

$x < 0$ or $x > \frac{2}{3}$ — $\frac{dy}{dx} > 0$ so find where the graph lies **above** the x-axis, giving two intervals.

y is increasing when $x < 0$ and $x > \frac{2}{3}$.

Exercise 9.1

1. For each of the following functions find the range of values of x for which $f(x)$ is increasing.

   a) $f(x) = x^2 - 6x$ b) $f(x) = 8 + 3x - 2x^2$ c) $f(x) = x^3 - 48x + 2$

   d) $f(x) = 2x^3 - 9x^2 + 12x - 3$ e) $f(x) = 3x^2 - 2x^3$ f) $f(x) = 3x^3 - 9x + 1$

2. Find the range of values of x for which y is decreasing.

   a) $y = x^2 - x + 2$ b) $y = 9 - 2x - x^2$ c) $y = 3x^2 + 6x + 5$

   d) $y = x^3 - 5x^2 + 3x - 4$ e) $y = 2x^3 - 54x - 1$ f) $y = x^3 - 6x^2 - 15x + 7$

3. Find the range of values of x for which y is increasing.

a) $y = 5x^2 + 5x$ b) $y = 3 + 16x - 4x^2$ c) $y = x^3 - 3x - 4$

d) $y = 6x^2 - 6x - 7$ e) $y = 4 - 12x + 9x^2 - 2x^3$ f) $y = 3x^3 - 18x^2 + 18x - 2$

4. Find the range of values of x for which $f(x)$ is decreasing.

a) $f(x) = 9x^2 - 2x^3$ b) $f(x) = 2 - 5x - 10x^2$ c) $f(x) = x^3 + 6x^2 + 12x - 1$

d) $f(x) = 5x^3 - 15x - 3$ e) $f(x) = 4x^2 - 12x^3$ f) $f(x) = x(10 - 2x)(16 - 2x)$

9.2 Stationary points

Any point on a curve where $\frac{dy}{dx} = 0$ is called a **stationary point**.

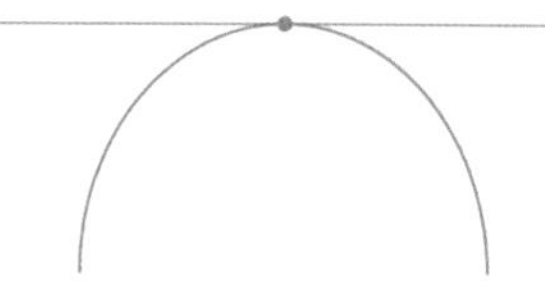

At a stationary point, the gradient of the curve and the gradient of the tangent = 0.

There are three types of stationary points: **minimum** point, **maximum** point, and **point of inflexion**.

Questions involving points of inflexion have not been included in this chapter as they will not be set in Cambridge International AS & A Level 9709 examination papers.

Minimum point

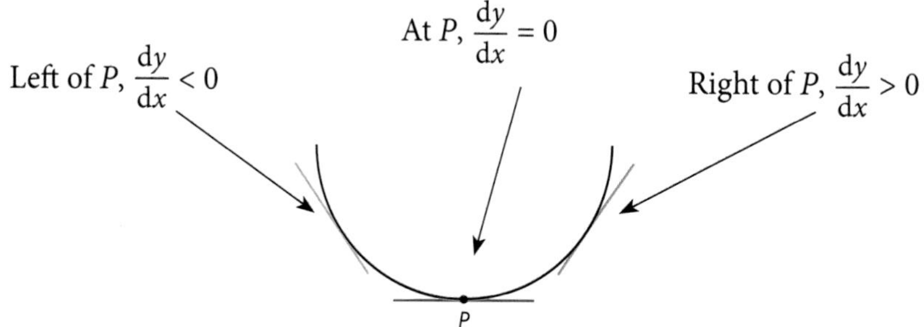

Note: At a minimum point, the sign of $\frac{dy}{dx}$ changes from – to + so the gradient of $\frac{dy}{dx}$ is +, i.e. $\frac{d^2y}{dx^2} \geq 0$.

Maximum point

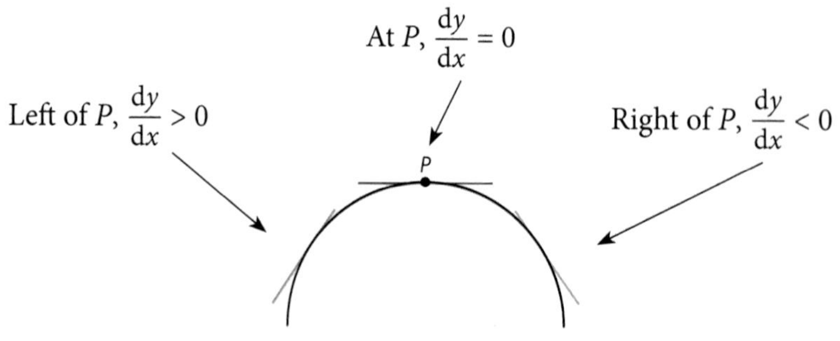

Note: At a maximum point $\frac{d^2y}{dx^2} \leq 0$.

Stationary points which are either a maximum point or a minimum point are often called turning points.

Example 3

Find the coordinates of the stationary points of the curve $y = 6x - 2x^3$ and **determine** their nature. **Sketch** the curve.

$\frac{dy}{dx} = 6 - 6x^2 = 0$ at stationary points

Differentiate and put $\frac{dy}{dx}$ equal to 0.

$6 = 6x^2$

$x = 1$ and $x = -1$

Solve for x.

$y = 4$ and $y = -4$

Substitute x in to $y = 6x - 2x^3$ to find y.

x	−2	−1	0	1	2
$\frac{dy}{dx}$	−18	0	6	0	−18

Use values for x less than −1, between 0 and 1, and greater than 1.

Draw the positive and negative gradients.

(1, 4) is a maximum turning point.

(−1, −4) is a minimum turning point.

Write down if it is a maximum or minimum.

Note: $\frac{d^2y}{dx^2} = -12x$

This is an alternative method. It is useful when $\frac{dy}{dx}$ is simple to differentiate again.

When $x = 1$, $\frac{d^2y}{dx^2} = -12$, $\frac{d^2y}{dx^2} < 0$, so maximum.

When $x = -1$, $\frac{d^2y}{dx^2} = 12$, $\frac{d^2y}{dx^2} > 0$, so minimum.

Sketch:

$y = 6x - 2x^3 = 2x(3 - x^2)$

Cuts x-axis at $0, \sqrt{3}, -\sqrt{3}$

Find when $y = 0$ for points where the graph crosses the x-axis.

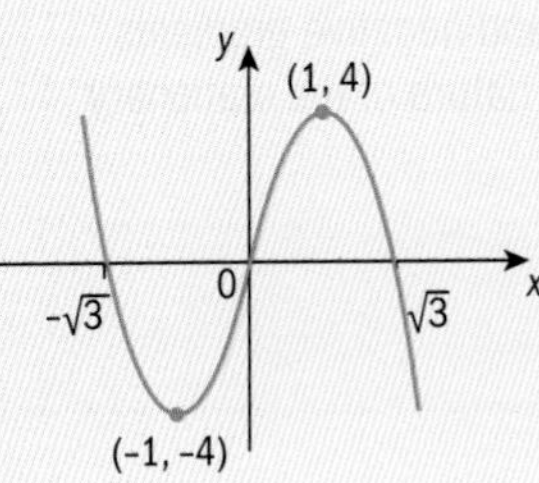

Example 4

Find the coordinates of the stationary points of the curve $y = \frac{2}{2x-1} + x$ and determine their nature.

Let $A = \frac{2}{2x-1}$ and let $u = 2x - 1$

Use the chain rule to differentiate $\frac{2}{2x-1}$.

$\frac{du}{dx} = 2$

Let $A = \frac{2}{u} = 2u^{-1}$ $\quad \frac{dA}{du} = -2u^{-2} = \frac{-2}{(2x-1)^2}$

$\frac{dA}{dx} = \frac{-2}{(2x-1)^2}(2) = \frac{-4}{(2x-1)^2}$

$\frac{dy}{dx} = \frac{-4}{(2x-1)^2} + 1 = 0$ at stationary points

Put $\frac{dy}{dx} = 0$.

$(2x - 1)^2 = 4$

$4x^2 - 4x - 3 = 0$

Multiply each term by $(2x - 1)^2$ and rearrange.

$(2x + 1)(2x - 3) = 0$

$x = -\frac{1}{2}, y = -\frac{3}{2}$ or $x = \frac{3}{2}, y = \frac{5}{2}$

x	-1	$-\frac{1}{2}$	0	$\frac{3}{2}$	2
$\frac{dy}{dx}$	$\frac{5}{9}$	0	-3	0	$\frac{5}{9}$
/	—	\	—	/	

Draw up a table to show the gradient of the tangent (**or** we could find $\frac{d^2y}{dx^2}$ etc.).

$\left(-\frac{1}{2}, -\frac{3}{2}\right)$ is a maximum turning point.

$\left(\frac{3}{2}, \frac{5}{2}\right)$ is a minimum turning point.

Draw the positive and negative gradients and determine the nature of the stationary points.

Exercise 9.2

1. For each of the following functions, find the coordinates of the stationary points and determine their nature.

   a) $y = x^2 - 2x + 5$ b) $y = 3 + x - x^2$

   c) $y = 2x^2 + 4x - 3$ d) $y = x^3 - 5x^2 + 3x + 1$

   e) $y = x^4 - 2x^2 - 2$ f) $y = 9 + 24x - 2x^3$

2. For each of the following functions, find the coordinates of the stationary points and determine their nature. Sketch the curve.

   a) $y = x^2 - 8x + 12$ b) $y = x^2 - x^3$

   c) $y = 4x^3 - x^4$ d) $y = x^2 - 2x + 7$

   e) $y = x^3(x - 2)$ f) $y = 20 + 15x - x^2 - \frac{x^3}{3}$

3. For each of the following functions, find the coordinates of the stationary points and determine their nature.

   a) $f(x) = 2x^3 - 3x^2$ b) $f(x) = 2x^3 - 9x^2 + 12x + 4$

   c) $f(x) = x + \frac{1}{x}, x \neq 0$ d) $f(x) = \frac{x^2 + 9}{2x}$

4. Find the coordinates of the stationary points of the curve $y = x^{\frac{1}{2}}\left(\frac{3}{4} - x\right)$ and determine their nature.

5. Find the coordinates of the stationary points of the curve $y = (3x - 2)^3 - 9x$ and determine their nature.

6. Find the coordinates of the stationary points of the curve $y = \frac{1 + 54x^3}{x^2}$ and determine their nature.

7. Find the coordinates of the stationary points of the curve $y = x^4 - 2x^3 + x^2 - 2$ and determine their nature. Sketch the curve.

8. Show that the curve $y = 3x^3 - 5x^2 + 3x + 4$ has no stationary points.

9. $f(x) = 3x^2 + 2x + 5$
   $g(x) = x^3 - 4x^2 - 3x + 6$

   a) Show that there is one value of x for which $f(x)$ and $g(x)$ both have the same stationary value.

   b) On the same axes sketch the graphs of $f(x)$ and $g(x)$.

10. The curve $y = ax^2 + bx + c$ has a minimum when $x = \frac{1}{2}$ and passes through the points $(2, 0)$ and $(1, -3)$.
    Find the values of a, b and c.

9.3 Problems involving maximum and minimum values

We can use differentiation to solve problems involving maximising or minimising quantities.

Example 5

The sum of two numbers is 36. Find the maximum product of the two numbers.

Let x and y be the two numbers.

$x + y = 36$

$y = 36 - x$ ← Write one variable in terms of the other.

Product $= xy$ ← Substitute for this variable to find the product in terms of one variable only.

$P = x(36 - x) = 36x - x^2$

$\frac{dP}{dx} = 36 - 2x = 0$ ← Use $\frac{dP}{dx} = 0$ for maximum/minimum.

$x = 18$

Maximum product $= 36 \times 18 - 18^2$

$= 324$

OR when $x = 18$, $y = 36 - 18 = 18$

Maximum product $= 18 \times 18 = 324$

Note: $\frac{d^2P}{dx^2} = -2 < 0$

so $x = 18$ gives a maximum, but this is obvious in these cases as the minimum product would be 0.

We can use these steps to help us solve a problem:

- Draw a diagram where possible.
- Choose letters to represent the unknown quantities.
- Write an expression for the quantity we are trying to maximise/minimise – it may contain two unknowns.
- Use a given condition from the question to write an equation connecting two variables.
- Rearrange the equation to express one variable in terms of the other.
- Use substitution to find an expression in terms of one variable only.
- Use differentiation to find maximum/minimum values.

Example 6

A closed box with a square base has a total surface area of $36\,\text{m}^2$.

Find the maximum possible volume of the box.

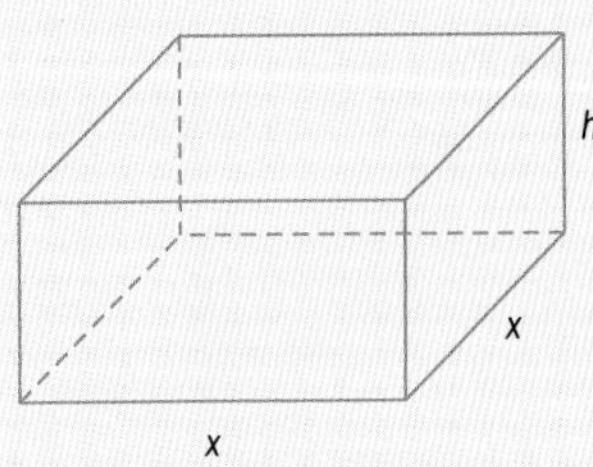

Always draw a diagram and use letters for the lengths of the sides.

Volume, $V = x^2h$

Surface area $= 2x^2 + 4xh$

$= 36$

$x^2 + 2xh = 18$

$h = \dfrac{18 - x^2}{2x}$

Using the given condition.

Rearrange to give h in terms of x.

$V = x^2\left(\dfrac{18 - x^2}{2x}\right)$

$= 9x - \dfrac{1}{2}x^3$

Substitute for h in the volume formula so that we only have one variable.

$\dfrac{dV}{dx} = 9 - \dfrac{3}{2}x^2 = 0$

For maximum/minimum.

$18 = 3x^2$

$x^2 = 6$

$x = \pm\sqrt{6}$

However, x is a length so $x > 0$.

$V = 9 \times \sqrt{6} - \dfrac{1}{2} \times (\sqrt{6})^3$

$= 6\sqrt{6}\,\text{m}^3$ (3 s.f.)

Substitute $x = \sqrt{6}$ into the expression for V.

Exercise 9.3

1. A farmer has a rectangular piece of land for pigs. One of the sides of the rectangle is a wall.
   The other three sides have fencing. The fencing is 80 m in length.
   Find the maximum possible area of this rectangular piece of land.

2. An open box with a square base has a total surface area of 300 cm^2.
   Find the greatest possible volume of the box.

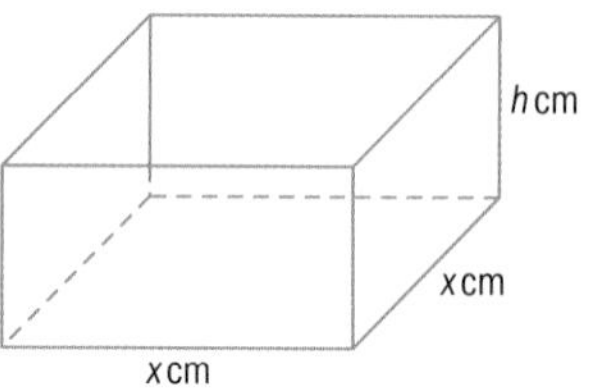

3. Fig. 1 shows a square sheet of metal of side 40 cm.
   A square x cm by x cm is cut from each corner.
   The sides are then bent upwards to form an open box as shown in Fig. 2.
   Find the value of x that maximises the volume of the box.

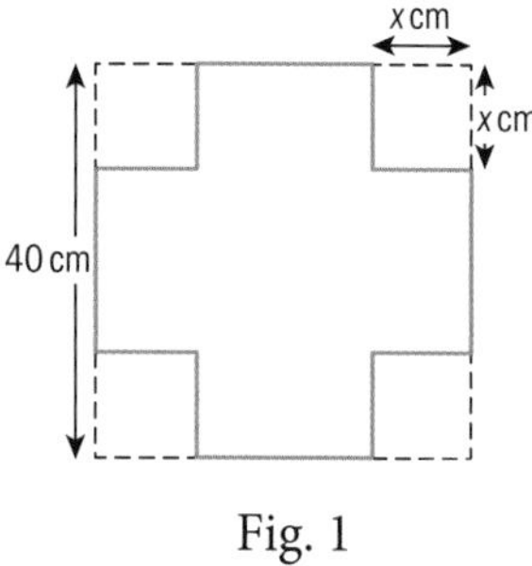

Fig. 1

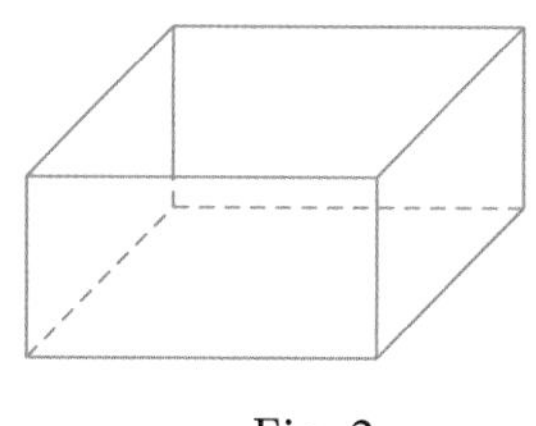

Fig. 2

4. Given that $x + y = 3$, find the least possible value of $x^2 + 14y$.

5. The diagram shows a sporting track made up of a rectangle with semicircles at each end.

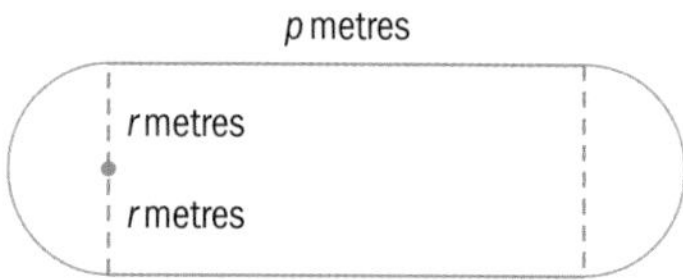

   The rectangle has dimensions p metres by $2r$ metres where r is the radius of each semicircle.
   The perimeter of the track is 1400 m. The track has a maximum area for this perimeter.
   Find the value of p and the value of r.

6. Find the maximum possible value of x^2y if $x + 2y = 8$.

7. A cylinder with an open top has radius r cm and a volume of $512\pi\,\text{cm}^3$.

   a) Write down the surface area of the cylinder in terms of r.

   b) Find the minimum surface area. Leave your answer in terms of π.

   c) Find the value of r and the height of the cylinder.

8. A sector of a circle, radius r has a perimeter of 20 cm. The angle of the sector is θ radians and the area is $A\,\text{cm}^2$.
   Find the maximum possible area of the sector.

9. A tank in the shape of a right circular cylinder with no top has a surface area of $3\pi\,\text{m}^2$.
   What height and base radius will maximise the volume of the cylinder?

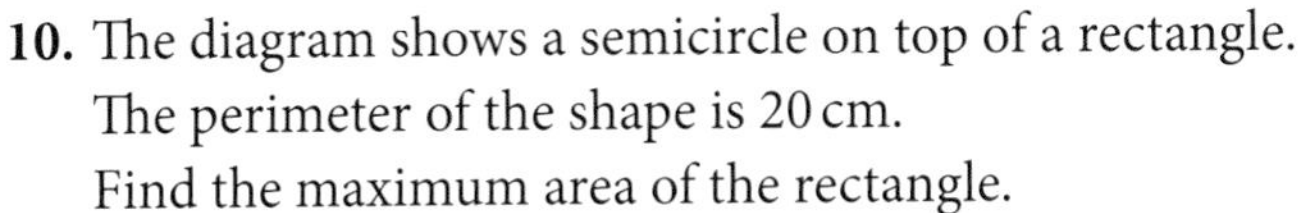

10. The diagram shows a semicircle on top of a rectangle.
    The perimeter of the shape is 20 cm.
    Find the maximum area of the rectangle.

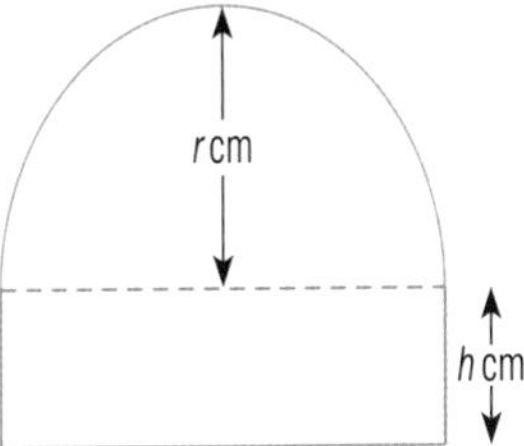

11. $W = pq$ and $2p + 5q = 100$. Find the maximum value of W.

12. An open tank made of metal is in the shape of a cuboid with a square base.
    The volume of the tank is $13.5\,\text{m}^3$.
    Find the dimensions of the tank that uses as little metal as possible.

9.4 Connected rates of change

There are times when we encounter problems which require us to find a rate of change of one variable with respect to another.

> Remember $\frac{dy}{dx}$ gives us the rate of change of x with respect to y.

In such cases, we often need to use the chain rule which states

$$\frac{dy}{dx} = \frac{dy}{du} \times \frac{du}{dx}$$

This rule may be used if we know y in terms of u (and so can find $\frac{dy}{du}$) and we know u in terms of x (and so can find $\frac{du}{dx}$), and we are required to find $\frac{dy}{dx}$ (the rate of change with respect to x).

We sometimes have more than three variables involved. We can use the chain rule to form other identities such as:

$$\frac{dy}{dx} = \frac{dy}{du} \times \frac{du}{dt} \times \frac{dt}{dx}$$

Example 7

The radius, r, of a circle is increasing at the rate of $\frac{2}{r^2}$ m s^{-1}.

Find the rate at which the area, A, is increasing when $r = 8$.

Follow the steps on the right of the examples for problems involving connected rates of change.

We want to find $\frac{dA}{dt}$. ← Write down the rate of change we want to find.

$\frac{dA}{dt} = \frac{dA}{dr} \times \frac{dr}{dt}$ ← Use the chain rule to write down an identity for $\frac{dA}{dt}$.

$\frac{dr}{dt} = \frac{2}{r^2}$ ← Write the information given as the rate of change of r with respect to t.

But, $A = \pi r^2$ ← Write down the link between A and r.

$\frac{dA}{dr} = 2\pi r$ ← Differentiate to the rate of change of A with respect to r.

So, $\frac{dA}{dt} = 2\pi r \times \frac{2}{r^2} = \frac{4\pi}{r}$ ← Substitute $\frac{dA}{dr}$ and $\frac{dr}{dt}$.

When $r = 8$, $\frac{dA}{dt} = \frac{4\pi}{8}$ ← We can now substitute $r = 8$.

$\frac{dA}{dt} = \frac{\pi}{2}$

$= 1.57$ m^2 s^{-1} (3 s.f.) ← Do not forget the units (area/time).

Example 8

A spherical balloon is being blown up so that its volume increases at a rate of 3 cm^3 s^{-1}.

Find the rate of increase of the radius of the balloon when the volume of the balloon is 60 cm^3.

We want to find $\frac{dr}{dt}$. ← Write down the rate of change we want to find.

$\frac{dr}{dt} = \frac{dr}{dV} \times \frac{dV}{dt}$ ← Write down the chain rule for $\frac{dr}{dt}$.

$\frac{dV}{dt} = 3$ ← Write the information given as a rate of change.

$V = \frac{4}{3}\pi r^3$, $\frac{dV}{dr} = 4\pi r^2$ ← Differentiate to get the rate of change of V with respect to r.

$\frac{dr}{dt} = \frac{1}{4\pi r^2} \times 3 = \frac{3}{4\pi r^2}$ ← $\frac{dr}{dV} = 1 \div \frac{dV}{dr}$

When $V = 60$, $60 = \frac{4}{3}\pi r^3$ ← Substitute for these rates of change.

$r^3 = \frac{45}{\pi}$

$r = 2.42859\ldots$

$\frac{dr}{dt} = \frac{3}{4\pi r^2} = \frac{3}{4\pi(2.42859\ldots)^2}$ ← We can now substitute for $r = 2.42859\ldots$

$= 0.0405$ cm s^{-1} (3 s.f.) ← Do not forget the units (radius/time).

Example 9

The surface area, A, of a cube is increasing at the rate of $12\,\text{cm}^2\,\text{s}^{-1}$.

Find the rate of increase of the volume, V, of the cube when each edge of the cube is 10 cm.

Working	Explanation
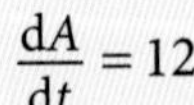 $\frac{dA}{dt} = 12$	Write the information given as a rate of change.
We want to find $\frac{dV}{dt}$.	Write down the rate of change we want to find.
$A = 6x^2$	We need to make the edge of the cube x.
$\frac{dA}{dx} = 12x$ $V = x^3$	Write down the surface area and volume in terms of the edge, say x cm.
$\frac{dV}{dx} = 3x^2$	Differentiate to get $\frac{dA}{dx}$ and $\frac{dV}{dx}$.
But, $\frac{dV}{dt} = \frac{dV}{dx} \times \frac{dx}{dA} \times \frac{dA}{dt}$	We need to express $\frac{dV}{dt}$ in terms of the other rates.
So, $\frac{dV}{dt} = 3x^2 \times \frac{1}{12x} \times 12 = 3x$	Substitute these three rates of change.
When $x = 10$, $\frac{dV}{dt} = 30\,\text{cm}^3\,\text{s}^{-1}$	We can now substitute $x = 10$.

Exercise 9.4

1. A spherical balloon is being blown up so that its radius increases at a rate of $0.4\,\text{cm}\,\text{s}^{-1}$.
   Find the rate of increase of the surface area of the balloon when the radius is 20 cm.

2. The radius of a circular ink blob is increasing at a rate of $5\,\text{cm}\,\text{s}^{-1}$.
   Find the exact rate of increase of the circumference of the circle.

3. The side of a cube is increasing at $0.2\,\text{cm}\,\text{s}^{-1}$.
   Find the rate of increase of the volume when the length of the side is 4 cm.

4. A spherical balloon is inflated so that its volume increases at a rate of $50\,\text{cm}^3\,\text{s}^{-1}$.
   Find the rate of increase of the radius of the balloon when the radius is 12 cm.

5. The side of a cube is decreasing at a rate of $0.4\,\text{cm}\,\text{s}^{-1}$.
   Find the rate of decrease of the surface area when the length of the side is 3 cm.

6. A cone has a height of 7 cm. The radius of the base of the cone is increasing at a rate of $8\,\text{cm}\,\text{s}^{-1}$.
   Find the rate of change of the volume of the cone when the base radius is 5 cm.

7. The volume of a cube is increasing at the rate of $12\,\text{cm}^3\,\text{s}^{-1}$.
   Find the rate of increase of the surface area of the cube when the side of the cube is 7 cm.

8. The surface area of a cube is increasing at $0.3\,\text{m}^2\,\text{s}^{-1}$.
   Find the rate of increase of the volume of the cube when the length of the side is 5 m.

9. A cuboid has a square base. The height of the cuboid is twice the length of the side of the base.
   The surface area of the cuboid is increasing at a rate of $10\,\text{cm}^2\,\text{s}^{-1}$.
   Find the rate of increase of the volume of the cuboid when the height of the cuboid is 12 cm.

10. Paint is poured onto a table, forming a circle which increases at a rate of $2.5\,\text{cm}^2\,\text{s}^{-1}$.
    Find the rate the radius is increasing when the area of the circle is $20\pi\,\text{cm}^2$.

Summary exercise 9

EXAM-STYLE QUESTIONS

1. Find the values of x for which $y = 3x^2 - 2x + 7$ is an increasing function.

2. Find the range of values of x for which $y = 3x^2 - 2x^3$ is decreasing.

3. Find the range of values of x for which $f(x) = x^3 + x^2 - x + 5$ is decreasing.

4. $f(x) = x^2 + px + 3$. $f(x)$ has a turning point when $x = -3$. Find the value of p.

5. Find the turning point of the curve $y = \frac{x^3+2}{x}$ and determine its nature.

6. a) Find the coordinates of the turning points on the curve $y = 2x^3 + 3x^2 - 12x + 6$ and determine their nature.

   b) Find the range of values of x for which y is increasing.

   c) Sketch the graph of y.

7. Find the coordinates of the stationary points of the curve $y = (x + 1)(2x - 1)^2$ and determine their nature. Sketch the curve.

8. The diagram shows a shape made from a rectangle and a semicircle.

   $y + \frac{1}{4}x(\pi + 2) = 25$

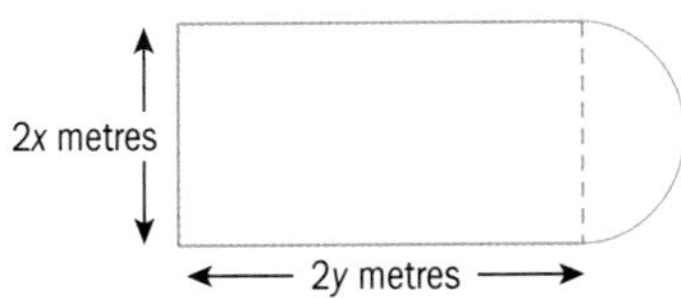

   a) Show that the area A square metres of the shape is given by $A = 100x - 2x^2 - \frac{1}{2}\pi x^2$.

   b) Find the maximum value of A and explain why this value is a maximum.

9. Triangle ABC is a right-angle triangle with angle $ACB = 90°$. $AC + BC = 6$ cm. Show that the maximum area of the triangle is 4.5 cm^2.

10. The diagram shows a cuboid with a square base.

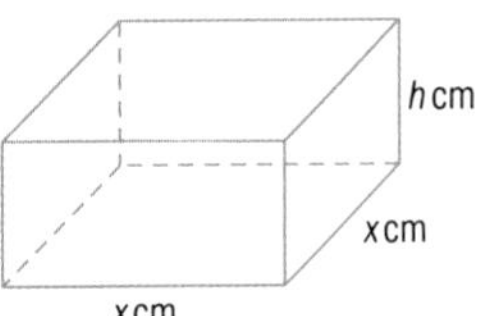

   Find the maximum volume of the cuboid if the sum of the height and the length of the base is 12 cm.

11. A solid cylinder with radius x cm has a volume of 1000 cm^3.

   a) Show that the total surface area of the cylinder is given by $A = 2\pi x^2 + \frac{2000}{x}$.

   b) Find the minimum total surface area. Give your answer to 1 decimal place.

12. The curve with equation $y = (2x + 1)(x^2 - k)$ has a stationary point where $x = 1$.

   a) Find k.

   b) Find the coordinates of the stationary points and determine their nature.

13. The radius of a sphere is increasing at a rate of 2 m s^{-1}.

   Find the exact rate of increase of the volume when the radius is equal to 4 m.

14. The volume of a cube is increasing at the rate of 10 cm^3 s^{-1}.

   Find the rate of increase of the surface area of the cube when the side of the cube is 8 cm.

15. A spherical balloon is being blown up so that its volume increases at a constant rate of 1.5 cm^3 s^{-1}.

   Find the rate of increase of the radius of the balloon when the volume is 56 cm^3.

16. A sector of a circle, radius r cm, has an area of 100 cm^2.

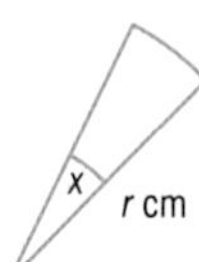

Find the minimum value for the perimeter.

17. The equation of a curve is $y = x^3 - 5x^2 + 7x - 14$.
    i) Find the coordinates of the stationary points on the curve.
    ii) Find the equation of the tangent to the curve at the point where the curve intersects the y-axis.
    iii) Find the set of values for x for which $x^3 - 5x^2 + 7x - 14$ is an increasing function of x.

18. The equation of a curve is $y = 4x^3 + px^2 + qx$ where p and q are positive constants.
    i) Show that the curve has only one stationary point when $p = 2\sqrt[3]{q}$.
    ii) In the case where $p = \frac{1}{2}$ and $q = -6$, find the set of values of x for which y is a decreasing function of x.

19. The equation of a curve is $y = 5x^2 + \frac{3}{x}$.
    i) Obtain an expression for $\frac{dy}{dx}$.
    ii) A point is moving along the curve in such a way that the x-coordinate is increasing at a constant rate of 0.04 units per second. Find the rate of change of the y-coordinate when $x = 2$.

20. The function f(x) is defined for $x \le \frac{4}{3}$ by $f(x) = \sqrt[3]{(4-3x)}$.
    i) Find $f'(x)$ and $f''(x)$.

    The first, second and third terms of a geometric progression are f(1), f′(1) and $kf''(1)$.

    ii) Find the value of the constant k.

Chapter summary

Increasing and decreasing functions

- When the gradient of a function is positive we say the function is an **increasing function**. A function $f(x)$ is increasing for $a < x < b$ if $f'(x) > 0$ for $a < x < b$.
- When the gradient of a function is negative we say the function is a **decreasing function**. A function $f(x)$ is decreasing for $a < x < b$ if $f'(x) < 0$ for $a < x < b$.

Stationary points

- Any point on a curve where $\frac{dy}{dx} = 0$ is called **a stationary point**.
- There are three types of stationary point: minimum points, maximum points and points of inflexion.
- At a minimum point, $\frac{d^2y}{dx^2} > 0$.
- At a maximum point, $\frac{d^2y}{dx^2} < 0$.
- At $\frac{d^2y}{dx^2} = 0$, the point may be a minimum, or a maximum, or a point of inflexion, and so we must look at the sign of $\frac{dy}{dx}$ either side of the stationary point to determine its nature.

Points of inflexion are not included in Cambridge International AS & A Level 9709 examination papers.

Problems involving maximum and minimum values

- We can use these steps to solve problems:
  - Draw a diagram where possible.
  - Choose letters to represent the unknown quantities.
  - Write an expression for the quantity we are trying to maximise/minimise – it may contain two unknowns.
  - Use a given condition from the question to write an equation connecting two variables.
  - Rearrange the equation to express one variable in terms of the other.
  - Use substitution to find an expression in terms of one variable only.
  - Use differentiation to find maximum/minimum values.

Connected rates of change

- Always derive a formula related to the rates of change given

  e.g. $\dfrac{dA}{dt} = \dfrac{dA}{dr} \times \dfrac{dr}{dt}$ e.g. $\dfrac{dV}{dt} = \dfrac{dV}{dr} \times \dfrac{dr}{dt}$

  e.g. $\dfrac{dV}{dt} = \dfrac{dV}{ds} \times \dfrac{ds}{dt}$ e.g. $\dfrac{dy}{dx} = \dfrac{dy}{du} \times \dfrac{du}{dt} \times \dfrac{dt}{dx}$.

- You can use these steps to solve problems:
  - Write the information given as a rate of change.
  - Write down the rate of change you want to find.
  - Write down the link between the two variables.
  - Differentiate to get another rate of change.
  - You may also need to use another formula and then differentiate this to get another rate of change.
  - Substitute these rates of change.
  - You can then substitute any given values to get the answer.

10 Integration

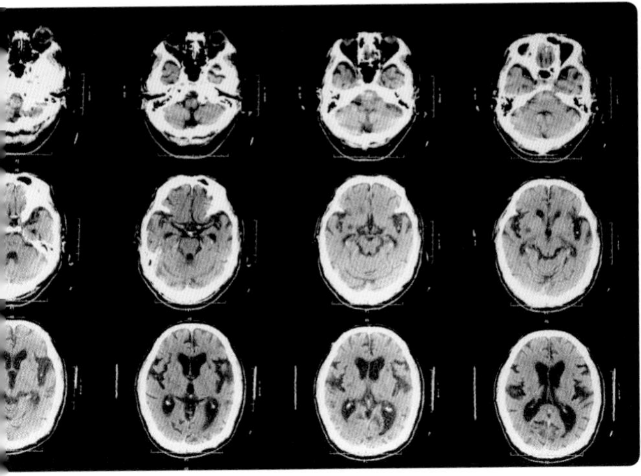

Integration is involved in practically every physical theory in some way. The Radon transform uses integration to reconstruct images from medical CT scans. As well as applications in modern day medicine, CT scanners have also been used to examine Egyptian mummies to find out information about their diets, health and lifestyle.

Objectives

- Understand integration as the reverse process of differentiation.
- Integrate $(ax + b)^n$ for any rational value of n, $n \neq -1$.
- Solve problems involving evaluating the constant of integration.
- Evaluate definite integrals.
- Evaluate definite integrals, including simple cases of 'improper' integrals, such as $\int_0^1 x^{-\frac{1}{2}}\,dx$ and $\int_1^{\infty} x^{-2}\,dx$.
- Use integration to find the area bounded by a curve and lines parallel to the axes.
- Use integration to find the area between a curve and a line, or between two curves.
- Use integration to work out the volume of revolution about one of the axes.

Before you start

You should know how to:

1. Find the value of a constant, c, given a curve and a point on the curve,

   e.g. $y = 3x + c$ $(-2, 5)$

   $5 = 3(-2) + c$ $c = 11$

   e.g. $y = x^2 - 2x + c$ $(3, -1)$

   $-1 = 3^2 - 2(3) + c$ $c = -4$

Skills check:

1. Find the value of c in each case where the equation of the curve and point on the curve are given.

   a) $y = -5x + c$ $(7, -2)$

   b) $y = 6x + c$ $(-4, -3)$

   c) $y = 2x^2 + 3x + c$ $(-1, -2)$

   d) $y = -x^2 - 4x + c$ $(5, 9)$

   e) $y = 3x^3 - x^2 + 7x + c$ $(-2, 8)$

2. Find the x-coordinates of the points where a curve crosses the x-axis,

e.g. $y = x^2 + 3x$ crosses the x axis where
$x^2 + 3x = 0$
$\Rightarrow x(x + 3) = 0$
$\Rightarrow x = 0$ or $x = 3$

e.g. $y = x^3 - x^2 - 6x$ crosses the x axis where
$x^3 - x^2 - 6x = 0$
$x(x^2 - x - 6) = 0$
$\Rightarrow x(x - 3)(x + 2) = 0$
$x = -2$ or $x = 0$ or $x = 3$

2. Find the x-coordinates of the points where each curve crosses the x-axis.

a) $y = x^2 - 2x$

b) $y = x^2 - 3x - 10$

c) $y = 12 + x - x^2$

d) $y = x^3 + 5x^2 - 6x$

e) $y = (x - 1)^2(3x + 4)$

10.1 Integration as the reverse process of differentiation

We know that if $y = 3x^2 - 5x$, then $\frac{dy}{dx} = 6x - 5$.

We also know that if, for example, $y = 3x^2 - 5x + 7$ or $y = 3x^2 -5x -2$, then $\frac{dy}{dx} = 6x - 5$.

Thus, if we are given $\frac{dy}{dx} = 6x - 5$ then we can calculate that $y = 3x^2 - 5x + c$, where c is a constant.

This process is the reverse of differentiation and is called integration.
The constant, c, is known as the constant of integration.

We write this process as follows: $\int (6x - 5)\,dx = 3x^2 - 5x + c$

We say: the integral of $6x - 5$ with respect to x is equal to $3x^2 - 5x + c$.

We can use these rules:

If $\frac{dy}{dx} = x^n$, then $y = \int x^n\,dx = \frac{x^{n+1}}{n+1} + c \ (n \neq -1)$

If $\frac{dy}{dx} = ax^n$, then $y = \int ax^n\,dx = \frac{ax^{n+1}}{n+1} + c \ (n \neq -1)$

Example 1

Given $\frac{dy}{dx} = 9x^2 - 4x + 3x^{\frac{1}{2}} - 7$, find y.

$y = \int (9x^2 - 4x + 3x^{\frac{1}{2}} - 7)\,dx$ — We need to integrate $\frac{dy}{dx}$ to get y.

$= \frac{9x^3}{3} - \frac{4x^2}{2} + \frac{3x^{\frac{3}{2}}}{\frac{3}{2}} - 7x + c$ — Use $\int x^n\,dx = \frac{x^{n+1}}{n+1}$ for each term.

$= 3x^3 - 2x^2 + 2x^{\frac{3}{2}} - 7x + c$ — Simplify where possible and include '+ c' in your answer.

Example 2

Given $f'(x) = x(3x - 4)^2$, find $f(x)$.

$f'(x) = x(3x - 4)^2 = x(9x^2 - 24x + 16)$ ← Expand the brackets.

$= 9x^3 - 24x^2 + 16x$

$f'(x) = \int (9x^3 - 24x^2 + 16x)\,dx$ ← Integrating $f'(x)$ gives $f(x)$.

$= \frac{9x^4}{4} - \frac{24x^3}{3} + \frac{16x^2}{2} + c$ ← Integrate the expression term by term.

$= \frac{9}{4}x^4 - 8x^3 + 8x^2 + c$ ← Simplify where possible and include '+ c' in your answer.

Exercise 10.1

1. Find y given $\frac{dy}{dx}$ in each case.

a) $2x$ **b)** x^7 **c)** $3x^2$ **d)** -15

e) $x - x^3$ **f)** $10x + 8x^7$ **g)** $5 - \frac{1}{2}x$ **h)** $x(x^4 - 6)$

i) $\sqrt{x}$ **j)** $(2x - 3)^2$ **k)** $\frac{2x^5 + 7x}{x}$ **l)** $\frac{12}{x^2} - \frac{6}{x^3}$

2. Find $f(x)$ given $f'(x)$ in each case.

a) $\frac{9}{8}x^{\frac{1}{2}}$ **b)** $-2x$ **c)** $\frac{\sqrt{x}}{3}$ **d)** $12x^5$

e) $3 - x - 2x^5$ **f)** $\left(\frac{1}{x^2} + 5\right)^2$ **g)** $\frac{4}{\sqrt[3]{x}}$ **h)** $\frac{24x^3 - 8x}{x}$

i) $(x - 4)(x + 7)$ **j)** $-\frac{5}{6}$ **k)** $x^2 + x^{-2}$ **l)** $\frac{14x^8 - 3}{x^2}$

3. Find

a) $\int 6x\,dx$ **b)** $\int 4x + 1\,dx$ **c)** $\int 4x^{-\frac{1}{2}}\,dx$

d) $\int 7x^{-8}\,dx$ **e)** $\int (x + 4)^2\,dx$ **f)** $\int \frac{4x^{-\frac{4}{3}}}{3}\,dx$

g) $\int (9 - 6x)\,dx$ **h)** $\int \frac{2x + 5x^3}{x}\,dx$ **i)** $\int 2x(1 - x)^2\,dx$

j) $\int \frac{(2x+1)^2}{\sqrt{x}}\,dx$ **k)** $\int \left(\frac{3}{\sqrt{x}} - \sqrt{x^3}\right)dx$ **l)** $\int \sqrt{x} - (\sqrt{x} + 5)^2\,dx$.

10.2 Finding the constant of integration

We can find the constant of integration when we know the equation of a curve and the coordinates of a given point on the curve.

Example 3

Find the equation of the curve that passes through the point (4, −1) and where $\frac{dy}{dx} = \frac{5x^2+1}{\sqrt{x}}$.

$\frac{dy}{dx} = \frac{5x^2+1}{\sqrt{x}} = 5x^{\frac{3}{2}} + x^{-\frac{1}{2}}$ ← Divide each term by $x^{\frac{1}{2}}$.

$y = \frac{5x^{\frac{5}{2}}}{\frac{5}{2}} + \frac{x^{\frac{1}{2}}}{\frac{1}{2}} + c$ ← Integrate each term and include '+ c' at the end.

$= 2x^{\frac{5}{2}} + 2x^{\frac{1}{2}} + c$ ← Dividing by $\frac{5}{2}$ is the same as multiplying by $\frac{2}{5}$.

As (4, −1) lies on the curve,

$-1 = 2(4)^{\frac{5}{2}} + 2(4)^{\frac{1}{2}} + c$ ← Substitute $x = 4$, $y = -1$.

$c = -1 - 64 - 4 = -69$ ← Find the value of c.

$y = 2x^{\frac{5}{2}} + 2x^{\frac{1}{2}} - 69$ ← Write down the equation of the curve.

Note: You can check you have integrated correctly by differentiating,

e.g. $\frac{d}{dx}\left(2x^{\frac{5}{2}} + 2x^{\frac{1}{2}} + c\right) = 5x^{\frac{3}{2}} + x^{-\frac{1}{2}}$

Example 4

A curve is such that $f'(x) = \frac{10}{x^3} - 4$ and the point $(-1, 2)$ lies on the curve.

Find the equation of the curve.

$f(x) = \int\left(\frac{10}{x^3} - 4\right)dx = \int(10x^{-3} - 4)\,dx$ ← Integrate $f'(x)$ to find $f(x)$ by expressing each term as a power of x.

$= \frac{10x^{-2}}{-2} - 4x + c$ ← Integrate term by term.

$= -\frac{5}{x^2} - 4x + c$ ← Simplify and write your negative powers as fractions.

$2 = -\frac{5}{(-1)^2} - 4(-1) + c$ ← Substitute for $x = -1$ and $y = 2$.

$c = 3$ ← Find the value of c.

$f(x) = -\frac{5}{x^2} - 4x + 3$ ← Write down the equation of the curve.

Example 5

A curve is such that $\frac{d^2y}{dx^2} = -6x$ and the curve has a maximum point at $(1, 2)$.

Find the equation of the curve.

$\frac{dy}{dx} = \int -6x\,dx$ ← Integrate $\frac{d^2y}{dx^2}$ to find $\frac{dy}{dx}$.

$= -3x^2 + c$

$0 = -3(1)^2 + c$ ← When $x = 1$ and $y = 2$ we have a maximum point, and so $\frac{dy}{dx} = 0$.

$c = 3$

$\frac{dy}{dx} = -3x^2 + 3$ ← We now have an equation for $\frac{dy}{dx}$.

$y = -x^3 + 3x + k$ ← Integrate to find y.

$2 = -(1)^3 + 3(1) + k$ ← $y = 2$ when $x = 1$.

$k = 0$

$y = -x^3 + 3x$ ← We can now write an equation for the curve.

Exercise 10.2

1. Find the equation of each curve, given $\frac{dy}{dx}$ and a point on the curve.

a) $\frac{dy}{dx} = 3x^2 - 6x + 2$ point = (−2, −10)

b) $\frac{dy}{dx} = (1 - 2x)^2$ point = (1, 8)

c) $\frac{dy}{dx} = x(2x + 5)$ point = (5, −1)

d) $\frac{dy}{dx} = \sqrt{x}\left(\sqrt{x} - 3\right)$ point = (9, 12)

e) $\frac{dy}{dx} = \frac{9x^3 - 3x}{x}$ point = (−5, 4)

f) $\frac{dy}{dx} = (3x - 1)(5x + 2)$ point = (−4, −6)

2. A curve is such that $\frac{dy}{dx} = \frac{5}{\sqrt{x}} - 10\sqrt{x^3}$ and the point (1, −6) lies on the curve. Find the equation of the curve.

3. A curve passes through the point (7, 10) and its gradient function is $\frac{6}{x^3} + 2$. Find the equation of the curve.

4. The curve C, with equation $y = f(x)$ passes through the point (−2, −1) and $f'(x) = x(3 - x)$. Find the equation of C in the form $y = f(x)$.

5. A curve is such that $\frac{d^2y}{dx^2} = -8x$. The curve has a maximum point when $x = 1$, and the point (2, −1) lies on the curve. Find the equation of the curve.

6. $f'(x) = 8x^3 - 4 + 3x^{\frac{-1}{2}}$ and $f(4) = 3$, find $f(x)$.

7. Given that $\frac{d^2y}{dx^2} = -3x + 2$ and that when $x = -1$, $\frac{dy}{dx} = 5$ and $y = 0$, find y in terms of x.

8. The curve C passes through the point (3, 10) and its gradient at any point is given by $\frac{dy}{dx} = 6x^2 - 4x + 3$.

a) Find the equation of the curve C.

b) Show that the point (2, −21) lies on the curve.

9. A curve is such that $\frac{d^2y}{dx^2} = 6x$. The curve has a maximum point when $x = -1$, and the point (3, −2) lies on the curve.
Find the equation of the curve.

10. The gradient of a curve is given by $\frac{dy}{dx} = ax + b$.
Given that the curve passes through (0, 0), (1, 1) and (−2, 16), find the equation of the curve.

10.3 Integrating expression of the form $(ax + b)^n$

We can integrate **linear** expressions that have been raised to a power by using the result

$$\int (ax+b)^n \, dx = \frac{(ax+b)^{n+1}}{a(n+1)} + c, \text{ provided } n \neq -1$$

Note: This result is valid for all values of $n \neq -1$, as $\frac{1}{n+1}$ is not defined for $n = -1$.

Note: Proof of this result is beyond the scope of this unit.

Example 6

Find $\int (4x-3)^5 \, dx$.

$$\int (4x-3)^5 \, dx = \frac{1}{4}\frac{(4x-3)^{5+1}}{(5+1)} + c$$

Use the standard result with $a = 4$, $b = -3$, $n = 5$.

$$= \frac{(4x-3)^6}{24} + c$$

Example 7

Find $\int \frac{4}{(1-2x)^7} \, dx$.

$$\int \frac{4}{(1-2x)^7} \, dx = 4\int (1-2x)^{-7} \, dx$$

We can take the constant factor 4 outside the integral.

$$= 4 \times \frac{1}{-2} \times \frac{(1-2x)^{-6}}{-6} + c$$

$a = 1$, $b = -2$, $n = -7$

$$= \frac{(1-2x)^{-6}}{3} + c$$

$$= \frac{1}{3(1-2x)^6} + c$$

Example 8

$\int \frac{6}{\sqrt{(4x+9)}} \, dx$

$$\int \frac{6}{\sqrt{(4x+9)}} \, dx = \int 6(4x+9)^{-\frac{1}{2}} \, dx$$

Write in a form using negative powers.

$$= 6 \times \frac{1}{4}\frac{(4x+9)^{\frac{1}{2}}}{\frac{1}{2}} + c$$

Use the standard result with $a = 4$, $b = 9$, $n = -\frac{1}{2}$.

$$= 3\sqrt{(4x+9)} + c$$

Simplify your answer.

Exercise 10.3

1. Find these integrals.

a) $\int (2x-1)^6\,dx$ b) $\int (4-3x)^8\,dx$ c) $\int (5x+2)^5\,dx$

d) $\int \frac{1}{(3x+5)^5}\,dx$ e) $\int \frac{15}{(1-3x)^6}\,dx$ f) $\int \frac{2}{(5+2x)^9}\,dx$

g) $\int \frac{3}{\sqrt{7x+1}}\,dx$ h) $\int \frac{6}{\sqrt{(6x-5)^3}}\,dx$ i) $\int \frac{1}{\sqrt[3]{(7-x)}}\,dx$

j) $\int 3\sqrt{(1-x)}\,dx$ k) $\int \frac{4}{(1-2x)^7}\,dx$ l) $\int \left(\sqrt{(2+3x)}\right)^5 dx$

2. A curve passes through the point (1, 5) and its gradient function is $(3x-4)^5$. Find the equation of the curve.

3. A curve is such that $\frac{dy}{dx} = (7-x)^4$ and the point (5, –3) lies on the curve. Find the equation of the curve.

4. $f'(x) = \frac{1}{(5x-3)^4}$ and $f(1) = -90$. Find $f(x)$.

5. $\frac{d^2y}{dx^2} = \left(\frac{1}{4}x+1\right)^7$. When $x = 4$, $\frac{dy}{dx} = 6$ and when $x = 4$, $y = 0$.
Find y in terms of x.

10.4 The definite integral

So far we have dealt with indefinite integrals. Indefinite integrals are integrals which include '+c', where c can be any constant number.
We can find a numerical value for an integral if we are given the upper limit and the lower limit.
This is called a definite integral and is defined as follows:

$$\int_a^b f'(x)dx = [f(x)]_a^b = f(b) - f(a)$$

where the constant a is the lower limit and the constant b is the upper limit, i.e. $b > a$.

Example 9

a) Find $\int (3x-2)^8 \mathrm{d}x$. b) Hence find $\int_0^1 (3x-2)^8 \mathrm{d}x$.

a) $\int (3x-2)^8 \mathrm{d}x = \frac{1}{3}\frac{(3x-2)^9}{9} + c$

$= \frac{(3x-2)^9}{27} + c$

Also divide by the coefficient of x.

b) $\int_0^1 (3x-2)^8 \mathrm{d}x = \left[\frac{(3x-2)^9}{27}\right]_0^1$

We use square brackets to show we have integrated and we write the limits in the positions shown.

$= \frac{(3(1)-2)^9}{27} - \frac{(3(0)-2)^9}{27}$

Calculate $f(b) - f(a)$.

$= \frac{1}{27} + \frac{512}{27} = \frac{513}{27} = 19$

Example 10

Find $\int_1^4 \frac{x^2 + x^3}{\sqrt{x}} \mathrm{d}x$.

$\int_1^4 \frac{x^2 + x^3}{\sqrt{x}} \mathrm{d}x = \int_1^4 \left(x^{\frac{3}{2}} + x^{\frac{5}{2}}\right) \mathrm{d}x$

Divide each term in the numerator by $x^{\frac{1}{2}}$.

$= \left[\frac{2}{5}x^{\frac{5}{2}} + \frac{2}{7}x^{\frac{7}{2}}\right]_1^4$

$= \left(\frac{2}{5}4^{\frac{5}{2}} + \frac{2}{7}4^{\frac{7}{2}}\right) - \left(\frac{2}{5}1^{\frac{5}{2}} + \frac{2}{7}1^{\frac{7}{2}}\right)$

Substitute $x = 4$ and $x = 1$.

$= \left(\frac{2}{5} \times 32 + \frac{2}{7} \times 128\right) - \left(\frac{2}{5} + \frac{2}{7}\right)$

Evaluate as separate brackets.

$= 12\frac{2}{5} + 36\frac{2}{7} = 48\frac{24}{35}$

Exercise 10.4

1. Find

a) $\int_0^1 6x \,\mathrm{d}x$ b) $\int_2^5 x^2 \mathrm{d}x$ c) $\int_4^9 \sqrt{x} \,\mathrm{d}x$

d) $\int_{-1}^1 \frac{x}{2} \mathrm{d}x$ e) $\int_1^2 \frac{3}{x^2} \mathrm{d}x$ f) $\int_1^4 \frac{1}{\sqrt{x}} \mathrm{d}x$.

2. Find

a) $\int_{-1}^{2}(6x^2 - 2x)\,dx$ b) $\int_{1}^{2}(9x^2 - 6x + 5)\,dx$ c) $\int_{-1}^{1}(2x^4 - 3x^3)\,dx$

d) $\int_{-3}^{-2}(3x^2 + 8x + 1)\,dx$ e) $\int_{1}^{3}(3x - 2)(3x + 2)\,dx$ f) $\int_{0}^{1}(2x + 1)^5\,dx.$

3. Evaluate

a) $\int_{0}^{2}\frac{x^2 + x^3}{x}\,dx$ b) $\int_{2}^{3}\frac{4x + 1}{x^3}\,dx$ c) $\int_{0}^{1}\frac{4\sqrt{x} - x}{\sqrt{x}}\,dx$

d) $\int_{1}^{9}\left(2 + \sqrt{x}\right)^2\,dx$ e) $\int_{2}^{3}\sqrt{(3 - x)}\,dx$ f) $\int_{2}^{3}\frac{4x - x^2}{2\sqrt{x}}\,dx.$

10.5 Finding area using definite integration

The diagram shows the area (A) bounded by the curve $y = f(x)$, the lines $x = a$, $x = b$ and the x-axis.

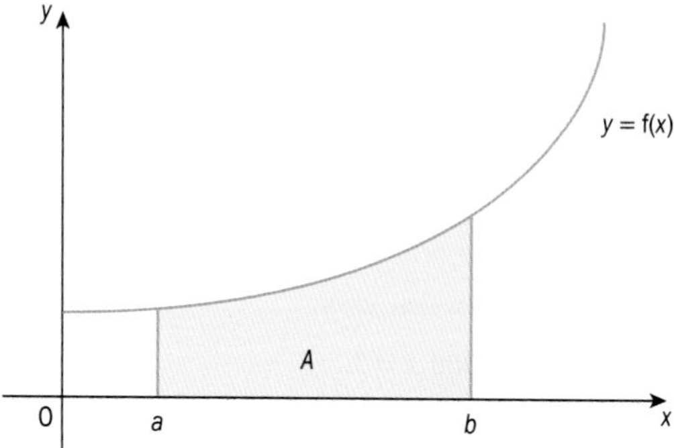

We can find an estimate of this area by dividing the required area into thin strips of equal width. The thinner the strips the better the approximation to the actual area A.

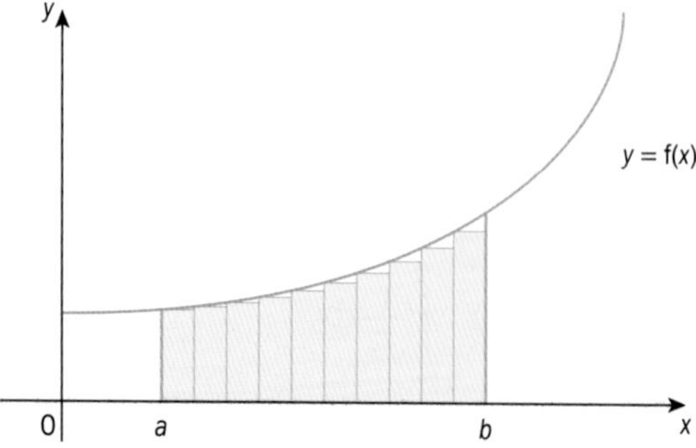

Consider the area of one of the rectangles:

- Its width is a small increase in x so we can say the width of the rectangle is δx.
- Similarly its area is a small increase in the value of A so we can say the area of the rectangle is δA.

As each strip is approximately a rectangle of width δx and height y, we can now say that $\boxed{\delta A \approx y \times \delta x}$ and $y \approx \dfrac{\delta A}{\delta x}$

The total area of $A = \sum_{x=a}^{x=b} \delta A = \sum_{x=a}^{x=b} y \delta x$

As δx gets smaller the accuracy of the area increases until, in the limiting case,

the total area $= \lim_{\delta x \to 0} \sum_{x=a}^{x=b} \dfrac{\delta A}{\delta x} = y$

But, $\lim_{\delta x \to 0} \dfrac{\delta A}{\delta x} = \dfrac{dA}{dx}$ and so, $\dfrac{dA}{dx} = y$

Hence $A = \int y \, dx$

> The area between the curve $y = f(x)$ and the x-axis, bounded by the line $x = a$ and the line $x = b$ is given by
>
> $$\text{Area} = \int_a^b f(x)\, dx = \int_a^b y \, dx$$

Note: The proof of this formula is not within the Cambridge 9709 syllabus.

Example 11

Find the area bounded by the curve with equation $y = (2 - x)(3 + x)$, the positive x-axis and the y-axis.

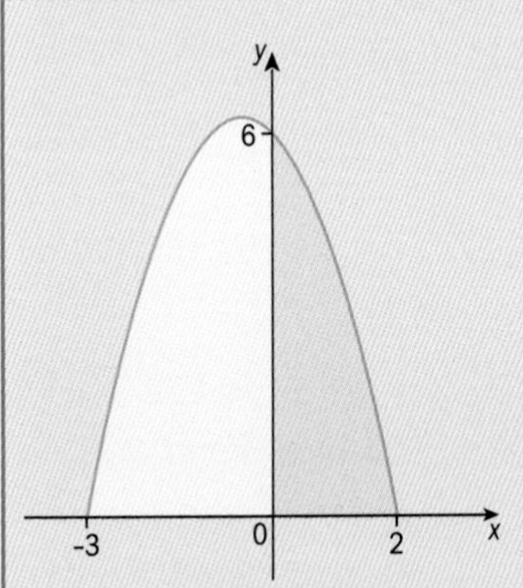

Draw a sketch to show the required region.

$$\text{Area} = \int_0^2 (6 - x - x^2)\, dx$$

Write the area as a definite integral.

$$= \left[6x - \frac{1}{2}x^2 - \frac{1}{3}x^3\right]_0^2$$

$$= \left(12 - 2 - \frac{8}{3}\right) - (0 - 0 - 0) = 7\frac{1}{3}$$

Work out the value of the integral.

Example 12

Find the area bounded by the curve with equation $y = x(x - 4)$ and the x-axis.

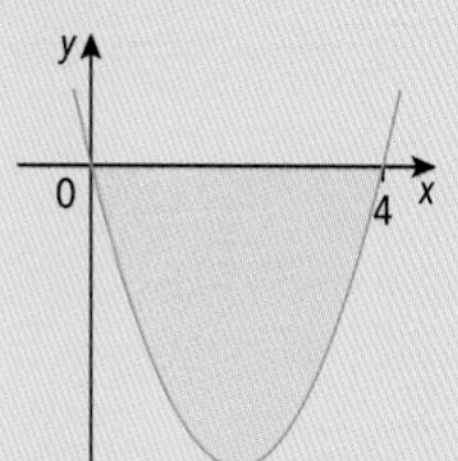

First draw a sketch.

$$\text{Area} = \int_0^4 (x^2 - 4x)\,dx$$

$$= \left[\frac{1}{3}x^3 - 2x^2\right]_0^4$$

$$= \left(\frac{64}{3} - 32\right) - (0 - 0) = -10\frac{2}{3}$$

Note: Where the value of the integral is negative this shows that the region is below the x-axis.

$$\text{Area} = 10\frac{2}{3}$$

Give the area as a positive value.

Example 13

Find the area bounded by the curve with equation $y = x(x + 2)(x - 2)$ and the x-axis.

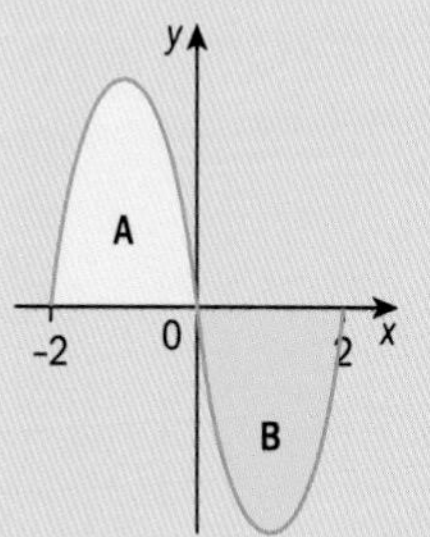

First draw a sketch.

We find the area by considering the two separate regions A and B.

$$\text{Area A} = \int_{-2}^{0} (x^3 - 4x)\,dx$$

Expand the brackets.

$$= \left[\frac{1}{4}x^4 - 2x^2\right]_{-2}^{0}$$

$$= (0 - 0) - (4 - 8) = 4$$

▶ Continued on the next page

$$\text{Area B} = \int_0^2 (x^3 - 4x)\,dx$$

$$= \left[\frac{1}{4}x^4 - 2x^2\right]_0^2$$

$$= (4 - 8) - (0 - 0) = -4$$

Total area = 4 + 4 = 8

We can use the result from region A above.

The negative value shows the area is under the x-axis.

Note that if we calculate $\int_{-2}^{2} (x^3 - 4x)\,dx$ we get 0 as the answer.

We can also use this technique to find the area bounded by a curve, $x = f(y)$, the y-axis and the lines $y = a$ and $y = b$.

The area between the curve $x = f(y)$, the **y-axis**, and the lines $y = a$ and $y = b$ is given by:

$$\text{Area} = \int_{y=a}^{y=b} x\,dy \text{ or } \int_{y=a}^{y=b} f(y)\,dy$$

Note: We usually write $\int_a^b$ instead of $\int_{y=a}^{y=b}$

Example 14

Find the area bounded by the curve with equation $x = y^2 - 2y$ and the y-axis as shown in the diagram.

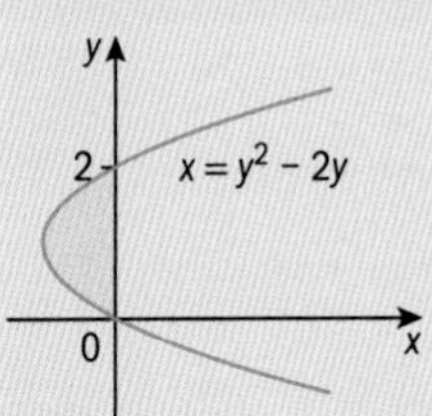

$$\text{Area} = \int_{y=a}^{y=b} x\,dy$$

Use the correct formula.

$$= \int_0^2 (y^2 - 2y)\,dy$$

Everything is using y, including the limits.

$$= \left[\frac{1}{3}y^3 - y^2\right]_0^2$$

$$= \left(\frac{8}{3} - 4\right) - (0 - 0) = -\frac{4}{3}$$

The negative value shows that the required area is **to the left** of the y-axis.

We write that the required area is $\frac{4}{3}$.

Write as a positive value.

Exercise 10.5

1. Find the area of each of the shaded regions.

a)

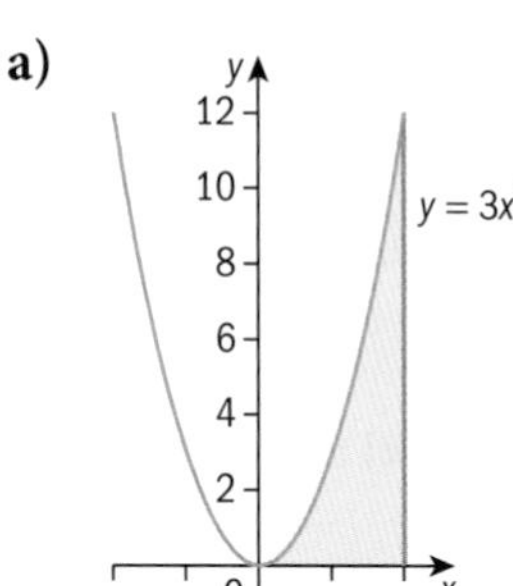

b)

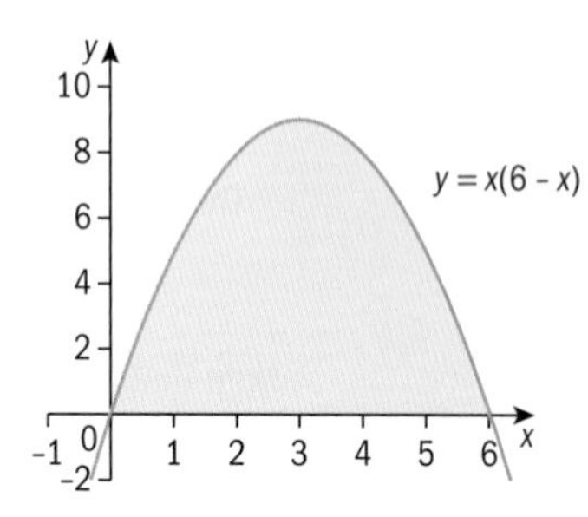

c)

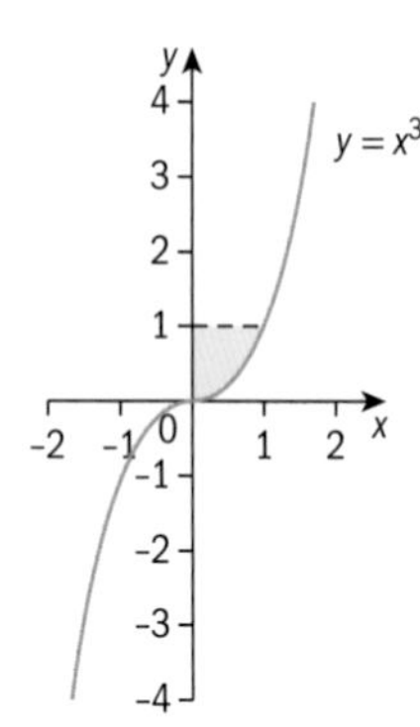

d)

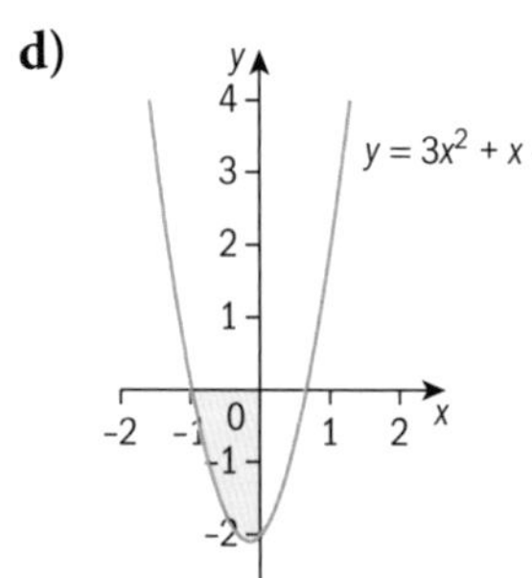

e)

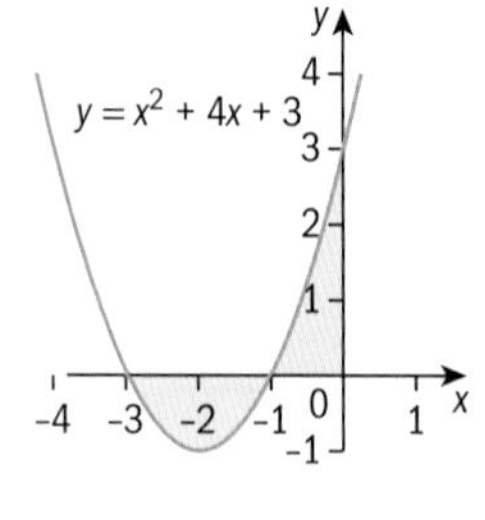

f)

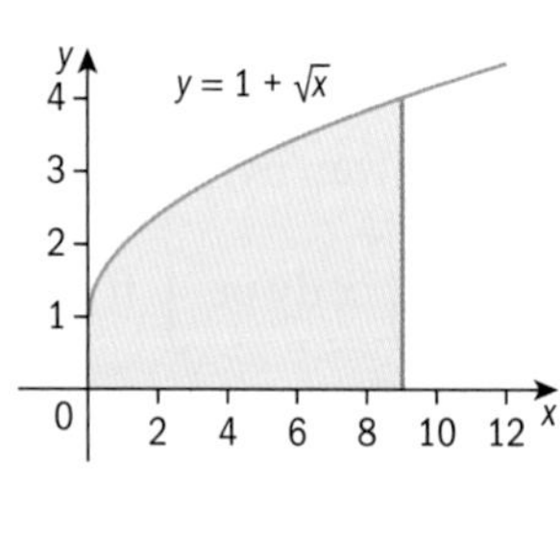

2. For each, find the area bounded by the curve and the x-axis.

   a) $y = 3x^2 - 3$ **b)** $y = x(x + 2)(x - 1)$

   c) $y = x^3 - 5x^2 - 6x$ **d)** $y = 1 - 4x^2$

First, sketch the curve.

3. Find the area bounded by the curve $y = \sqrt{x}$, the y-axis and the lines $y = 1$ and $y = 3$.

4. Find the area bounded by the curve $y = x^3$, the x-axis and the lines $x = -1$ and $x = 2$.

5. Find the area bounded by the curve $x = (y + 3)(y - 1)$ and the y-axis.

6. Find the area bounded by the curve $y = 2\sqrt{x}$, the x-axis and the lines $x = 1$ and $x = 9$.

7. **a)** Draw the graph of $x + y = 5$ and find the area bounded by the line $x + y = 5$, the x-axis and the y-axis.

   b) Check this area using integration.

8. The diagram shows part of the curve with equation $y = \frac{16}{\sqrt{x}} + \sqrt{x}$.

Find the area bounded by the curve, the lines $x = 1$, $x = 4$ and the x-axis.

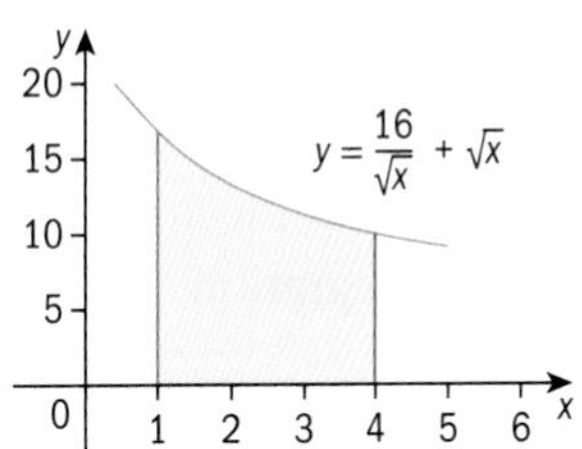

9.

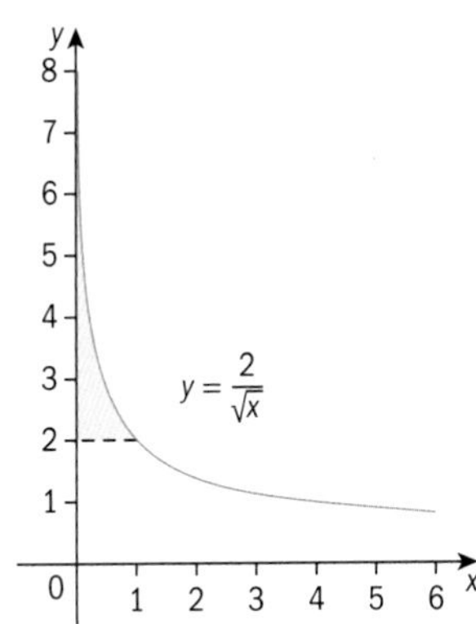

Find the area of the shaded region.

10. The diagram shows part of the curve with equation $y = (x - 5)(x + 2)(x - 3)$.

a) Write down the coordinates of A and B.

b) Show that the equation may be written as $y = x^3 - 6x^2 - x + 30$.

c) Find the total shaded area.

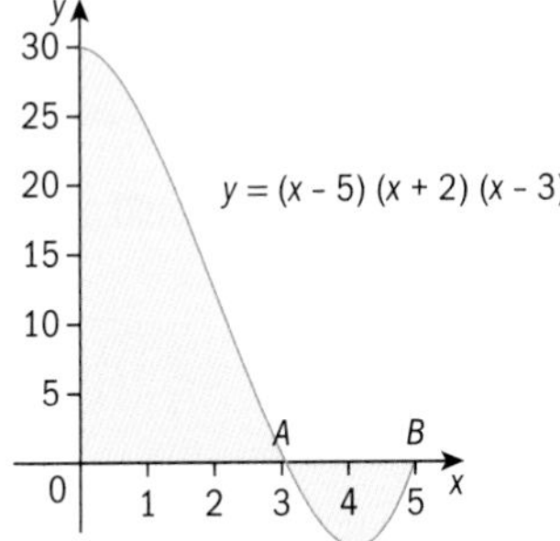

10.6 Area bounded by two curves or a curve and a line

We can find the area between two curves or between a curve and a line using integration.

Example 15

The diagram shows a sketch of the curve with equation $y = x(5 - x)$ and the line $y = 2x$.
Work out the shaded area.

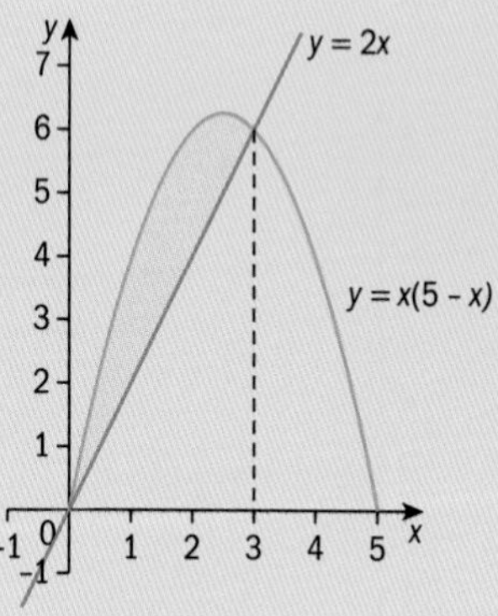

The curve meets the line when $x(5 - x) = 2x$

$5x - x^2 = 2x$

$x^2 - 3x = 0$

$x(x - 3) = 0$

$x = 0$ or $x = 3$ ← Solve the two equations simultaneously to find where the line and curve intersect.

When $x = 3$, $y = 6$

Area under the curve between $x = 0$ and $x = 3$

$$= \int_0^3 (5x - x^2)\,dx$$

$$= \left[\frac{5}{2}x^2 - \frac{1}{3}x^3\right]_0^3$$

$$= (22.5 - 9) - (0) = 13.5$$

Shaded area $= 13.5 - 9 = 4.5$ ← Area of triangle $= \frac{1}{2} \times 3 \times 6 = 9$. This could also be worked out using $\int_0^3 2x\,dx$.

Alternative method

Shaded area $= \int_0^3 (5x - x^2)\,dx - \int_0^3 2x\,dx$ ← Area under curve minus area under line

$$= \int_0^3 (5x - x^2 - 2x)\,dx$$ ← Combine this into one integral.

$$= \int_0^3 (3x - x^2)\,dx$$

$$= \left[\frac{3}{2}x^2 - \frac{1}{3}x^3\right]_0^3$$

$$= (13.5 - 9) - (0) = 4.5$$

Example 16

The diagram shows the curve $y = x^2 + 2x$ and the line $y = -x$.
Work out the area of the shaded region.

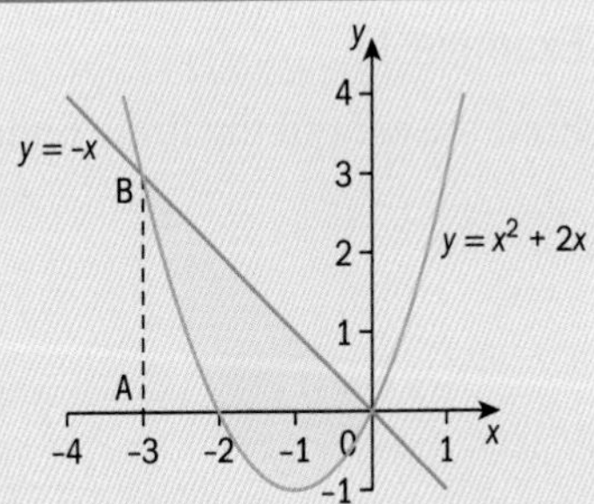

The curve meets the line when $x^2 + 2x = -x$

$x^2 + 3x = 0$

$x(x + 3) = 0$

Solve the two equations simultaneously to find where the line intersects the curve.

They meet at (0, 0) and (−3, 3)

Area of triangle 0AB $= \frac{1}{2}(3 \times 3) = \frac{9}{2}$

Area under the curve, between $x = -3$ and $x = -2$

$$= \int_{-3}^{-2} (x^2 + 2x)\,dx$$

The dotted line is at $x = -3$.

$$= \left[\frac{1}{3}x^3 + x^2\right]_{-3}^{-2}$$

$$= \left(-\frac{8}{3} + 4\right) - (-9 + 9)$$

$$= \frac{4}{3}$$

Area of shaded region above the x-axis $= \frac{9}{2} - \frac{4}{3} = \frac{19}{6}$

Area of shaded region below the x-axis

$$= \int_{-2}^{0} (x^2 + 2x)\,dx$$

This is the area between the curve and the x-axis.

$$= \left[\frac{1}{3}x^3 + x^2\right]_{-2}^{0}$$

$$= (0 + 0) - \left(-\frac{8}{3} + 4\right) = -\frac{4}{3}$$

The area is a negative value as it is below the curve.

Area required $= \frac{19}{6} + \frac{4}{3} = \frac{9}{2}$

Example 17

Find the area of the finite region bounded by $y = x^2$ and $y = x - x^2$.

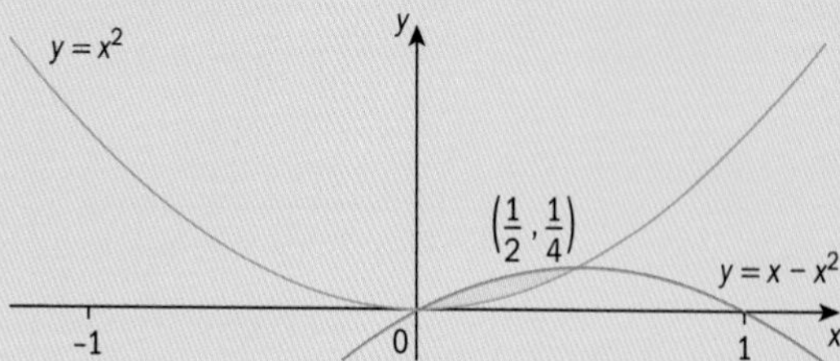

$$\text{Area} = \int_0^{\frac{1}{2}} (x - x^2)\,dx - \int_0^{\frac{1}{2}} x^2\,dx$$

Area under $y = x - x^2$ minus area under $y = x^2$

$$= \int_0^{\frac{1}{2}} (x - x^2 - x^2)\,dx$$

Combine the two integrals.

$$= \int_0^{\frac{1}{2}} (x - 2x^2)\,dx$$

$$= \left[\frac{1}{2}x^2 - \frac{2}{3}x^3\right]_0^{\frac{1}{2}}$$

$$= \left(\frac{1}{8} - \frac{1}{12}\right) - (0 - 0)$$

$$= \frac{1}{24}$$

Exercise 10.6

1. Find the shaded areas.

a)

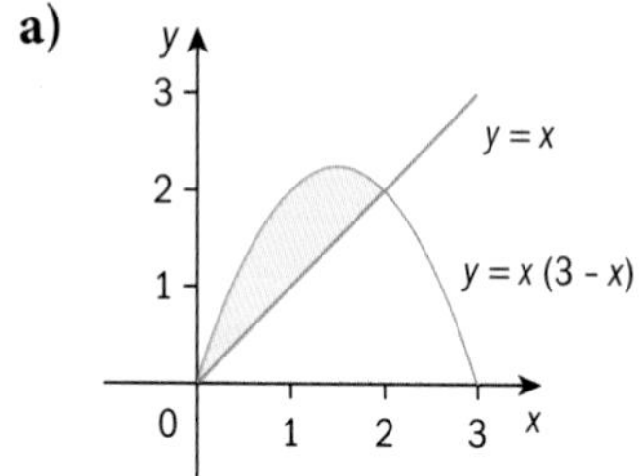

b)

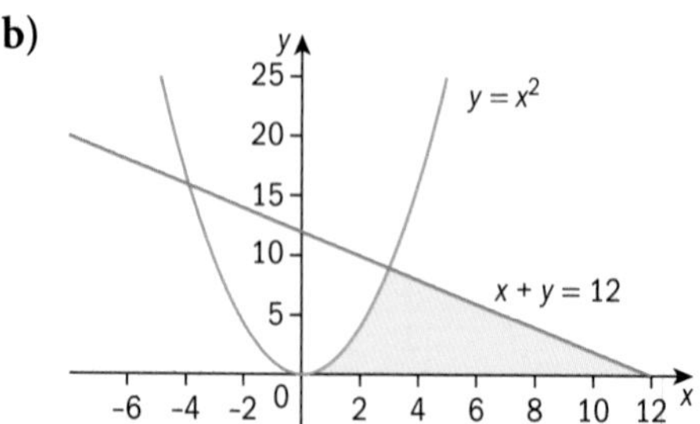

2. Find the area of the region bounded by the curve with equation $y = x^2$ and the curve with equation $y = 2x - x^2$.

3.

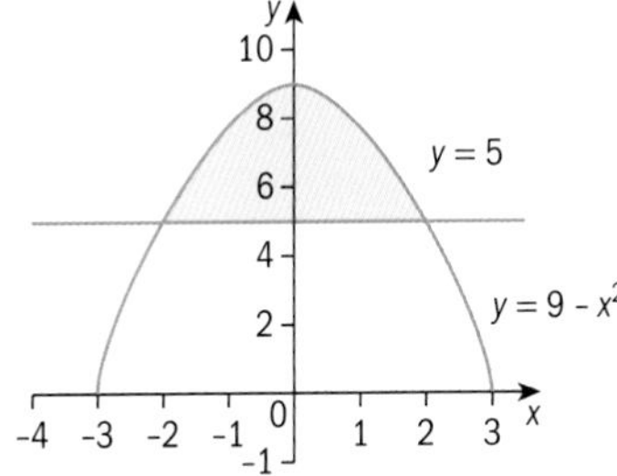

The diagram shows the curve $y = 9 - x^2$ and the line $y = 5$.
Find the shaded area.

4.

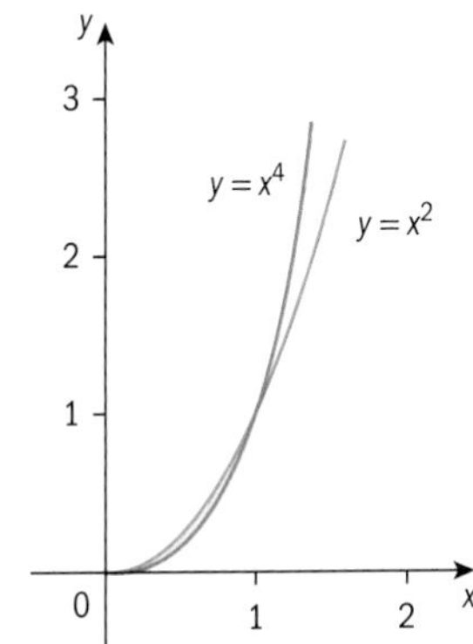

The diagram shows parts of the curves $y = x^2$ and $y = x^4$.
Find the shaded area.

5. Find the area between the curve with equation $y = x(x - 1)$ and the line $y = 3x$.

Find the coordinates of the points of intersection and sketch the graphs.

6. **a)** Show that the curve $y = 2x^2 + 3$ and the curve $y = 10x - x^2$ meet at the points where $x = 3$ and $x = \frac{1}{3}$.

b) Find the area between these two curves.

7.

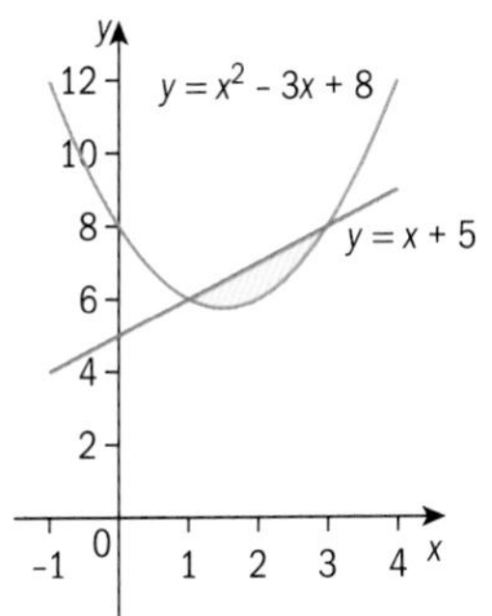

Find the area bounded by the curve $y = x^2 - 3x + 8$ and the line $y = x + 5$.

8.

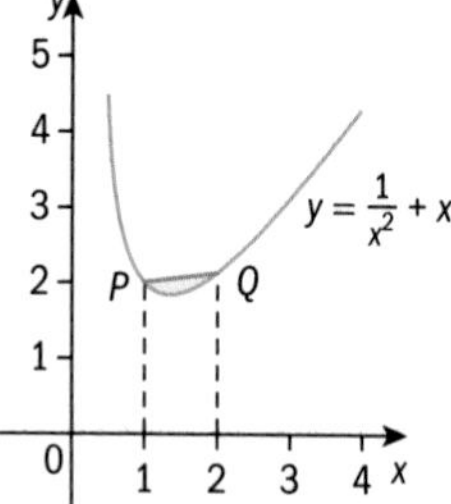

The diagram shows part of the curve with equation $y = \frac{1}{x^2} + x$.

At point P on the curve, $x = 1$ and at point Q on the curve, $x = 2$.

Find the shaded area.

9.

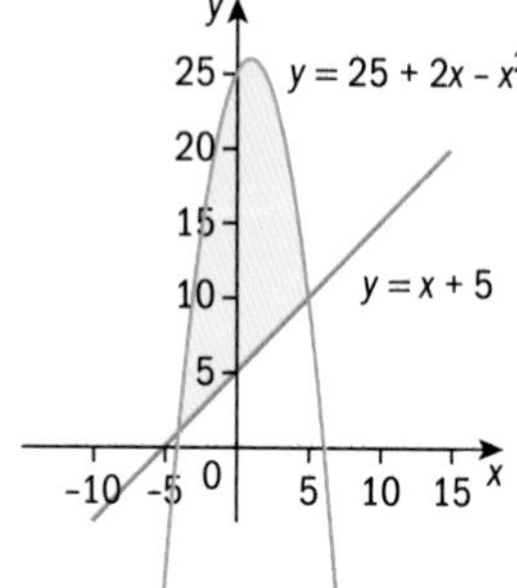

The diagram shows the curve with equation $y = 25 + 2x - x^2$.

The straight line with equation $y = x + 5$ cuts the curve in two places.

Find the exact area of the shaded region.

10.

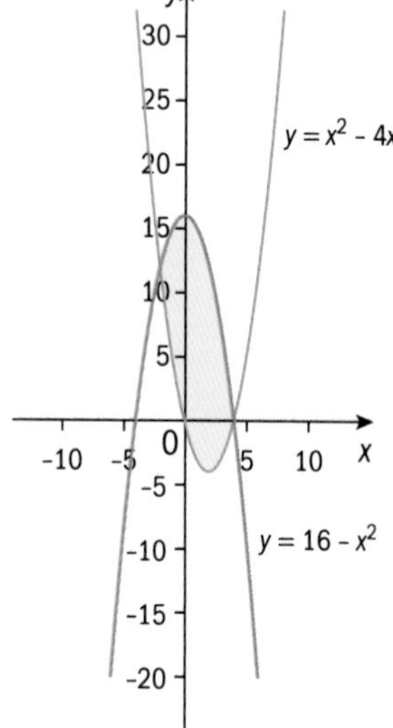

Find the area bounded by the two curves.

10.7 Improper integrals

Improper integrals can be one of two types:

(1) Integrals where at least one of the limits is infinite,

e.g. **i)** $\int_{1}^{\infty} \frac{1}{x^2\sqrt{x}}\,dx$, e.g. **ii)** $\int_{-\infty}^{-2} \frac{2}{x^5}\,dx$.

(2) Integrals where the function to be integrated is not defined at a point in the interval of integration.

In this case, we will restrict our consideration to examples where the function is not defined at one end of the interval, e.g. $\int_{-2}^{0} \frac{2}{x^5}\,dx$ where $\frac{2}{x^5}$ is not defined at $x = 0$.

For **(1) i)**

> We define the **improper integral** $\int_{a}^{\infty} f(x)dx$ as $\lim_{b\to\infty} \int_{a}^{b} f(x)dx$, provided the limit exists.

ii)

> We define the **improper integral** $\int_{-\infty}^{b} f(x)dx$ as $\lim_{a\to-\infty} \int_{a}^{b} f(x)dx$, provided the limit exists.

For **(2)**

> When f(x) is defined for $0 < x < b$, but f(x) is not defined when $x = 0$, then the **improper integral** $\int_{0}^{b} f(x)dx = \lim_{a\to 0^+} \int_{a}^{b} f(x)dx$, provided the limit exists.

Note: $\lim_{a\to 0^+}$ denotes 'a tends to 0 from just above 0'.

Example 18

Show that the improper integral $\int_{1}^{\infty} \frac{1}{x^2\sqrt{x}}\,dx$ has a value and find that value.

$$\int_{1}^{b} \frac{1}{x^2\sqrt{x}}\,dx = \int_{1}^{b} x^{-\frac{5}{2}}dx$$

Write the integral with an upper limit, say b (where b is a large number).

$$= \left[-\frac{2}{3x^{\frac{3}{2}}}\right]_1^b$$

Integrate the function in the normal way.

$$= \left(-\frac{2}{3b^{\frac{3}{2}}}\right) - \left(-\frac{2}{3}\right)$$

As $b \to \infty$, $\frac{2}{3b^{\frac{3}{2}}} \to 0$

As b tends to ∞, the denominator gets larger and the value of the fraction tends to 0.

Thus $\int_{1}^{\infty} \frac{1}{x^2\sqrt{x}}\,dx = -0 + \frac{2}{3}$

Hence the improper integral does exist (as it has a finite value) and has a value of $\frac{2}{3}$.

Example 19

Show that only one of the following improper integrals has a finite value and find that value.

a) $\int_{-2}^{0} \frac{2}{x^5}\,dx$ **b)** $\int_{-\infty}^{-2} \frac{2}{x^5}\,dx$

a) $\int_{-2}^{p} \frac{2}{x^5}\,dx = \left[-\frac{2}{4x^4}\right]_{-2}^{p}$

$= \left(-\frac{1}{2p^4}\right) - \left(-\frac{1}{2(-2)^4}\right)$

As the function to be integrated is not defined at $x = 0$, replace the upper limit 0 with a letter, say p ($-2 < p < 0$ and p just less than 0).

As $p \to 0$, $\frac{1}{2p^4}$ has no finite value

As $p \to 0$, $\frac{1}{2p^4} \to \infty$ and so has no finite value.

and so, $-\frac{1}{2p^4} + \frac{1}{32}$ has no finite value.

We can say the integral does not exist.

b) $\int_{q}^{-2} \frac{2}{x^5}\,dx = \left[-\frac{2}{4x^4}\right]_{q}^{-2}$

Start by replacing $-\infty$ with a variable, say q.

$= \left(-\frac{1}{2(-2)^4}\right) - \left(-\frac{1}{2q^4}\right)$

As $q \to -\infty$, $\frac{1}{2q^4} \to 0$

Thus $\int_{-\infty}^{-2} \frac{2}{x^5}\,dx = -\frac{1}{32} + 0$

Hence, this improper integral does exist and has a value of $-\frac{1}{32}$.

Exercise 10.7

1. Show that only one of the following improper integrals has a finite value and find that value.

   a) $\int_{0}^{1} \frac{4}{\sqrt{x}}\,dx$ b) $\int_{1}^{\infty} \frac{4}{\sqrt{x}}\,dx$

2. Find a) $\int_{2}^{p} x^3\,dx$ b) $\int_{2}^{p} x^{-3}\,dx$.

   Hence, find whether the following integrals exist and, if they do, find their value.

   c) $\int_{2}^{\infty} x^3\,dx$ d) $\int_{2}^{\infty} x^{-3}\,dx$.

3. a) Find, in terms of a and b, the value of the integral $\int_a^b \frac{6}{x^{\frac{5}{2}}}\,\mathrm{d}x$.

b) Show that only one of the following improper integrals has a finite value and find that value.

i) $\int_0^9 \frac{6}{x^{\frac{5}{2}}}\,\mathrm{d}x$ ii) $\int_9^\infty \frac{6}{x^{\frac{5}{2}}}\,\mathrm{d}x$

4. For each of the following improper integrals, find the value of the integral or explain briefly why it does not have a value.

a) $\int_{27}^\infty \sqrt[3]{x}\,\mathrm{d}x$ b) $\int_{27}^\infty \frac{1}{\sqrt[3]{x^4}}\,\mathrm{d}x$

5. Explain briefly why $\int_0^{100} 3x^{-\frac{1}{2}}\mathrm{d}x$ is an improper integral.

6. For each of the following improper integrals, find the value of the integral or explain briefly why it does not have a value.

a) $\int_0^{100} 3x^{-\frac{1}{2}}\,\mathrm{d}x$ b) $\int_0^{100} 3x^{-\frac{3}{2}}\,\mathrm{d}x$ c) $\int_{100}^\infty 3x^{-\frac{1}{2}}\,\mathrm{d}x$

10.8 Volumes of revolution

The diagram shows the area bounded by the curve $y = \mathrm{f}(x)$, the lines $x = a$ and $x = b$, and the x-axis.

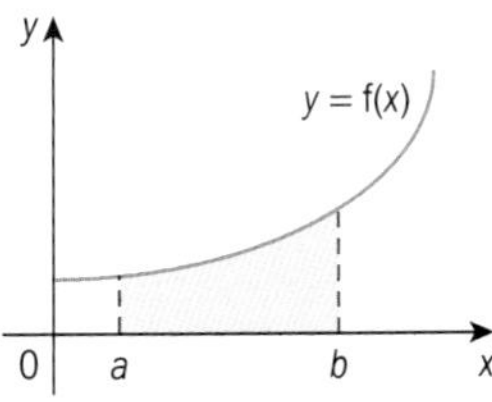

If we rotate this area about the x-axis, a solid of revolution is formed.

We can divide this three-dimensional shape into 'slices', each approximately in the shape of a cylinder, where the radius is the y value at that point and the thickness of each slice is δx.

Note: We use the symbol δx to represent a small increase in x.

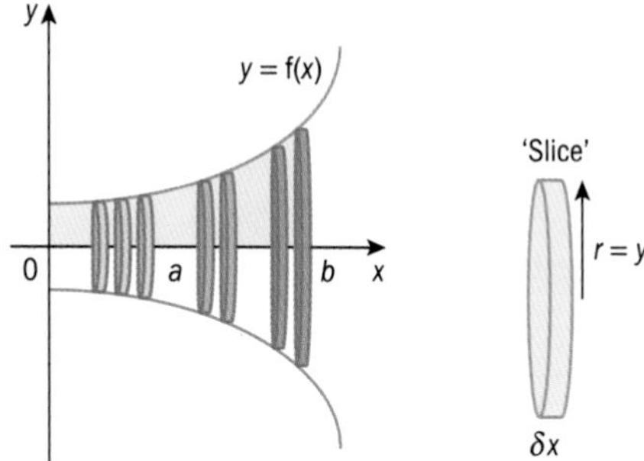

The volume of each circular 'slice' = $\pi y^2\ \delta x$ and the total volume of the solid = $\sum \pi y^2 \delta x$. As $\delta x \to 0$ we get closer and closer to the real volume V.

We can write $V = \lim_{\delta x \to 0} \sum \pi y^2 \delta x = \pi \int_a^b y^2 \mathrm{d}x$ where a and b are values of x.

The volume of a solid of revolution generated by rotating the curve $y = \mathrm{f}(x)$ between $x = a$ and $x = b$ through 360° about the x-axis is given by $V_x = \int_a^b \pi y^2 \,\mathrm{d}x$.

Similarly, the volume of a solid of revolution generated by rotating the curve $x = \mathrm{f}(y)$ between $y = a$ and $y = b$ through 360° about the y-axis is given by $V_y = \int_a^b \pi x^2 \,\mathrm{d}y$.

Example 20

Find the volume obtained when the shaded region is rotated through 360° about the x-axis.

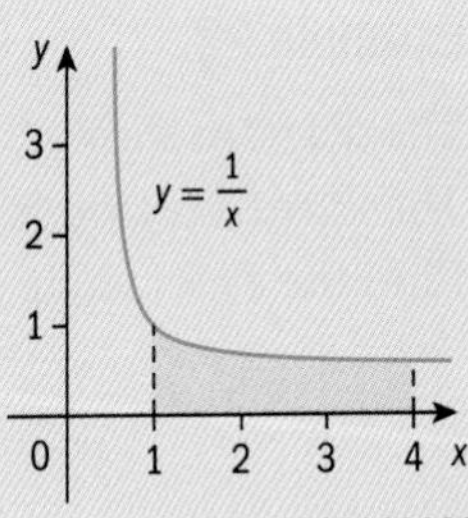

$$V = \int_1^4 \pi y^2 \, dx = \pi \int_1^4 \frac{1}{x^2} dx$$

$y^2 = \left(\frac{1}{x}\right)^2$ and $a = 1$, $b = 4$

$$= \pi \left[-\frac{1}{x} \right]_1^4$$

π is a constant so can be taken outside the integral.

$$= \pi \left[\left(-\frac{1}{4} \right) - (-1) \right]$$

$$= \frac{3}{4}\pi$$

You can leave your answer in terms of π.

Example 21

Find the volume obtained when the shaded region is rotated through 360° about the x-axis.

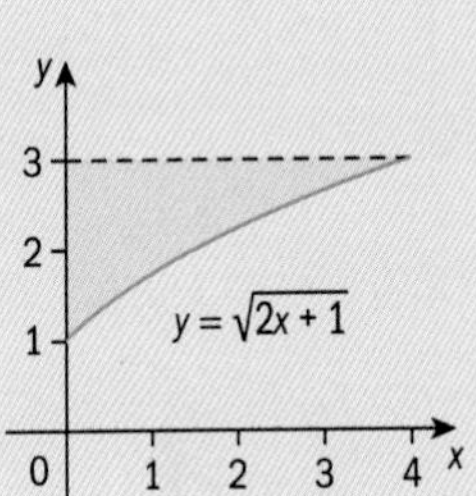

Volume required = volume of cylinder, with $r = 3$ and $h = 4$, minus volume of revolution about x axis.

Work out a strategy for finding the required volume.

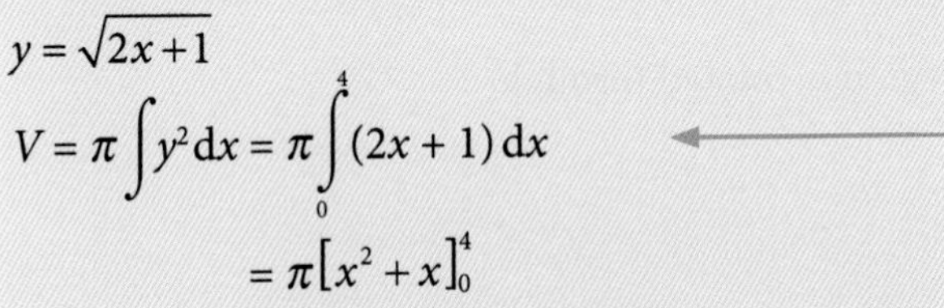

$y = \sqrt{2x+1}$

$$V = \pi \int y^2 dx = \pi \int_0^4 (2x + 1) \, dx$$

Don't forget to square y and put values of x for the bounds.

$$= \pi \left[x^2 + x \right]_0^4$$

$$= \pi \left[(4^2 + 4) - (0^2 + 0) \right]$$

Show the substitutions clearly.

$$= 20\pi$$

This gives us the volume when the area **below** the curve is rotated. The required volume is the volume of the cylinder found when rotating the line $y = 3$ about the x-axis with the volume of revolution removed. The volume of the cylinder is $\pi r^2 h$ where $r = 3$ and $h = 4$.

Required volume = $\pi \times 3^2 \times 4 - 20\pi = 16\pi$

Leave answer in terms of π unless told otherwise.

Example 22

The diagram shows the line $y = 1$, the line $y = 4$ and part of the curve $y = 3x^2$.
The shaded region is rotated through 360° about the y-axis.
Find the exact value of the volume of revolution obtained.

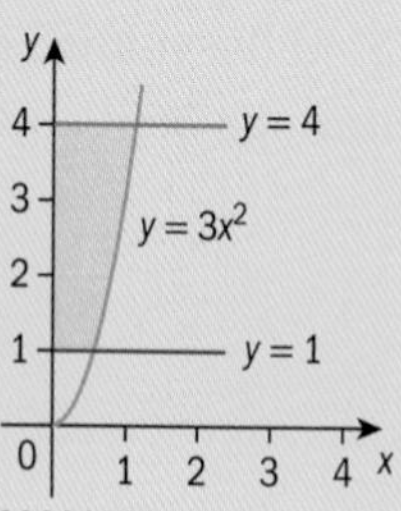

$y = 3x^2,\ x^2 = \frac{y}{3}$ — Rearrange to $x^2 = \ldots$

$V = \int_a^b \pi x^2\,dy = \pi \int_1^4 \frac{y}{3}\,dy$ — Use the correct formula.

$= \pi\left[\frac{y^2}{6}\right]_1^4$

$= \pi\left[\left(\frac{16}{6}\right) - \left(\frac{1}{6}\right)\right]$

$= \frac{15}{6}\pi = \frac{5\pi}{2}$ — The question asked for the exact answer so leave the answer in terms of π.

Exercise 10.8

1.

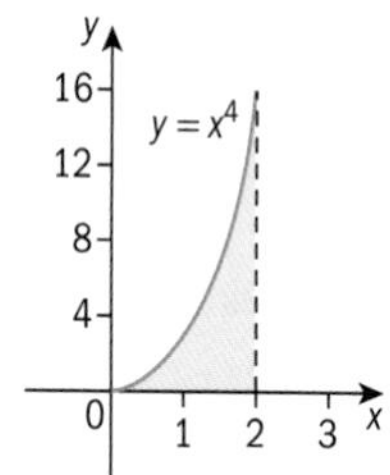

Find the volume obtained when the shaded region is rotated through 360° about the x-axis.

2.

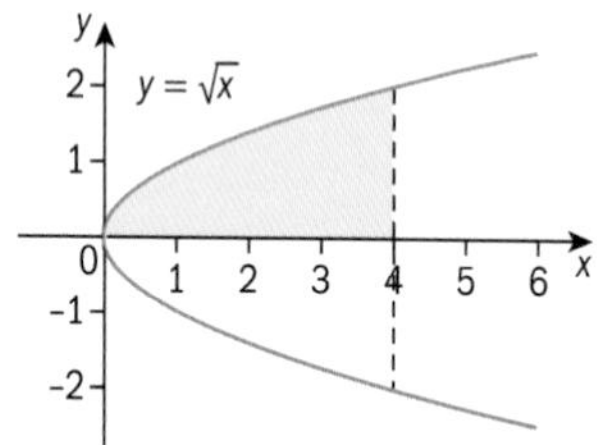

Find the volume obtained when the shaded region is rotated through 360° about the x-axis.

3.

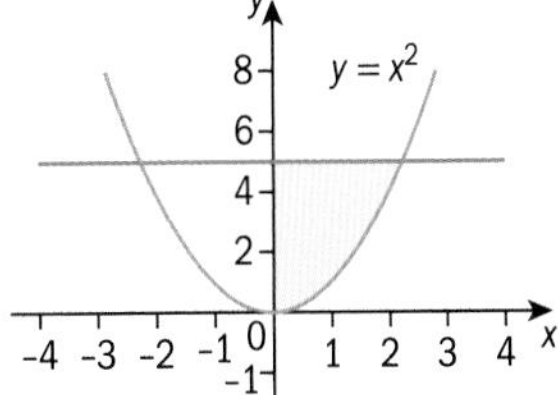

Find the volume obtained when the shaded region is rotated through 360° about the y-axis.

4.

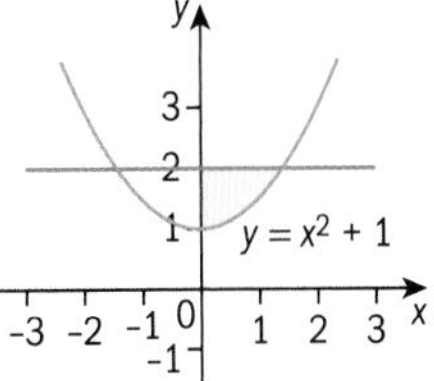

Find the volume obtained when the shaded region is rotated through 360° about the y-axis.

5.

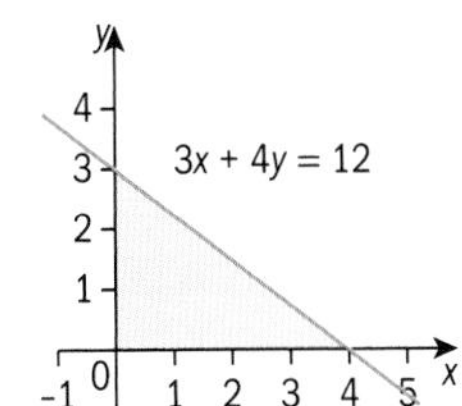

Find the volume obtained when the shaded region is rotated through 360° about the x-axis.

Find the volume of the cone formed.

6. Find the volume obtained when region bounded by the curve $y = x(3 - x)$ and the x-axis is rotated through 360° about the x-axis.

7.

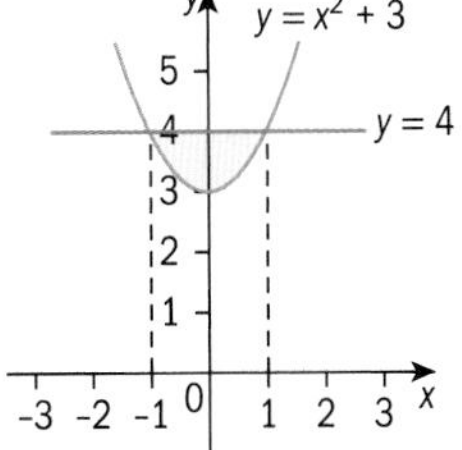

Find the volume obtained when the shaded region is rotated through 360° about the x-axis.

8. Find the volume obtained when the shaded region is rotated through 360° about the y-axis.

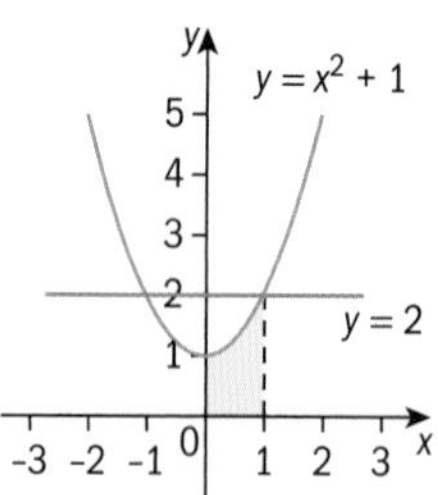

9. The curve $y^2 = x$ and the line $y = x$ meet at the points A and B.
   a) Find the coordinates of the points A and B.
   b) Sketch the curve $y^2 = x$ and the line $y = x$ on one set of axes.
   c) Find the volume obtained when the region between the line and the curve is rotated through 360° about the x-axis.

10. Find the volume obtained when the shaded region is rotated through 360° about the x-axis. Give your answer to the nearest whole number.

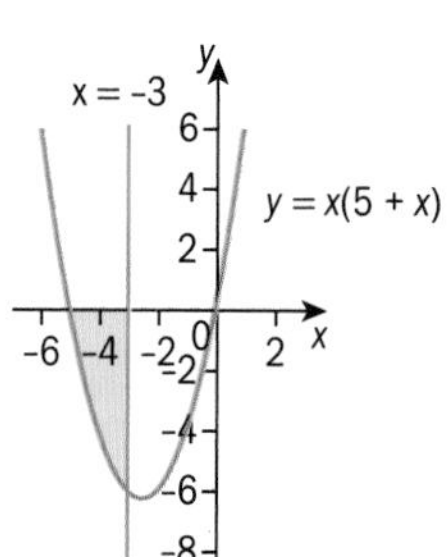

11.

The diagram shows the lines $y = 2$ and $y = 5$ and part of the curve $y = 2x^2 + 1$.
The shaded region is rotated through 360° about the y-axis.
Find the exact value of the volume of revolution obtained.

12. The area between curve $y^2 = x(4 - x)^2$ and the x-axis is rotated through 360° about the x-axis. Find the exact value of the volume of the solid of revolution formed by this rotation.

13. The diagram shows part of the curve $y^2 = 32x$ and part of the curve $y = x^3$.
    The shaded region is rotated through 360° about the x-axis.
    Find the exact value of the volume of revolution obtained.

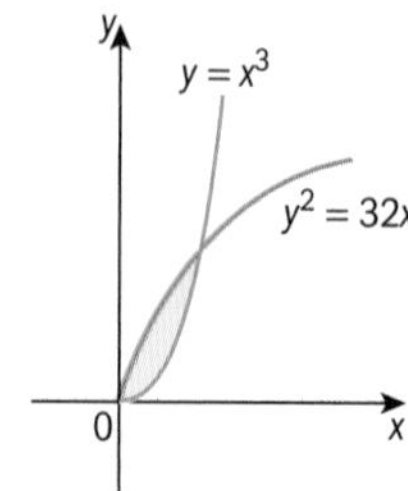

Summary exercise 10

1. Find these integrals.

   a) $\int 12x^2\,dx$ b) $\int\left(\frac{3}{x^2}-2x\right)dx$

   c) $\int(6-x)^2\,dx$ d) $\int 2\sqrt{x}\left(1+\sqrt{x}\right)^2\,dx$

   e) $\int\frac{x+5x^7}{x^3}\,dx$ f) $\int\left(\frac{2}{\sqrt[3]{x}}-14\sqrt{x^5}\right)dx$

2. Find the equation of the curve, given $\frac{dy}{dx}$ and a point on the curve.

   a) $\frac{dy}{dx}=9x^2-2x+1$ point = (−3, 5)

   b) $\frac{dy}{dx}=\frac{4x^4-5x}{x}$ point = (−1, −2)

   c) $\frac{dy}{dx}=\left(\sqrt{x}+3\right)^2$ point = (4, 2)

   d) $\frac{dy}{dx}=8x-\frac{7}{x^2}$ point = (1, −4)

3. Given that $f'(x)=2x^2-x+3\sqrt{x}$ and $f(4)=0$, find $f(x)$.

EXAM-STYLE QUESTIONS

4. Find an expression for y in terms of x if $\frac{dy}{dx}=(3x-5)(x-1)$ and $y=0$ when $x=5$.

5. Find

   a) $\int_1^3\left(x+\frac{1}{x}\right)^2dx$ b) $\int_{-2}^{-1}\frac{2x^4+3}{x^4}\,dx$

   c) $\int_{-2}^{2}\frac{x^2(x-1)^2}{x}\,dx$ d) $\int_3^6\sqrt{(7-x)}\,dx.$

6. Find $\int_{-2}^{-1}\frac{1}{(3x+2)^5}\,dx.$

7. Show that only one of the following improper integrals has a finite value and find that value.

   a) $\int_0^{16}\frac{1}{\sqrt{x}}\,dx$ b) $\int_{16}^{\infty}\frac{1}{\sqrt{x}}\,dx$

8. Find in terms of p and q the value of the integral $\int_p^q\frac{3}{x^4}\,dx.$

9. Show that only one of the following improper integrals has a finite value and find that value.

   a) $\int_0^1\frac{3}{x^4}\,dx$ b) $\int_1^{\infty}\frac{3}{x^4}\,dx$

10. For each of the following integrals explain briefly why it is an improper integral.

    a) $\int_0^{16}\frac{1}{x}\,dx$ b) $\int_0^3\frac{1}{2}x^{-2}\,dx$

    c) $\int_0^{27}4x^{-\frac{1}{3}}\,dx$

    d) Find whether each of the integrals has a finite value and, where possible, find its value.

11.

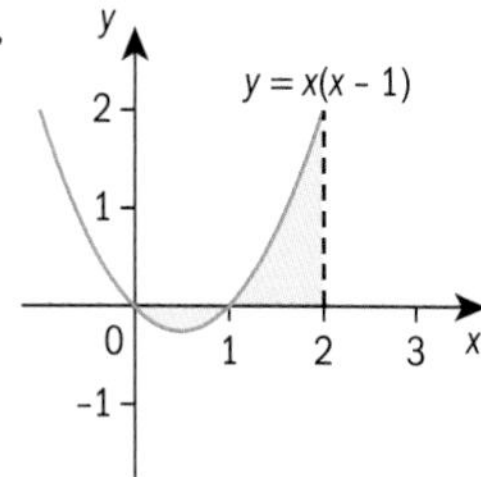

The diagram shows the curve $y = x(x - 1)$. Find the total shaded area.

12. Find the area bounded by the curve $y = 3x^2 - 2x + 1$, the x-axis and the lines $x = 1$ and $x = 2$.

First draw a sketch.

13. Find the area bounded by the curve $y = 2 + x - x^2$ and the line $y = x + 1$.

14.

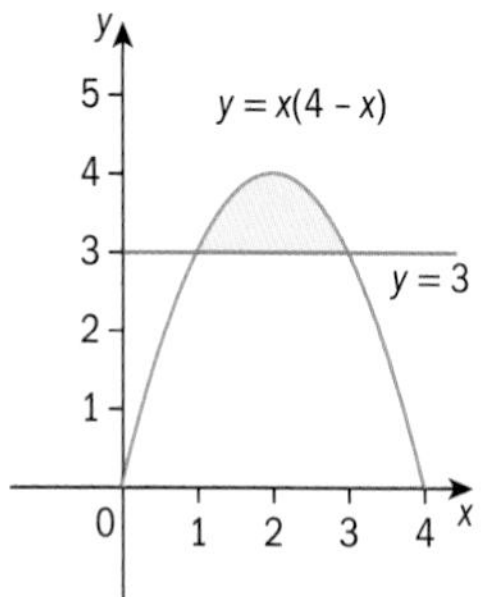

The diagram shows the curve $y = x(4 - x)$ and the line $y = 3$. Find the shaded area.

15.

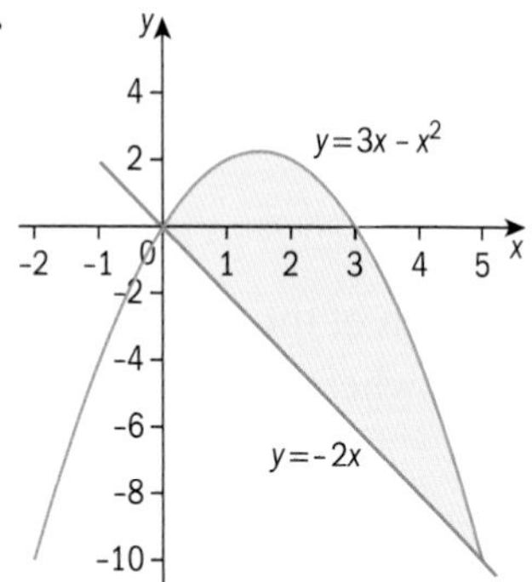

The diagram shows a sketch of the curve $y = 3x - x^2$ and the line $y = -2x$. Find the area of the shaded region.

16. Find the value of $\int_{-4}^{4} x^3\, dx$ and explain the significance of the answer.

17. a) Sketch the curve $y = x(x^2 - 1)$ showing clearly where the curve crosses the x-axis.

b) Find the area between the curve and the x-axis.

c) Find the volume obtained when the area bounded by the curve and the positive x-axis is rotated 360° about the x-axis.

18.

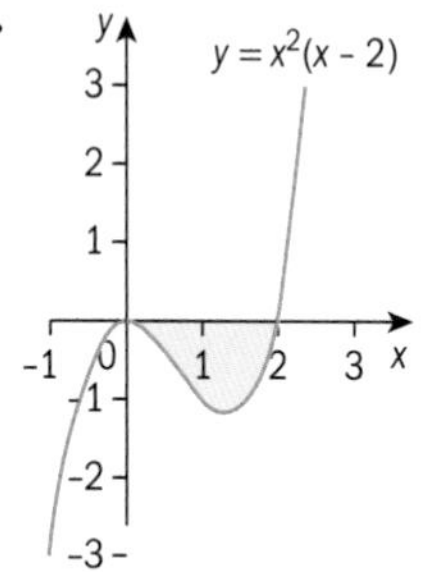

Find the volume generated when the shaded area is rotated through 360° about the x-axis.

19.

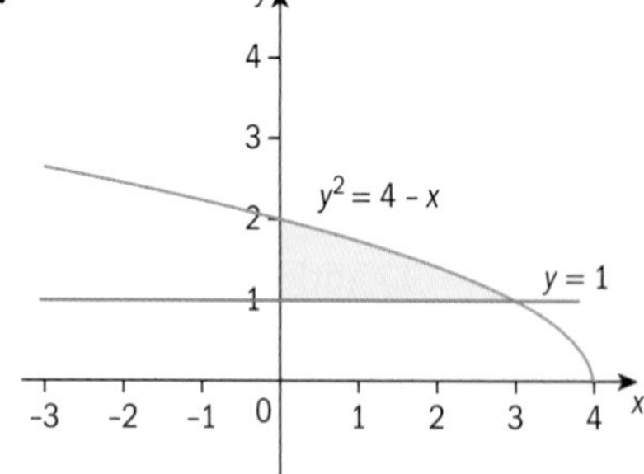

The diagram shows the line $y = 1$ and part of the curve with equation $y^2 = 4 - x$.

a) Find the volume obtained when the shaded region is rotated through 360° about the x-axis.

b) Find the volume obtained when the shaded region is rotated through 360° about the y-axis.

EXAM-STYLE QUESTIONS

20.

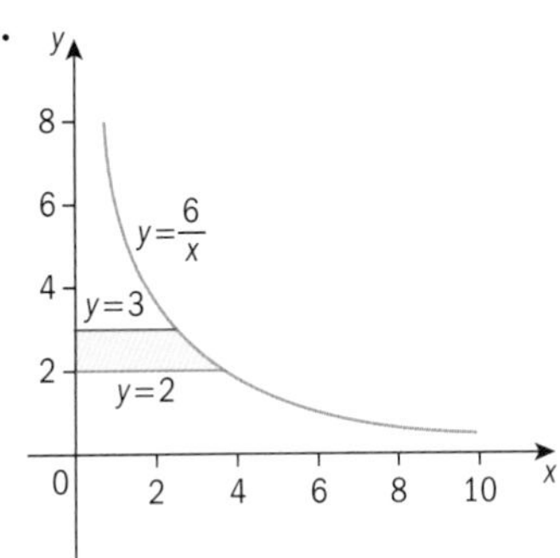

The diagram shows the region enclosed by the curve with equation $y = \frac{6}{x}$, the y-axis and the lines $y = 2$ and $y = 3$. Find, in terms of π, the volume obtained when this region is rotated through 360° about the y-axis.

21. Find the area enclosed by the curve $y = \frac{2x^3 + 3}{x^2}$, the line $x = \frac{1}{2}$ and the line $x = 4$.

22.

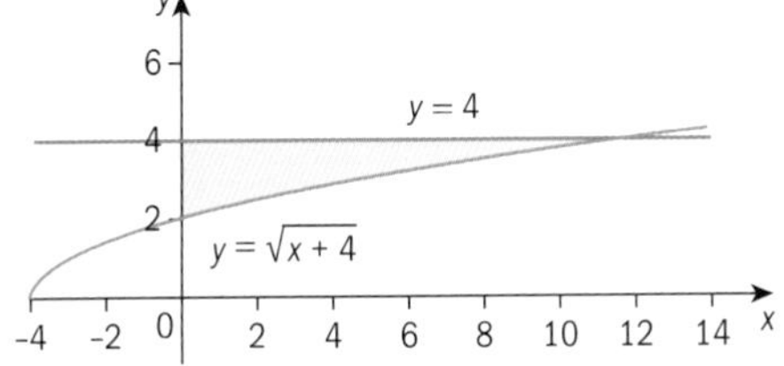

The diagram shows the line $y = 4$ and part of the curve with equation $y = \sqrt{x+4}$.

a) Show that the equation $y = \sqrt{x+4}$ can be written in the form $x = y^2 - 4$.

b) Find the area of the shaded region.

c) Find the volume obtained when the shaded region is rotated through 360° about the y-axis.

23. The region bounded by the curve $4y = x^2$, the line $x = 4$ and the line $y = 1$, is rotated through 360° about the x-axis. Find the volume of the solid formed.

24. The area bounded by the curve $y = 15kx - 15x^2$ and the x-axis is rotated through 360° about the x-axis, where k is an integer. The volume of the solid formed is 240π. Find the value of k.

Chapter summary

Integrating x^n and ax^n

- $\int x^n \, dx = \frac{x^{n+1}}{n+1} + c$, provided $n \neq -1$.
- $\int ax^n \, dx = \frac{ax^{n+1}}{n+1} + c$, provided $n \neq -1$.

Integrating $\int (ax + b)^n dx$

- $\int (ax + b)^n \, dx = \frac{1}{a}\frac{(ax+b)^{n+1}}{(n+1)} + c$, provided $n \neq -1$.

The definite integral

- $\int_a^b f'(x)dx = [f(x)]_a^b = f(b) - f(a)$

Improper integrals

- When we have a definite integral with an upper limit of ∞ or a lower limit of $-\infty$ then we have an **improper** integral.
- When we have an integral where the function to be integrated is not defined at a point in the interval of integration then we have an **improper** integral.
- We define the improper integral $\int_a^{\infty} f(x)\, dx$ as $\lim_{b \to \infty} \int_a^b f(x)\, dx$, provided the limit exists.
- We define the improper integral $\int_{-\infty}^{b} f(x)\, dx$ as $\lim_{a \to -\infty} \int_a^b f(x)\, dx$, provided the limit exists.
- When $f(x)$ is defined for $0 \le x < b$, but $f(x)$ is not defined when $x = 0$, then the **improper integral** $\int_0^b f(x)\, dx$ is defined as $\lim_{a \to 0^+} \int_a^b f(x)\, dx$, provided the limit exists.

Area under a curve

- The area under the curve $y = f(x)$ bounded by the x-axis, the line $x = a$ and the line $x = b$ is given by:

  $$\text{Area} = \int_a^b y\,dx = \int_a^b f(x)\,dx$$

  If the result is positive the region lies above the x-axis.

  If the result is negative the region lies below the x-axis.
- The area under the curve $x = f(y)$ bounded by the **y-axis**, the line $y = a$ and the line $y = b$ is given by:

  $$\text{Area} = \int_a^b x\,dy$$

  If the result is positive the region is to the right of the y-axis.

  If the result is negative the region is to the left of the y-axis.

Volume of revolution

- If an area bounded by a curve is rotated through 360° about the x-axis we can write the volume of the solid formed is $V = \int_a^b \pi y^2\,dx$, where a and b are the boundary values for x.
- If an area bounded by a curve is rotated through 360° about the y-axis we can write the volume of the solid formed is: $V = \int_a^b \pi x^2 dy$, where a and b are the boundary values for y.

Maths in real-life

Describing change mathematically

Calculus is the study of rates of change, so almost everything in the real-life application of mathematics involves calculus. The two major branches that we have been studying, differentiation and integration, are related to each other by the fundamental theorem of calculus which has many worldwide applications, from engineering and medicine, to business and space travel.

Isaac Newton (1643–1727) and Gottfried Wilhelm Leibniz (1646–1715) are credited for developing modern calculus in 17th century Europe, but many of the related ideas appeared in ancient Greece, China, India, the Middle East and medieval Europe.

Newton's discovery of differential calculus was perhaps ten years earlier than Leibniz', but Leibniz was the first to publish his account, written independently of Newton, in 1684. Soon after, he published an exposition of integral calculus that included the fundamental theorem of calculus. There was much controversy over which man deserved credit!

This is a hydroelectric power station in Canada. It provides clean, cost-efficient power.

In this type of power station, calculus is applied in many ways, for example:

- By engineers in designing and building the structure to make sure it can withstand the enormous forces generated.
- By the teams putting together the business case proposing it, calculating the amount of energy that would be generated from different flows of water through the system.
- By the environmentalists trying to model the effect building such a power station would have on the surrounding ecosystems.

Astronomers regularly use calculus to study the motion of planets, meteorites and spaceships.

The laws of planetary motion that are commonly used by astronomers to calculate orbits are derived using calculus. When sending a rocket into space, calculus is used to work out exactly how much fuel the rocket needs to accelerate to the correct velocity.

Along with calculating fuel requirements, calculus is used in calculating almost every aspect of space travel:

- Velocity
- The effects of gravity during take off
- Changes in pressure
- Oxygen expenditure.

We could use algebra in a huge number of problems which are static (i.e. not changing), however as most things in real-life are constantly changing we need calculus.

For this reason, calculus also has many applications in medicine.

Scientists can use calculus to model the growth rate of tumours or bacteria, as well as to model the relationship between related physical traits such as spine length and skull length.

The study of the movement of a drug into, through and out of the body is extremely important for the understanding of how best to administer medication. Scientists use calculus to determine the absorption, duration and intensity of a drug's effect as well as the excretion from the body. Formulae are used to summarise the behaviour of most drugs, which require calculus to calculate suitable doses of a specific drug and also to model how that specific drug is absorbed by the body.

Many drug effects occur primarily when the blood level of the drug is either increasing or decreasing. When the drug reaches steady state, these effects can be either reduced or completely absent. One way to understand this is that these effects only take place if there is a first derivative other than zero.

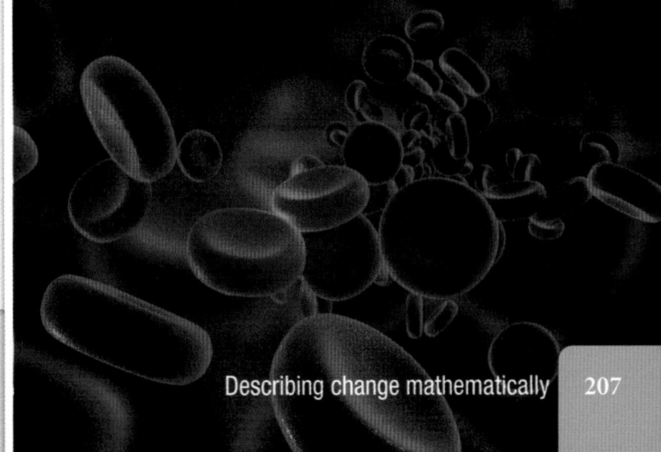

Exam-style paper A – Pure 1

75 Marks

1. A circle has equation $x^2 + y^2 - 6x + 4y - 3 = 0$.
   Find the centre and radius of the circle. [3]

2. Find the points of intersection of the graphs $2x - y = 4$ and $4x^2 + y^2 = 10$. [4]

3. The curve $y = f(x)$ has a maximum at $A(-2, 20)$, a minimum at $B(1, -7)$, and passes through the origin. The graph of $y = f(x)$ is shown below.

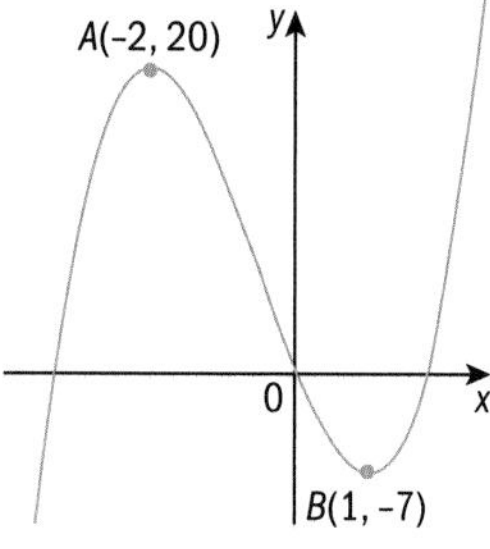

   On separate diagrams, sketch

   a) $y = f(x) + 4$

   b) $y = \frac{1}{2}f(x)$

   Show clearly the coordinates of the maximum and minimum points and where the curve cuts the y-axis. [4]

4. The first 3 terms of a geometric progression are x, $10 - x$ and $2x + 1$ where $x > 0$.
   Find the value of the common ratio, r. [5]

5. **a)** Find the first 3 terms in the expansion of $\left(3x - \frac{2}{x}\right)^6$ in descending powers of x. [3]

   b) Hence find the coefficient of x^2 in the expansion of $\left(1 + \frac{3}{x^2}\right)\left(3x - \frac{2}{x}\right)^6$. [2]

6. The function f is defined by $f : x \mapsto 7 - 5\cos 2x$ for all x.
   The function g is defined by $g : x \mapsto x + \frac{\pi}{2}$ for all x.

   a) State the range of f. [2]

   b) Sketch the graph of $y = f(x)$ for $0 \le x \le \pi$. [2]

   c) Find the function fg. [2]

7. 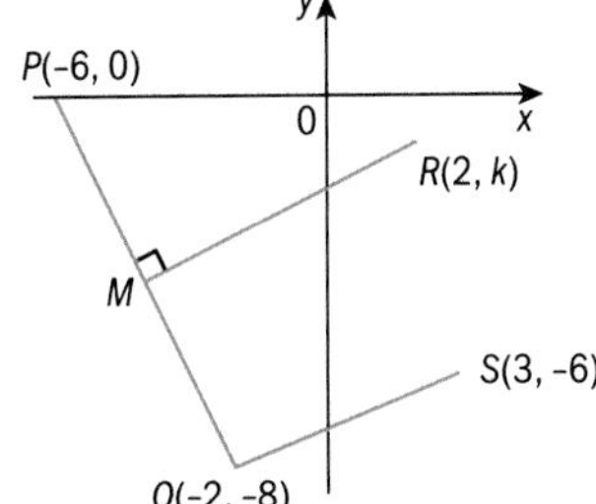

The diagram shows the line QS where Q has coordinates $(-2, -8)$ and S has coordinates $(3, -6)$.

a) Find the exact length of QS. [2]

P has coordinates $(-6, 0)$ and M is the mid-point of PQ. MR is perpendicular to PQ where R has coordinates $(2, k)$.

b) Find the value of k. [4]

8. 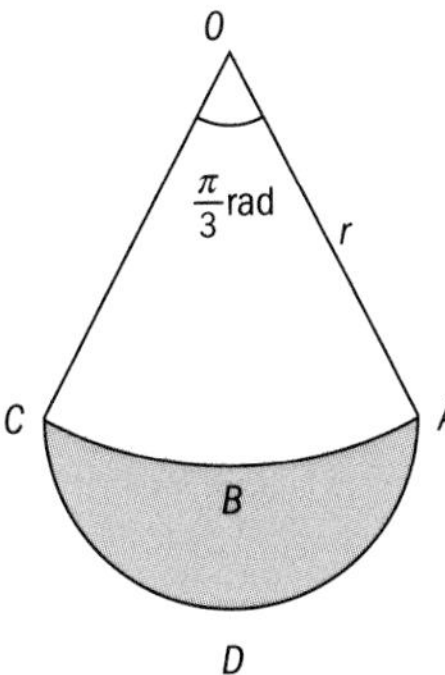

In the diagram, ADC is a semicircle. ABC is the arc of a circle, centre O and radius r. Angle $COA = \frac{\pi}{3}$ radians.

Find the area of the shaded region, leaving your answer in terms of $\sqrt{3}$, π and r. [6]

9. **a)** Solve the equation $4\sin^2 x + 7\cos x = 7$ for $0° \leq x \leq 360°$. [5]

b) Hence solve the equation $4\sin^2(\theta + 20)° + 7\cos(\theta + 20)° = 7$ for $0° \leq \theta \leq 360°$. [2]

10. 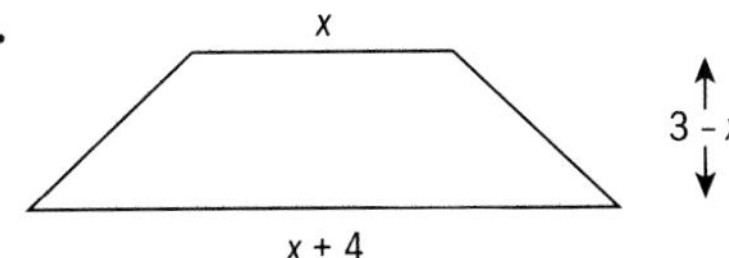

The diagram shows a trapezium. All measurements are in centimetres.

a) Show that the area of the trapezium is $(6 + x - x^2)$ cm^2. [2]

b) The area of the trapezium can be written in the form $a(x + b)^2 + c$.

i) Find the values of a, b and c.

ii) Hence find the maximum area of the trapezium and the value of x which gives the maximum area. [6]

11. a) Differentiate with respect to x $\quad 5x^4 + 6\sqrt{x} - \dfrac{3-x^2}{2x^2}$. [4]

b) Find $\displaystyle\int_{-4}^{1} \frac{1}{\sqrt{5-x}}\,\mathrm{d}x$. [5]

12.

The diagram shows part of the curve with equation $y = x^3 - 12x^2 + 36x$.

The x-axis is a tangent to the curve at P.

a) Find the coordinates of P. [3]

The curve has a maximum turning point at Q.

b) Find the coordinates of Q. [4]

The shaded area is bounded by the curve and the x-axis.

c) Find the area of the shaded region. [5]

Exam-style paper B – Pure 1

75 Marks

1. Solve $6x^2 - 7x - 5 > 0$. [2]

2. 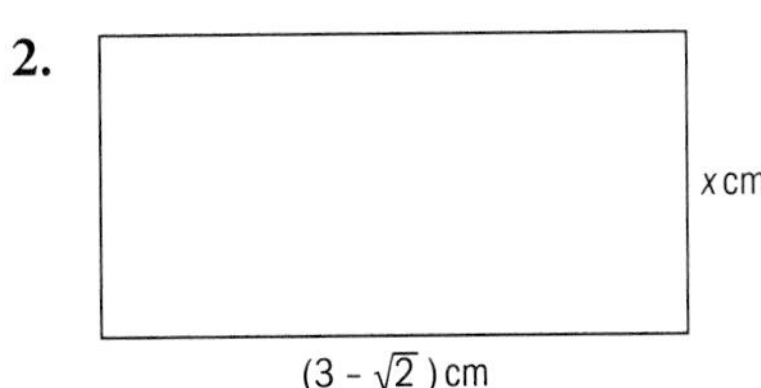

The area of a rectangle is $(5 - \sqrt{2})\text{ cm}^2$. The base of the rectangle is $(3 - \sqrt{2})$ cm and the height is x cm. Find the height of the rectangle. Give your answer in the form $a + b\sqrt{c}$. [3]

3. Describe the geometric transformation which maps $y = \sqrt{x^2 + 16}$ onto the graph of

   a) $-3 + y = \sqrt{x^2 + 16}$ [2]

   b) $y = 4\sqrt{x^2 + 1}$. [2]

4. The sum to n terms of an arithmetic progression is n^2. Find the nth term. [4]

5. A circle with centre $C(4, -2)$ passes through the point $P(7, 2)$.

   a) Find the equation of the circle in the form $(x - a)^2 + (y - b)^2 = k$. [2]

   b) Find the equation of the tangent to the circle at P. [5]

6. Find the values of k for which the line $y = 7x - 4$ is a tangent to the curve $y = 4x^2 - kx + k$. [5]

7. The coefficient of x^4 in the expansion of $(3 - x)^6 - (kx - 3)^5$ is 375. Find the possible values of k. [5]

8. **a)** Find, in terms of a and b, the value of the integral $\displaystyle\int_a^b \frac{3}{2x^{\frac{5}{2}}}\,dx$. [2]

   b) Show that only one of the following improper integrals has a finite value and find that value.

   i) $\displaystyle\int_0^4 \frac{3}{2x^{\frac{5}{2}}}\,dx$ **ii)** $\displaystyle\int_4^\infty \frac{3}{2x^{\frac{5}{2}}}\,dx$ [3]

9. **a)** Prove the identity $\cos^4\theta - \sin^4\theta = 2\cos^2\theta - 1$. [3]

b) Solve the equation $\cos^4\theta - \sin^4\theta = 0$ for $0° \leq \theta \leq 360°$. [3]

10. 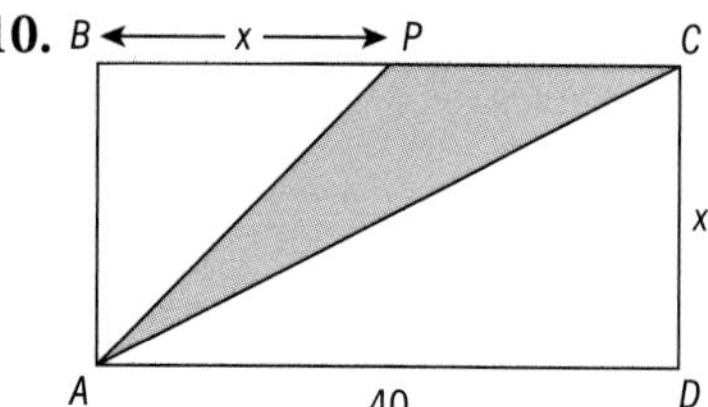

The diagram shows a plan for a rectangular piece of land $ABCD$. $AD = 40$ metres.
Vegetables are going to be planted in the triangular vegetable patch ACP.
$BP = CD = x$ metres.

a) Show that the area, A m^2 of the vegetable patch is given by

$$A = 20x - \frac{1}{2}x^2.$$ [2]

b) Given that x can vary, find the maximum area of the vegetable patch, showing clearly that A is a maximum not a minimum. [4]

11.

The diagram shows a sector OAB of a circle with centre O and radius r cm.
The angle AOB is θ radians and the perimeter of the sector is 30 cm.

a) Show that $\theta = \frac{30}{r} - 2$. [2]

b) Find the area of the sector in terms of r. [2]

c) Find the maximum possible area of the sector. [3]

12. 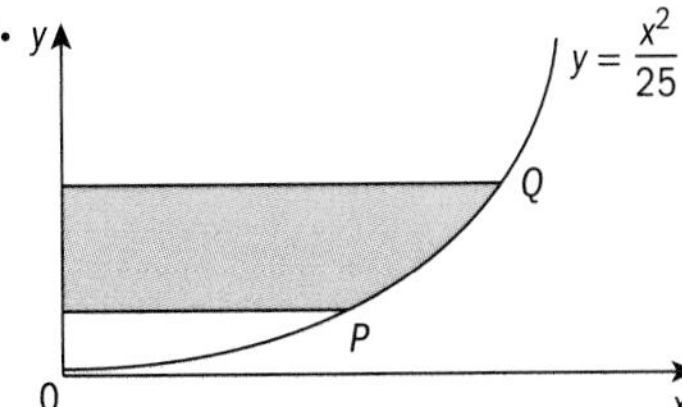

The diagram shows the graph of $y = \frac{x^2}{25}$, $x \geq 0$, which passes through the points P and Q.

The x-coordinate of P is 5 and the x-coordinate of Q is 10.

a) Find the y-coordinates of P and Q. [1]

b) Find the equation of the tangent to the curve at Q. [4]

c) Find the volume obtained when the region enclosed by the curve, the y-axis and the lines through P and Q parallel to the x-axis is rotated through 360° about the y-axis. [5]

13. Functions f and g are defined by

$\mathrm{f}: x \mapsto \frac{1}{2-x}, x \in \mathbb{R}, x \neq 2$ $\quad$ $\mathrm{g}: x \mapsto 3x + 4, x \in \mathbb{R}$

a) Express in terms of x

i) $\mathrm{f}^{-1}(x)$ **ii)** $\mathrm{g}^{-1}(x)$. [3]

b) Sketch in a single diagram the graphs of $y = \mathrm{g}(x)$ and $y = \mathrm{g}^{-1}(x)$, making clear the relationship between the graphs and writing down the coordinates of their intersection. [4]

c) Given that $\mathrm{fg}(-3) = \mathrm{f}^{-1}(k)$, find the value of k. [4]

Answers

The answers given here are concise. However, when answering exam-style questions, you should show as many steps in your working as possible.

1 Quadratics

Skills check page 2

1. **a)** 7 **b)** 13 **c)** 66 **d)** 200

2. **a)** $(3x+1)(2x-1)$ **b)** $(1+10x)(1-10x)$
c) $(7-x)(3+2x)$ **d)** $2(x+3)(x-3)$

3. **a)** $x \leq \frac{4}{5}$ **b)** $x < 1\frac{5}{12}$ **c)** $x < -\frac{9}{13}$

4. **a)** $x=-1 \quad y=\frac{1}{2}$ **b)** $x=4 \quad y=5$
c) $x=-2 \quad y=-1$ **d)** $x=1\frac{1}{2} \quad y=4$

Exercise 1.1 page 3

1. **a)** $x=5$ or $x=-7$ **b)** $x=5$ or $x=2$
c) $x=4$ or $x=-3$ **d)** $x=1$ or $x=-9$
e) $x=6$ or $x=-3$ **f)** $x=3$ or $x=-2$
g) $x=-6$ or $x=-2$ **h)** $x=3$ or $x=-8$
i) $x=0$ or $x=4$ **j)** $x=4$ or $x=-4$
k) $x=4$ or $x=\frac{1}{2}$ **l)** $x=-6$ or $x=-\frac{1}{3}$
m) $x=\frac{2}{5}$ or $x=-\frac{1}{3}$ **n)** $x=\frac{3}{2}$ or $x=\frac{3}{4}$
o) $x=0$ or $x=4$ **p)** $x=\frac{2}{5}$ or $x=-5$
q) $x=2$ or $x=-4$ **r)** $x=1$ or $x=-\frac{9}{2}$
s) $x=2$ or $x=-\frac{2}{5}$ **t)** $x=\frac{3}{10}$ or $x=-\frac{1}{10}$
u) $x=\frac{4}{3}$ or $x=\frac{5}{2}$ **v)** $x=0$ or $x=-2$
w) $x=-6$ or $x=\frac{3}{5}$ **x)** $x=\frac{3}{2}$ or $x=-7$
y) $x=1$ or $x=-1$ **z)** $x=-\frac{1}{6}$ or $x=-\frac{3}{2}$

2. $x=-\frac{1}{3}$

3. $49\,\text{cm}^2$

4. 3 and 7 or −7 and −3

5. $x=-\frac{5}{2}$

Exercise 1.2 page 5

1. **a)** $-2 \leq x \leq 5$ **b)** $x < 1$ or $x > 3$

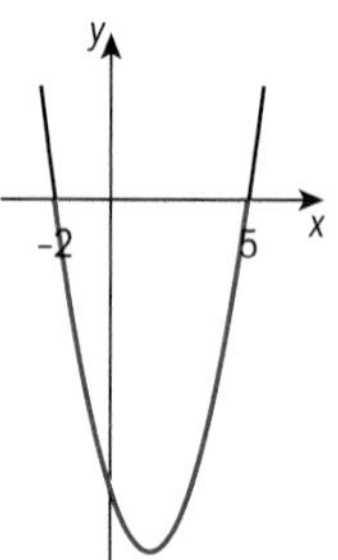

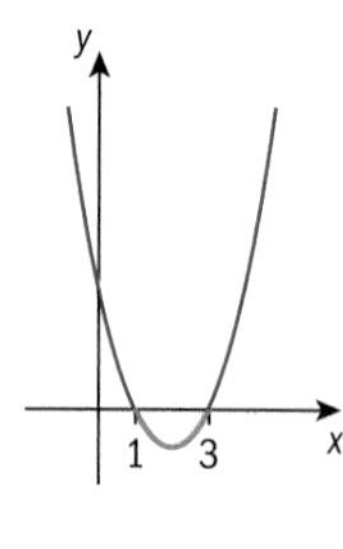

c) $x < -4$ or $x > -\frac{2}{3}$ **d)** $-10 < x < 7$

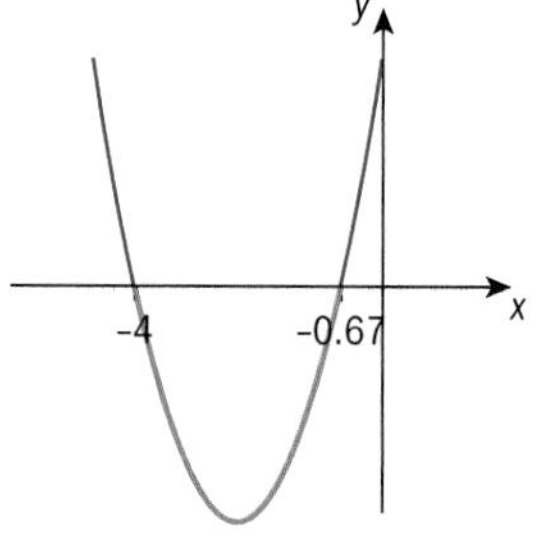

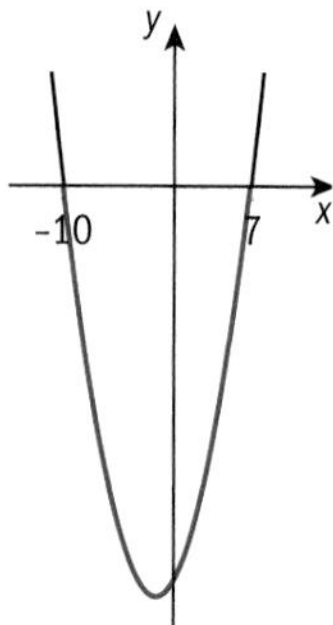

e) $3 \leq x \leq 6$ **f)** 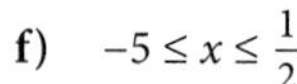 $-5 \leq x \leq \frac{1}{2}$

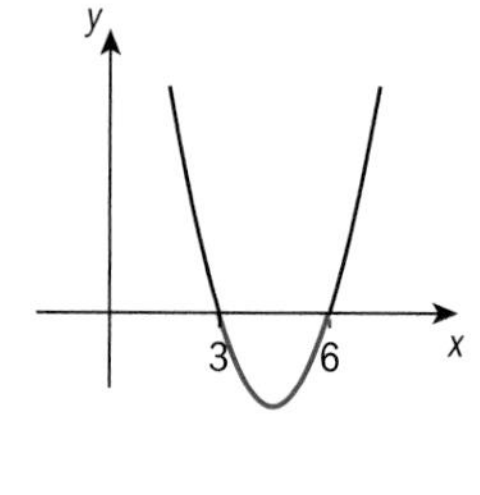

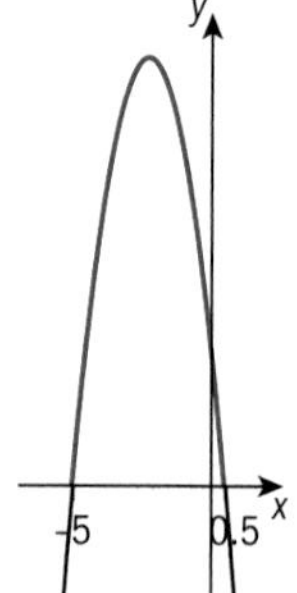

2. **a)** $-4 \leq x \leq -3$ **b)** $-5 < x < 6$

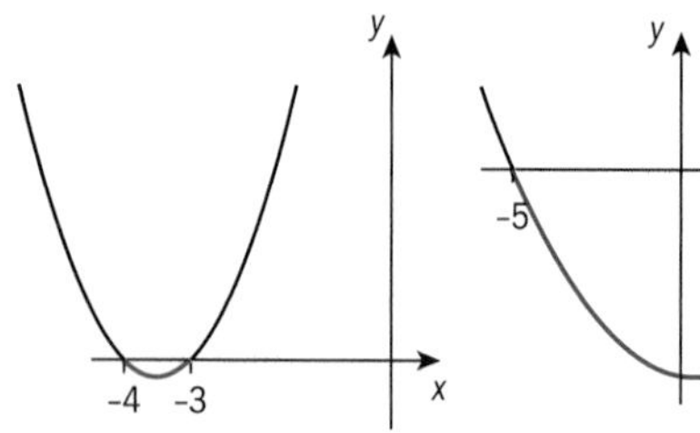

c) $x < -8$ or $x > 6$

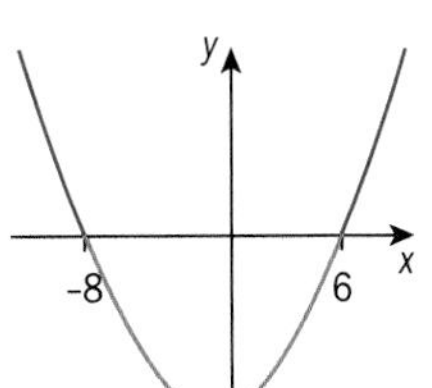

d) $x \le 2$ or $x \ge 3$

e) $x < \frac{3}{2}$ or $x > 4$

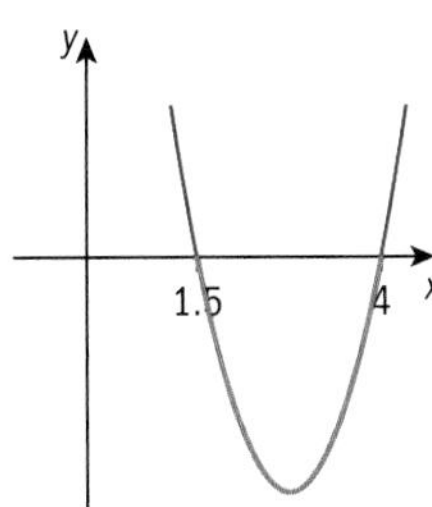

f) $-5 \le x \le \frac{1}{3}$

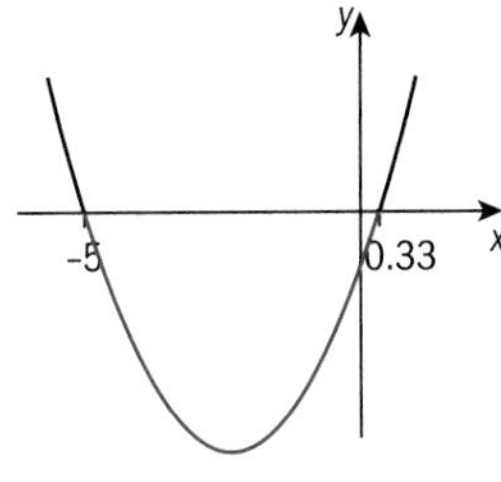

g) $x \le -3$ or $x \ge 5$

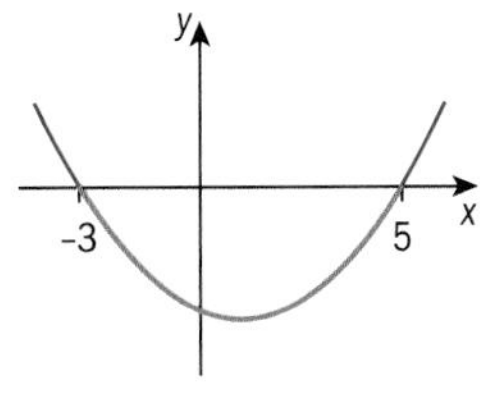

h) $x < -8$ or $x > 2$

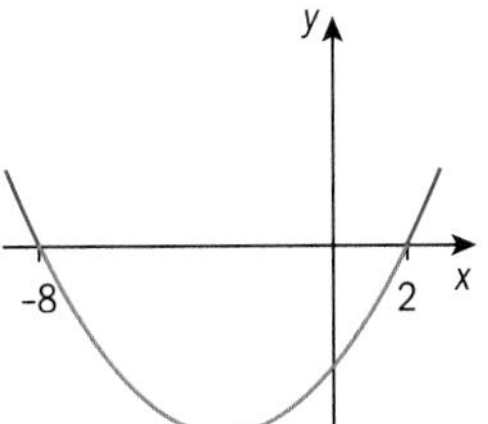

i) $-\frac{7}{2} < x < 3$

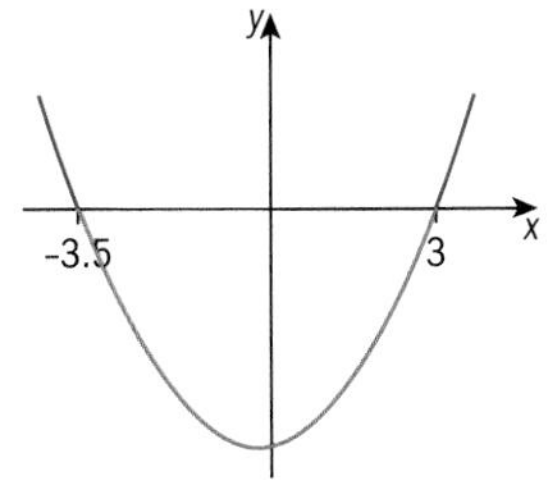

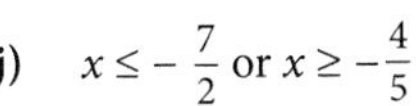

j) $x \le -\frac{7}{2}$ or $x \ge -\frac{4}{5}$

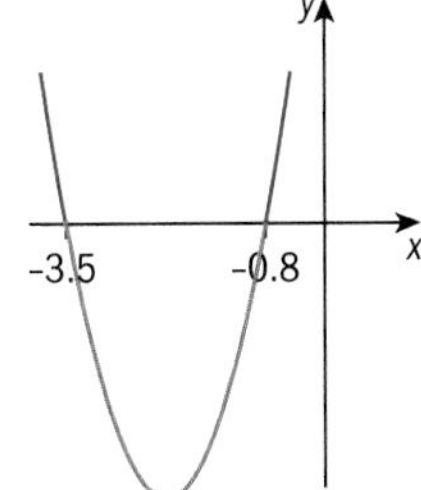

k) $x \le -3$ or $x \ge 3$

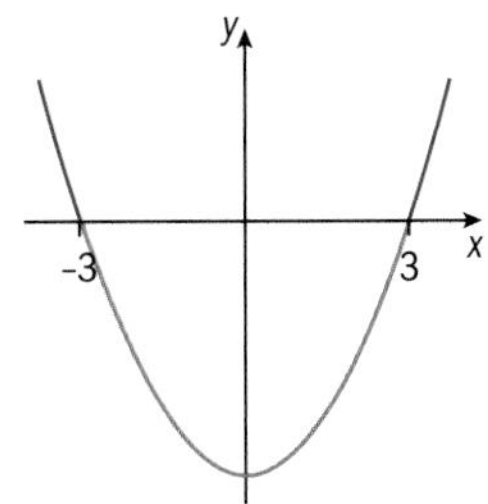

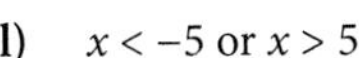

l) $x < -5$ or $x > 5$

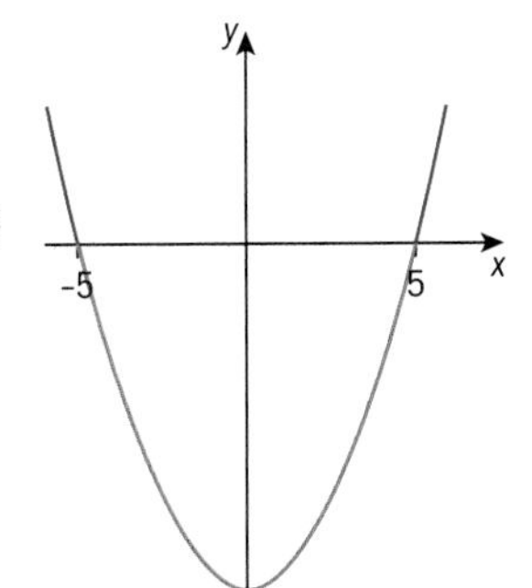

Note: In questions 2 and 3 all equations have been arranged so that x^2 is positive.

3. a) $x < -5$ or $x > 7$

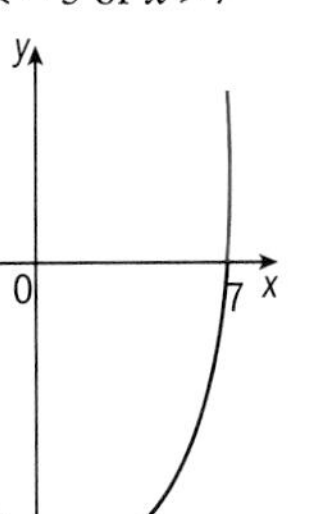

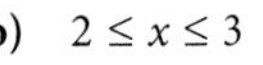

b) $2 \le x \le 3$

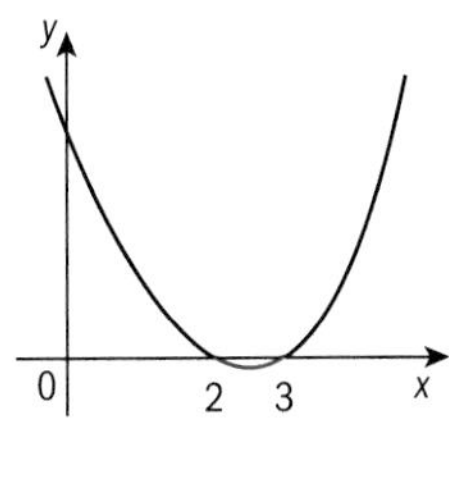
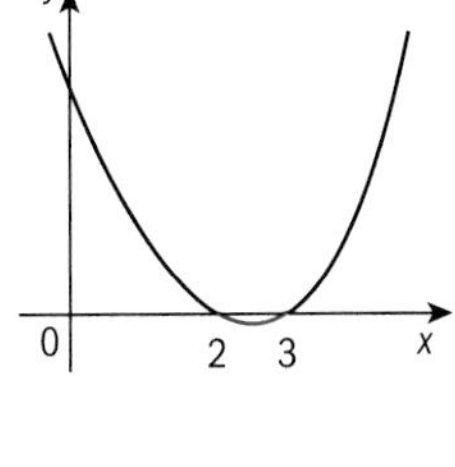

c) $-4 \le x \le 5$

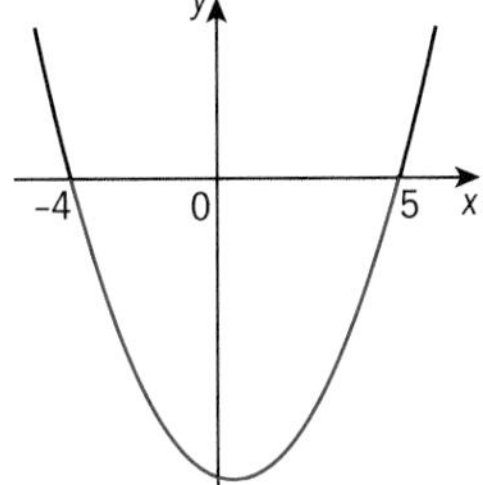

d) $x \le -5$ or $x \ge 2$

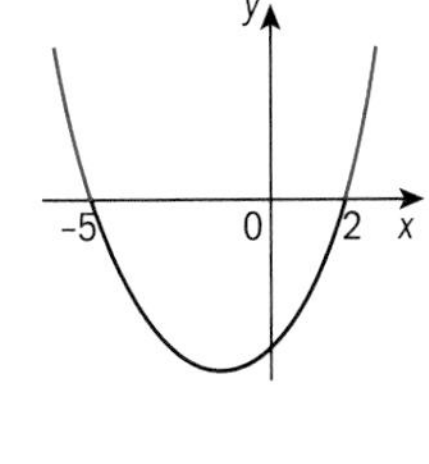

e) $-1 < x < 10$

f) $x > 9$ or $x < -5$

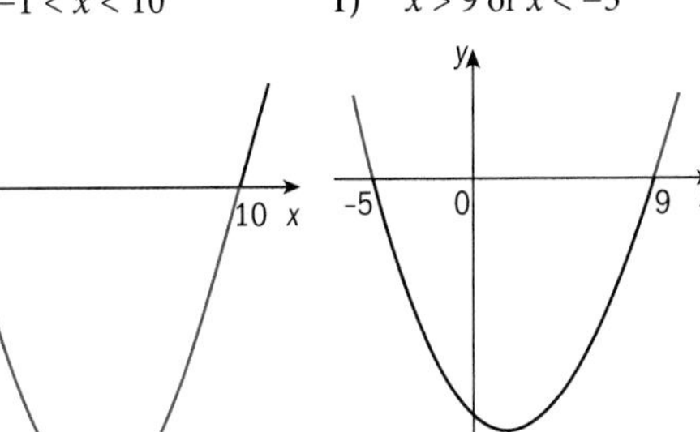

g) $x < 3$ or $x > 8$

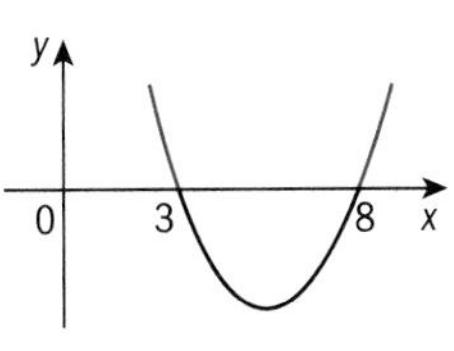

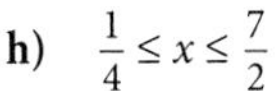

h) $\frac{1}{4} \le x \le \frac{7}{2}$

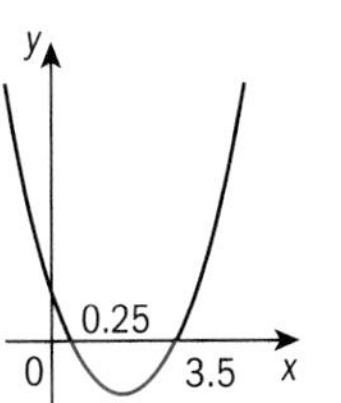

i) $x \le 2$ or $x \ge 6$

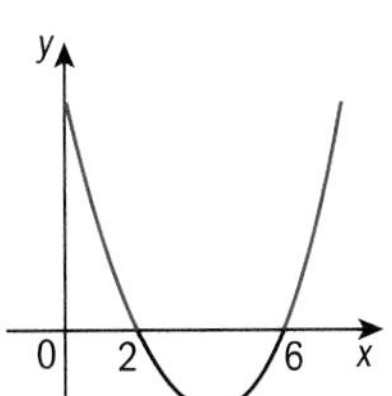

j) $x < -6$ or $x > -4$

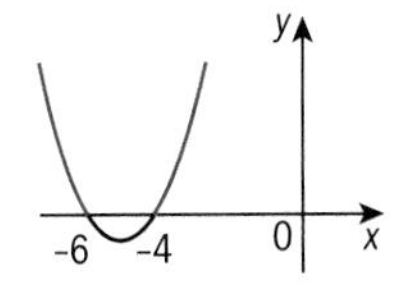

Exercise 1.3 page 8

1. a) $(x+1)^2-6$ b) $(x-5)^2-5$
   c) $(x-2)^2-3$ d) $22-(x+4)^2$
   e) $74-(x+8)^2$ f) $(x+10)^2-145$
   g) $(x+6)^2-20$ h) $12-(x+3)^2$

2. a) $x=-3\pm\sqrt{8}$ $\left(\text{or } x=-3\pm 2\sqrt{2}\right)$
   b) $x=2\pm\sqrt{12}$ $\left(\text{or } x=2\pm 2\sqrt{3}\right)$
   c) $x=-1\pm\sqrt{5}$
   d) $x=10\pm\sqrt{70}$
   e) $x=4\pm\sqrt{19}$
   f) $x=-7\pm\sqrt{50}$ $\left(\text{or } x=-7\pm 5\sqrt{2}\right)$
   g) $x=\frac{3\pm\sqrt{17}}{2}$
   h) $x=\frac{1}{2}\pm\sqrt{\frac{13}{4}}$ $\left(\text{or } x=\frac{1\pm\sqrt{13}}{2}\right)$

3. a) $6(x+1)^2-9$ b) $3(x-1)^2-18$
   c) $3(x-3)^2-23$ d) $4(x+3)^2-45$
   e) $\frac{11}{2}-2\left(x+\frac{1}{2}\right)^2$ f) $5(x-2)^2-18$
   g) $-2\left(x-\frac{3}{4}\right)^2+\frac{41}{8}$ h) $3\left(x+\frac{5}{6}\right)^2-\frac{13}{12}$

4. a) $x=-2\pm\sqrt{\frac{10}{3}}$ $\left(\text{or } -2\pm\frac{1}{3}\sqrt{30}\right)$
   b) $x=-1\pm\sqrt{\frac{8}{3}}$ $\left(\text{or } -1\pm\frac{2\sqrt{6}}{3}\right)$
   c) $x=-5\pm\sqrt{\frac{132}{5}}$
   d) $x=5\pm\sqrt{\frac{53}{2}}$ $\left(\text{or } 5\pm\frac{1}{2}\sqrt{106}\right)$
   e) $x=2\pm\sqrt{\frac{7}{2}}$ $\left(\text{or } 2\pm\frac{1}{2}\sqrt{14}\right)$
   f) $x=6\pm\sqrt{\frac{85}{2}}$ $\left(\text{or } x=6\pm\frac{1}{2}\sqrt{170}\right)$
   g) $x=\frac{1\pm\sqrt{97}}{4}$
   h) $x=\frac{1}{3}\pm\sqrt{\frac{28}{9}}$ $\left(\text{or } x=\frac{1\pm 2\sqrt{7}}{3}\right)$

5. a) $-2-\sqrt{2}<x<-2+\sqrt{2}$
   b) $x\le 3-\sqrt{12}$ or $x\ge 3+\sqrt{12}$
   c) $x<1-\sqrt{2}$ or $x>1+\sqrt{2}$
   d) $-5-\sqrt{18}\le x\le -5+\sqrt{18}$
   e) $x\ge 3+\frac{5\sqrt{2}}{2}$ or $x\le 3-\frac{5\sqrt{2}}{2}$
   f) $-1-\frac{\sqrt{6}}{3}<x<-1+\frac{\sqrt{6}}{3}$
   g) $-1-\frac{\sqrt{35}}{5}\le x\le -1+\frac{\sqrt{35}}{5}$
   h) $x<2-\frac{\sqrt{22}}{2}$ or $x>2+\frac{\sqrt{22}}{2}$

Exercise 1.4 page 10

1. a) $x=2.35$ or $x=-0.85$
   b) $x=1.74$ or $x=0.46$
   c) $x=-3.53$ or $x=-0.47$
   d) $x=-5.37$ or $x=0.37$
   e) $x=-1.40$ or $x=0.90$
   f) $x=-2.56$ or $x=1.56$
   g) $x=0.42$ or $x=1.58$
   h) $x=0.27$ or $x=1.23$
   i) $x=-4.35$ or $x=0.35$
   j) $x=-3.30$ or $x=0.30$

2. a) $x=\frac{1\pm\sqrt{41}}{4}$ b) $x=\frac{3\pm\sqrt{6}}{3}$
   c) $x=\frac{3\pm\sqrt{149}}{10}$ d) $x=\frac{1\pm\sqrt{97}}{12}$
   e) $x=-4\pm\sqrt{19}$ f) $x=\frac{-7\pm\sqrt{37}}{2}$
   g) $x=-5\pm\sqrt{30}$ h) $x=\frac{-5\pm\sqrt{57}}{4}$
   i) $x=\frac{-5\pm 3\sqrt{5}}{4}$ j) $x=2\pm\sqrt{3}$

3. a) $\frac{-10-\sqrt{40}}{6}<x<\frac{-10+\sqrt{40}}{6}$
   b) $x\le 3-\sqrt{2}$ or $x\ge 3+\sqrt{2}$
   c) $x<\frac{-3-\sqrt{41}}{8}$ or $x>\frac{-3+\sqrt{41}}{8}$
   d) $\frac{-1-\sqrt{17}}{4}\le x\le\frac{-1+\sqrt{17}}{4}$
   e) $x\le\frac{4-\sqrt{26}}{5}$ or $x\ge\frac{4+\sqrt{26}}{5}$
   f) $x\le\frac{3-\sqrt{3}}{6}$ or $x\ge\frac{3+\sqrt{3}}{6}$
   g) $x\le\frac{3-\sqrt{29}}{2}$ or $x\ge\frac{3+\sqrt{29}}{2}$

h) $\dfrac{-1-\sqrt{7}}{2} < x < \dfrac{-1+\sqrt{7}}{2}$

i) $\dfrac{1-\sqrt{3}}{2} < x < \dfrac{1+\sqrt{3}}{2}$

j) $\dfrac{-1-\sqrt{5}}{2} < x < \dfrac{-1+\sqrt{5}}{2}$

Exercise 1.5 page 12

1. **a)** $x = \pm\sqrt{7}$ **b)** $x = \pm\sqrt{\dfrac{2}{3}}$

c) $x = \sqrt[3]{-5}$ or $x = \sqrt[3]{-2}$ **d)** $x = \pm 2$

e) $x = \pm\sqrt[4]{\dfrac{2}{3}}$ or $x = \pm\sqrt[4]{\dfrac{3}{2}}$

f) $x = -2$ or $x = \sqrt[3]{5}$

2. **a)** $x = 1.99$ or $x = 0.50$

b) $x = \pm 0.68$

c) $x = \pm 1.52$ **d)** $x = 1.16$ or $x = -1.37$

e) $x = \pm 1.79$ **f)** $x = \pm 1.98$ or $x = \pm 0.29$

3. **a)** $x = \pm 1.09$

b) $x = 1.20$ or $x = 0.664$

c) $x = 1.37$ or $x = -0.636$ **d)** no solutions

e) $x = 0.901$ or $x = -0.966$ **f)** $x = 1.13$

Exercise 1.6 page 14

1. **a)** -23; no real roots

b) 41; two distinct roots

c) 0; equal roots

d) $+17$; two distinct roots

e) -7; no real roots

f) 0; equal roots

g) -24; no real roots

h) 8; two distinct roots

i) 33; two distinct roots

j) 12; two distinct roots

k) 0; equal roots

l) 8; two distinct roots

2. Proof

3. Proof

4. $-8 < a < 8$

5. $k = -\dfrac{1}{6}$

6. $p = 1 \quad q = -6$

7. Discriminant = 76; two distinct roots

8. $k < \dfrac{25}{8}$ **9.** $p = \dfrac{2}{5}$

10. Proof **11.** $k = 6$ or -6

12. $p = 8, \quad q = -2, \quad r = -3$

Exercise 1.7 page 16

1. $(-1, 0), (2, 3)$

2. $(-1, 3), (3, 19)$

3. $(1, 0), (6, 5)$

4. $(3, 4), (-3, 10)$

5. $(1, 2), (2, 4)$

6. $(-2, 3), (5, 10)$

7. $(-3, -3), (2, 2)$

8. $(5, 25), (-1, 1)$

9. $(2, 1)$

10. $(6, 17), (-1, 3)$

11. $(2, -1), (1, 2)$

12. $(-3, -7), (2, -2)$

13. $(-4, -6), (12, 2)$

14. $\left(-\frac{1}{3}, 3\right), \left(\frac{5}{4}, \frac{25}{2}\right)$

15. $(-3, -4), (1, 0)$

16. $(-\frac{2}{5}, -\frac{14}{5}), (2, 2)$

17. $\left(2, \frac{9}{4}\right), \left(3, \frac{3}{2}\right)$

18. $(-1, -1), \left(-\frac{1}{2}, 0\right)$

19. $(-2, -3), (0, 1)$

20. $(-1, -2), (3, 4)$

21. Proof

22. 2.5 metres by 1.4 metres

23. **a)** $2xy = 54$; $4x + 4y + 2xy = 96$ (or $2x + 2y + xy = 48$)

b) $x = 6, y = 4.5$

24. **a)** $x + y = 9 \quad 2x + 6 = y^2$

b) $x = 5, y = 4$ perimeter = 50 cm

25. $x = 2, y = 3$

Exercise 1.8 page 20

1. $y = (x + 1)^2 - 1$; minimum at $(-1, -1)$

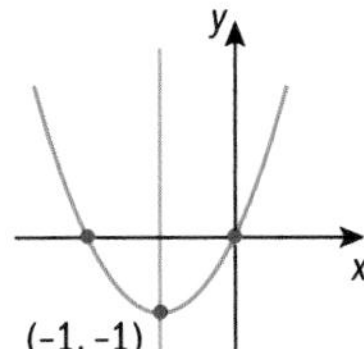

2. $y = (x - 3)^2 + 1$; minimum at $(3, 1)$

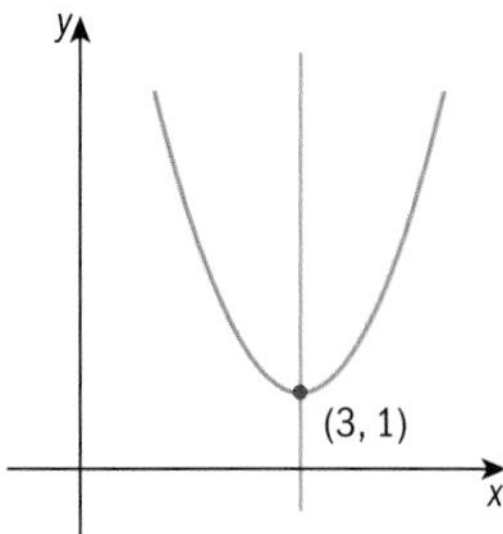

3. $y = -4\left(x + \frac{1}{4}\right)^2 + 3\frac{1}{4}$; maximum at $\left(-\frac{1}{4}, 3\frac{1}{4}\right)$

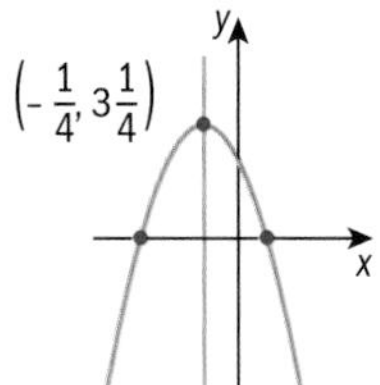

4. $y = (x + 2)^2 - 7$; minimum at $(-2, -7)$

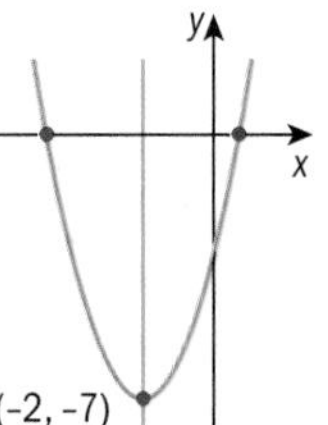

5. $y = -(x - 1)^2 + 6$; maximum at $(1, 6)$

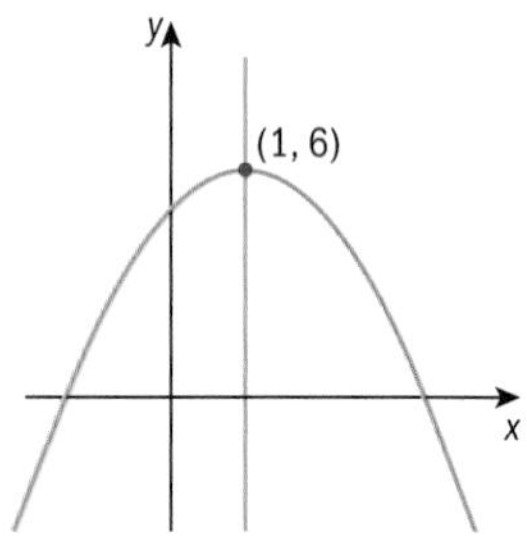

6. $y = -4\left(x - \frac{1}{2}\right)^2 + 4$; maximum at $\left(\frac{1}{2}, 4\right)$

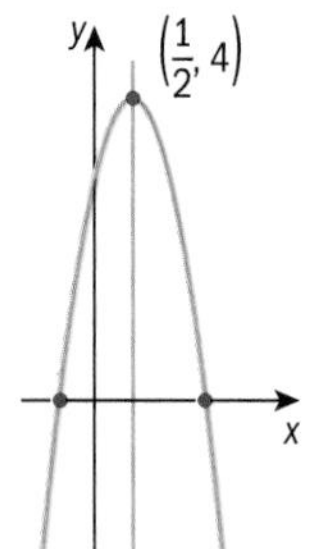

7. $y = (x - 2)^2 - 12$; minimum at $(2, -12)$

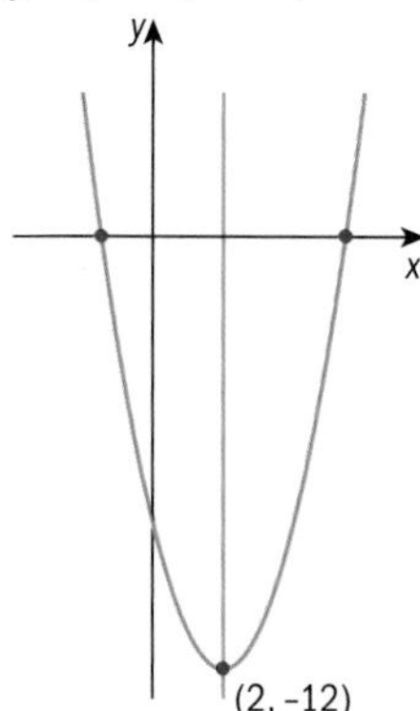

8. $y = (x + 3)^2 - 2$; minimum at $(-3, -2)$

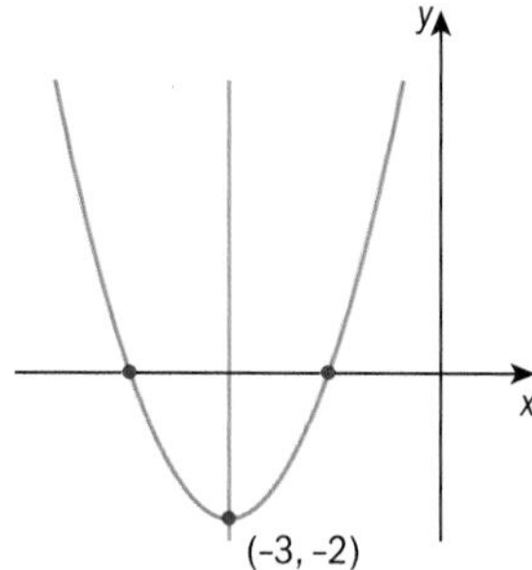

9. $y = 5\left(x - \frac{2}{5}\right)^2 - \frac{14}{5}$; minimum at $\left(\frac{2}{5}, -\frac{14}{5}\right)$

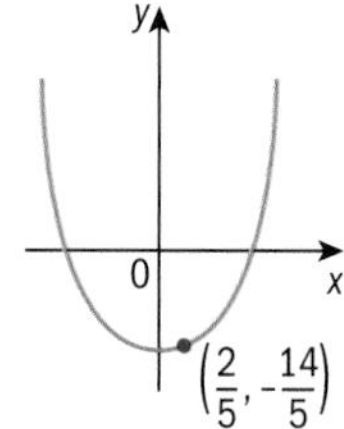

10. $y = 2\left(x - \frac{5}{4}\right)^2 - \frac{49}{8}$; minimum at $\left(\frac{5}{4}, -\frac{49}{8}\right)$

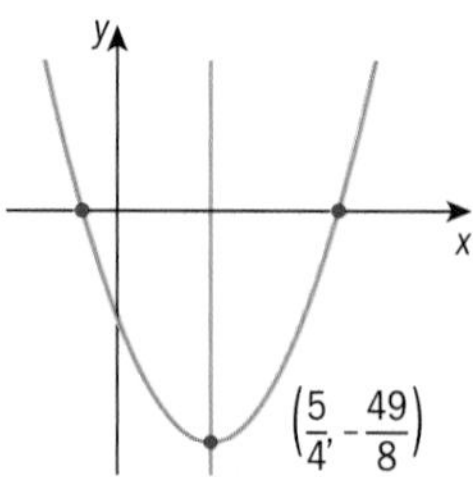

11. $24.5\,\text{cm}^2$

12. $64\,\text{cm}^2$

13. $\frac{81}{16}\,\text{cm}^2$ when $x = 1\frac{3}{4}$

14. **a)** $-(x - 10)^2 + 100$

b) 10 years

Summary exercise 1 page 21

1. $x = \frac{5}{6}$ or $x = \frac{4}{3}$
2. a) $(2x - 7)(5x + 3)$ b) $x = \frac{7}{2}$ or $x = -\frac{3}{5}$
3. $y = \sqrt[3]{6}$ or $y = \sqrt[3]{3}$
4. $x = \frac{1}{4}$ or $x = 4$
5. $x = 2.02$ or $x = -0.568$
6. $x = \pm 0.91$
7. 10 cm by 10 cm
8. $-5 < x < 4$
9. $-\frac{4}{3} \leq x \leq \frac{5}{2}$
10. $x = -0.219$ (3 s.f.) or $x = -2.28$ (3 s.f.)
11. $\frac{3-\sqrt{149}}{14} \leq x \leq \frac{3+\sqrt{149}}{14}$
12. $x = 3$ or $x = -1$
13. $q^2 = 4p$
14. a) No real roots b) Equal roots
    c) Two distinct roots
15. $k < -12$ or $k > 12$
16. a) $p = -3$ $q = -4$ b) $r = 6.25$
17. $x = 3, y = -2$
18. $x = \sqrt{2}, y = -\sqrt{2}$ or $x = -\sqrt{2}, y = \sqrt{2}$
19. $x = 5, y = -2$ or $x = 1, y = 2$
20. $x = 3, y = -1$ or $x = 1, y = 5$
21. a) $3(x + 2)^2 - 7$ b) -7
22. $x = 2 + \sqrt{8}$ or $\left(x = 2 + 2\sqrt{2}\right)$
23. $-4 \leq x \leq \frac{7}{2}$
24. $x > 1$
25. $k > 4$ or $k < -4$
26. $k = -8$
27. $x = \pm\sqrt{\frac{8}{3}}$

2 Functions and transformations

Skills check page 24

1. a) 7 b) −26 c) −2050
   d) 1064 e) −69 f) −7
2. a)

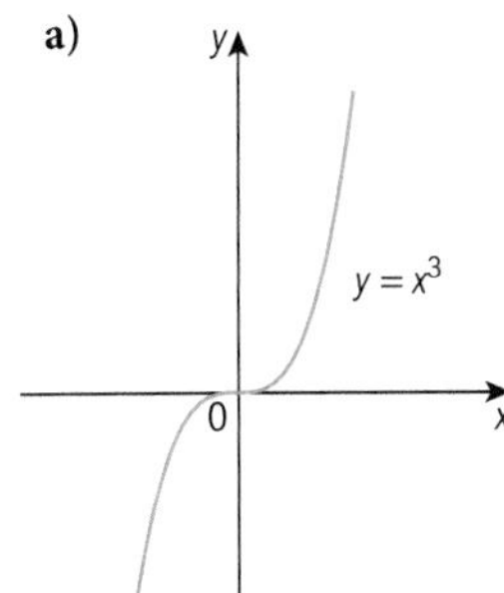

b)

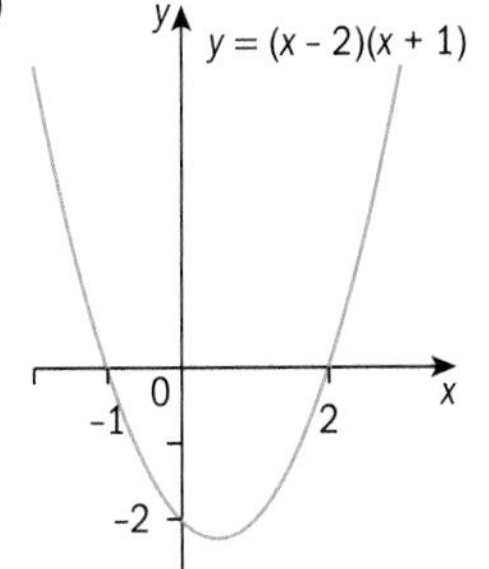

c)

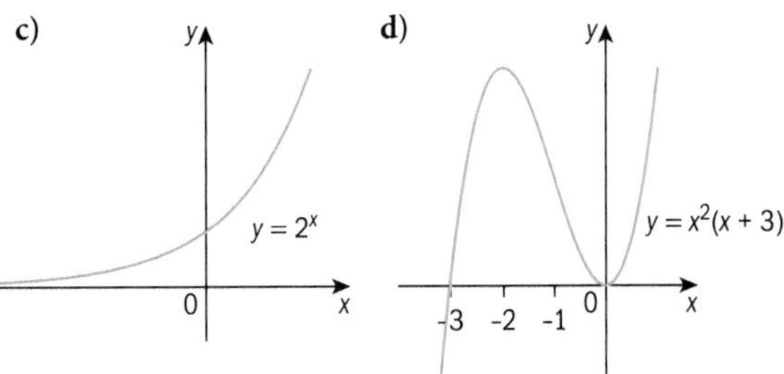

e)

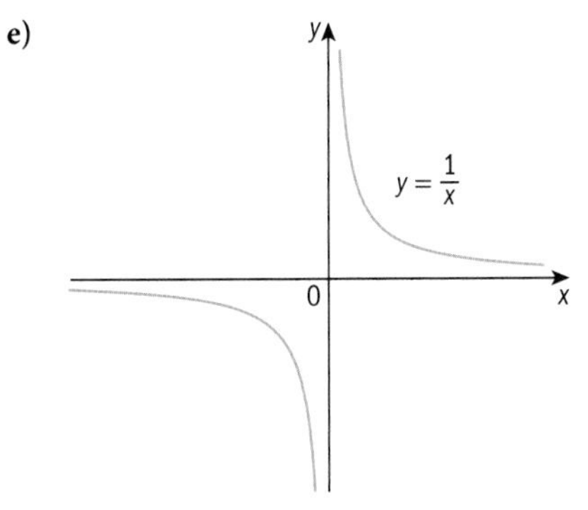

f)

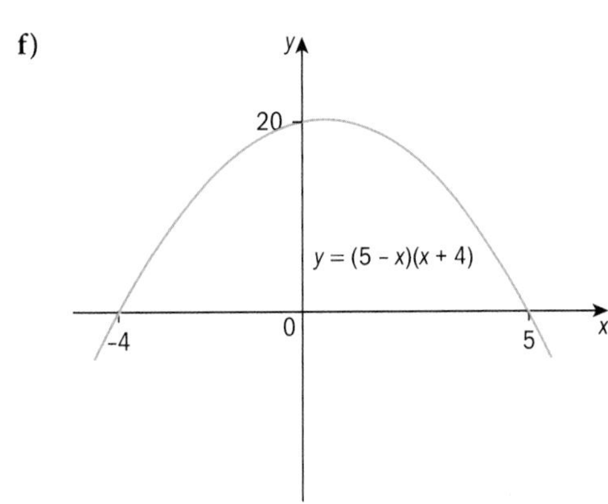

Exercise 2.1 page 27

1. a) $f(x) \leq 1$ b) $0 \leq g(x) \leq 18$
   c) $h(x) > -4$ d) $0 < f(x) \leq 1$
   e) $-6 \leq g(x) \leq 22$ f) $h(x) \geq 0$
2. a) Range: $f(x) \in \mathbb{R}$; one-to-one function

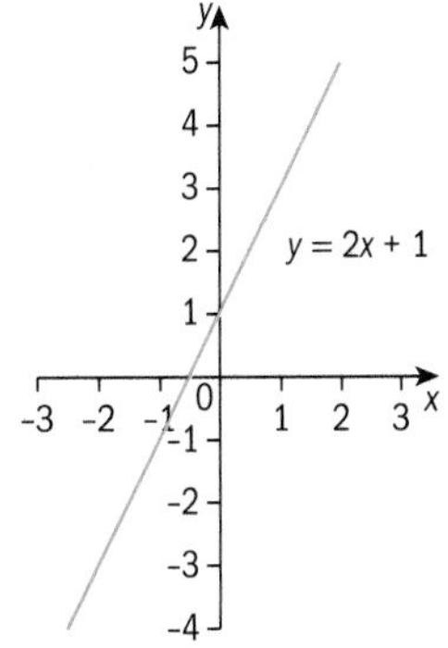

b) Range: $-4 \le g(x) < 5$; many-to-one function

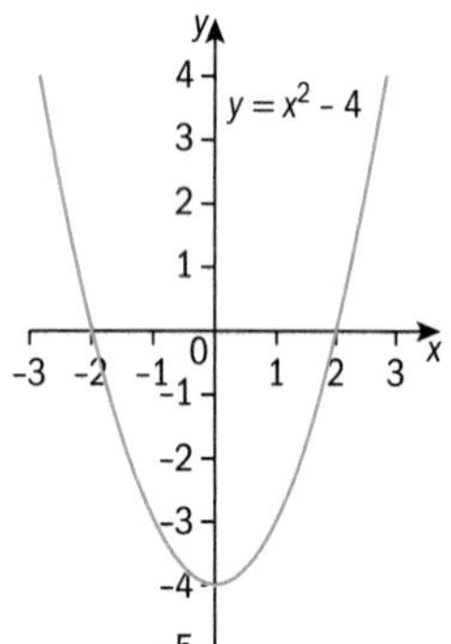

c) Range: $-1 \le h(x) < 8$; one-to-one function

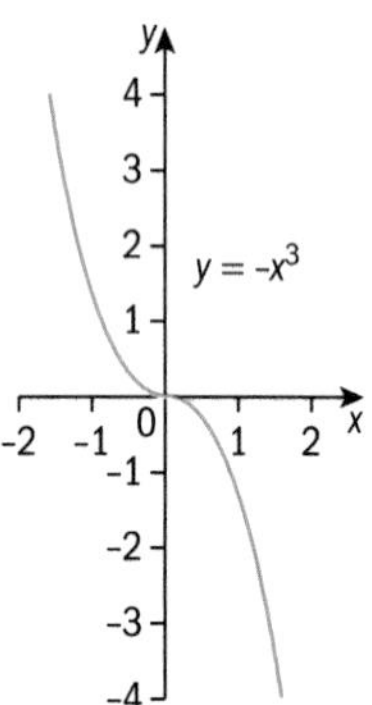

d) Range: $f(x) \ge 1$; one-to-one function

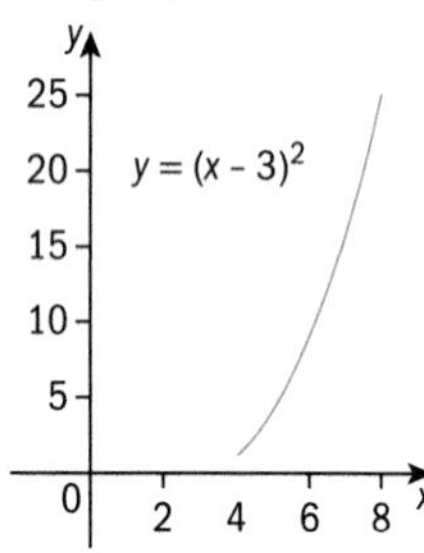

e) Range: $-5 \le g(x) \le -1$; one-to-one function

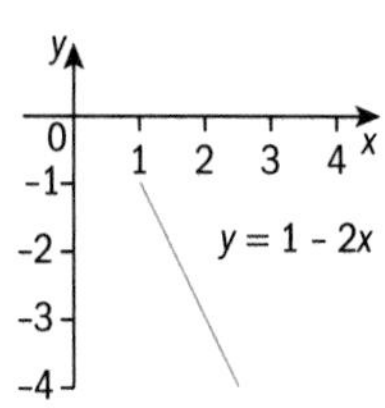

f) Range: $h(x) \ge 0$; one-to-one function

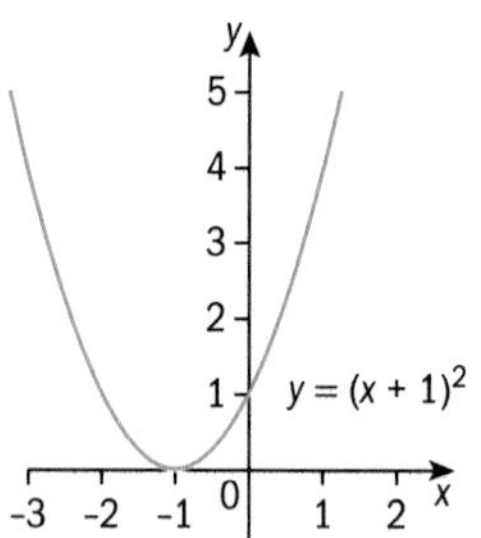

3. **a)**

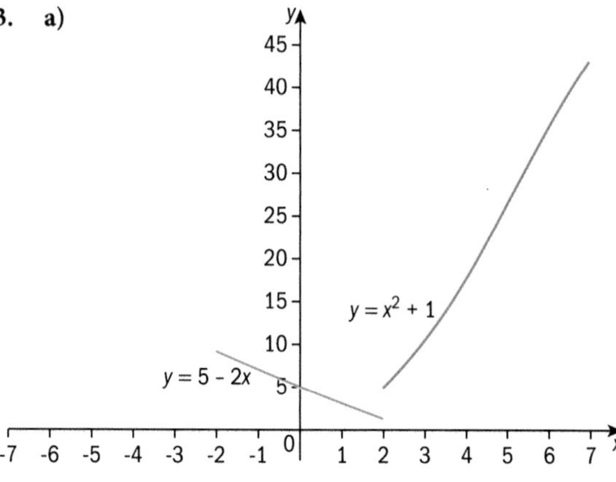

b) Range: $f(x) > 1$

4. **a)** $3(x+1)^2 - 21$ **b)** Range: $f(x) \ge -21$

5. $x < 4 - \sqrt{6}$ or $x > 4 + \sqrt{6}$ **6.** Range: $h(x) \le 6$

7. f is a function as it is one-to-one. g is not a function as it is one-to-many, as $g(2) = 6$ and 3.

8. Range: $0 \le f(x) \le 18$

9. **a)** **i)** $x \in \mathbb{R}$ **ii)** $f(x) \in \mathbb{R}$

iii) One-to-one

b) **i)** $x \in \mathbb{R}, x \ge 0$ **ii)** $g(x) \ge 0$

iii) One-to-one

c) **i)** $x \in \mathbb{R}, x \ne 0$ **ii)** $h(x) \in \mathbb{R}, h(x) > 0$

iii) Many-to-one

10. **a)** $g(x) = (x+5)^2 - 25 + p$ **b)** Range: $g(x) \ge p - 25$

Exercise 2.2 page 31

1. **a)** $5x^2 - 1$ **b)** $25x - 6$ **c)** 256 **d)** 1

2. **a)** 51 **b)** $2 - 2x^2$

3. **a)** $\frac{1}{x-2}$ **b)** $\frac{x-2}{x-3}$ **c)** $-\frac{5}{16}$

4. **a)** $2x^2 + 1$ **b)** $2 - 2x$

c) -1 **d)** 10

5. $k = -\frac{1}{3}$ **6.** $x = \frac{1}{2}$

7. $x = 6$ or $x = -3$

8. $x = \pm\frac{1}{2}$

9. Proof

10. $fg(x) \geq 6$

11. $x = \frac{3}{2}$

12. $a = -2, b = -4$ or $a = -\frac{1}{2}, b = -1$

13. Proof

14. $x \leq 1$ or $x \geq 2$

15. $4\left(x - \frac{1}{4}\right)^2 - \frac{1}{4}$

Exercise 2.3 page 35

1. a) $f^{-1}(x) = \frac{x-3}{2}, x \in \mathbb{R}$
   b) $f^{-1}(x) = \sqrt{x+2}, x \in \mathbb{R}, x \geq -2$
   c) $f^{-1}(x) = \frac{1-x}{4}, x \in \mathbb{R}$
   d) $f^{-1}(x) = 2x + 3, x \in \mathbb{R}$
   e) $f^{-1}(x) = \frac{1}{x}, x \in \mathbb{R}, x \neq 0$
   f) $f^{-1}(x) = \frac{x}{2} + 1, x \in \mathbb{R}$
2. a) $f^{-1}: x \mapsto 3 - x, x \in \mathbb{R}$
   b) $g^{-1}: x \mapsto \frac{x+6}{24}, x \in \mathbb{R}$
   c) $h^{-1}: x \mapsto \sqrt{\frac{x-2}{3}}, x \in \mathbb{R}, x \geq 2$
   d) $f^{-1}: x \mapsto \frac{2}{x} - 5, x \in \mathbb{R}, x \neq 0$
   e) $g^{-1}: x \mapsto \frac{2}{x} - 1, x \in \mathbb{R}, x \neq 0$
   f) $h^{-1}: x \mapsto x^2 + 3, x \in \mathbb{R}, x \geq 0$
3. $-\frac{13}{3}$
4. $f^{-1}: x \mapsto \frac{x+1}{x-2}, x \in \mathbb{R}, x \neq 2$
5. a) $f(x) = (x + 1)^2 - 2$ b) $-1 \leq f(x) \leq 7$
   c) $f^{-1}: x \mapsto -1 + \sqrt{x+2}, x \in \mathbb{R}, -1 \leq x \leq 7$
   d) Proof
   e)

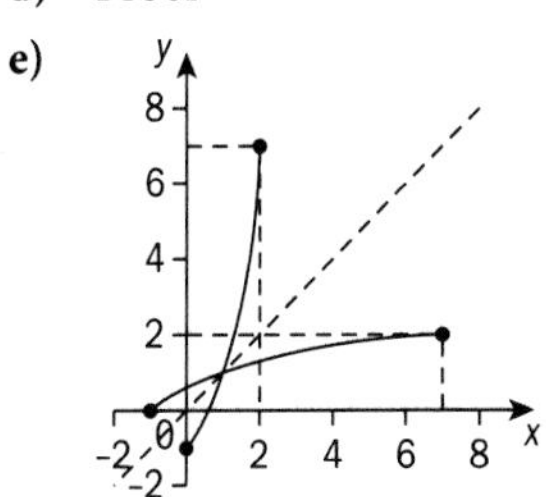

6. $1 - \frac{1}{x-2}, x \in \mathbb{R}, x \neq 2$
7. Proof
8. a) $\frac{5}{2}$ b) Proof
9. Proof
10. a) Self-inverse
    b) Not self-inverse
    c) Self-inverse
11. a) i) $2(x - 3)^2 - 10$ ii) $f(x) \geq -10$
    b) i) 3
    ii) $g^{-1}: x \mapsto 3 + \sqrt{\frac{x+10}{2}}, x \in \mathbb{R}, x \geq -10$
12. a) $4 - (x - 2)^2$
    b) $f^{-1}: x \mapsto 2 + \sqrt{4-x}, x \in \mathbb{R}, x \leq 4$
13. a) $f^{-1}: x \mapsto 3x^2 - 12x + 13$ b) $x \in \mathbb{R}, x \geq 2$
14. $a = 7, b = 20$
15. a) $f^{-1}: x \mapsto \frac{2x+1}{1-x}, x \in \mathbb{R}, x \neq 1$ b) Proof

Exercise 2.4 page 38

1. a)

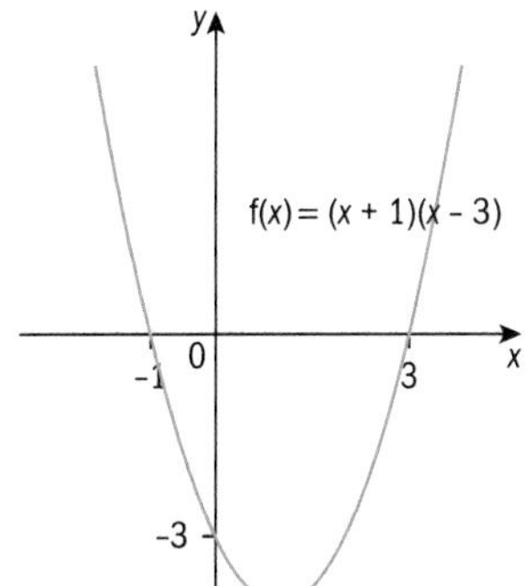

b) i)

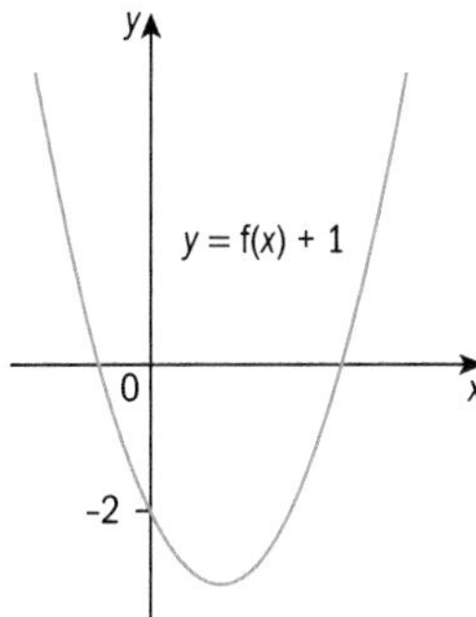

ii)

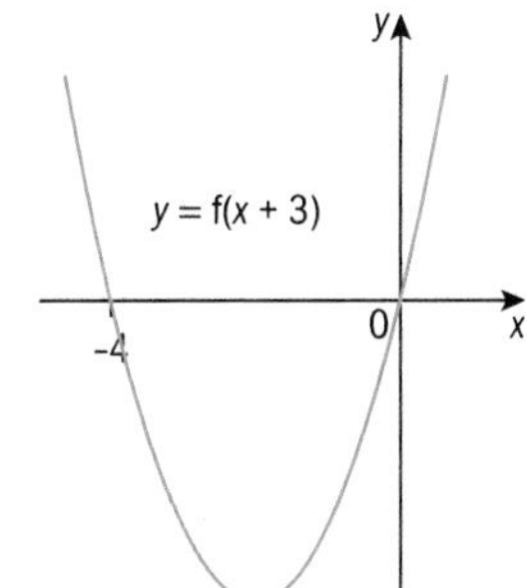

iii)

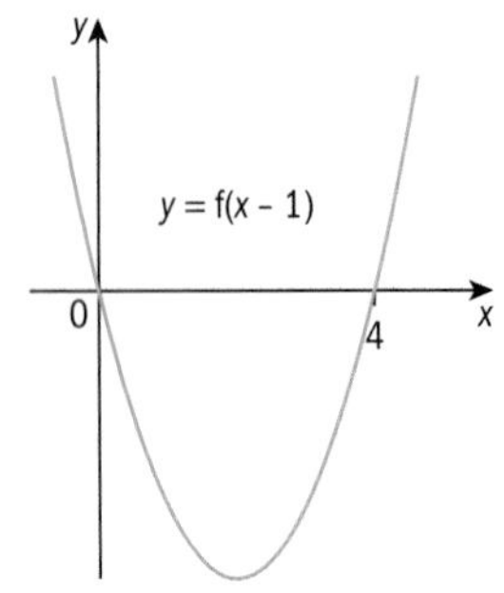

2. a)

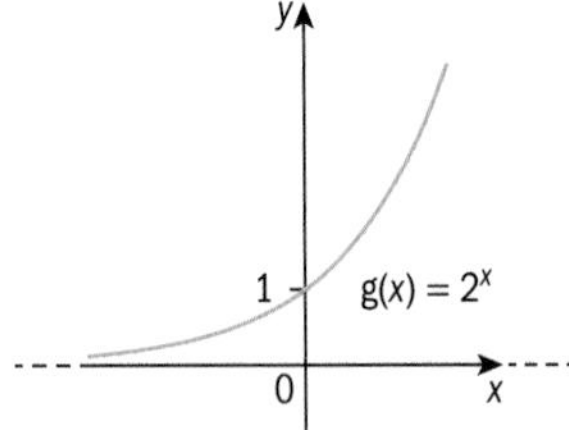

b) i)

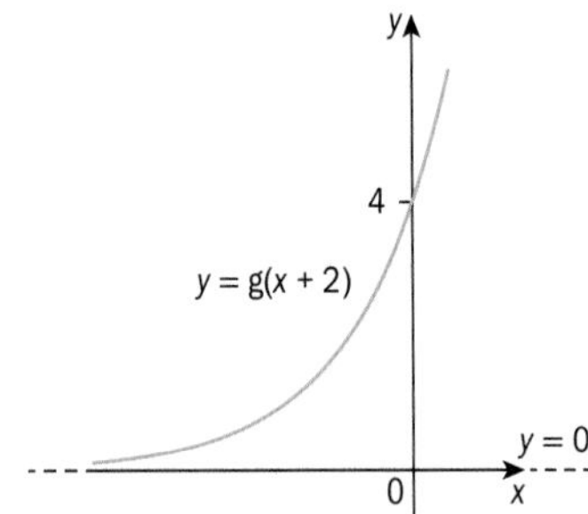

ii)

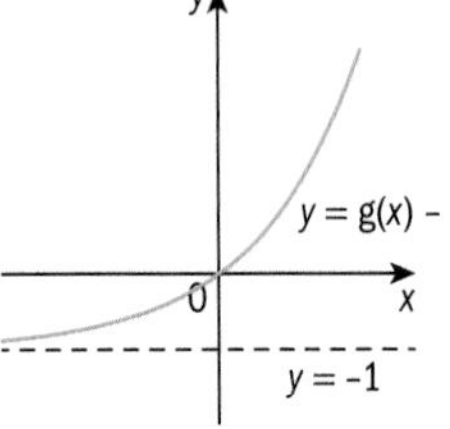

iii)

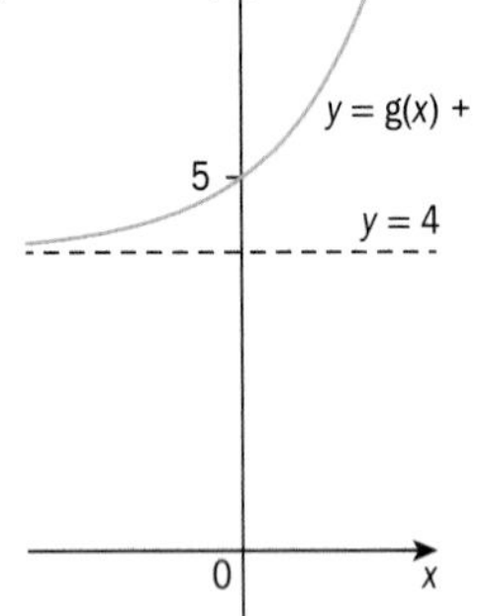

3. Translation by the vector $\begin{pmatrix} -5 \\ 0 \end{pmatrix}$

4. a) $y = x^2 + x + 1$

b) Translation by the vector $\begin{pmatrix} 1 \\ 3 \end{pmatrix}$

5. $y = x^2 - x + 4$

6. $y = x^2 + 2x - 2$

Exercise 2.5 page 40

1. a)

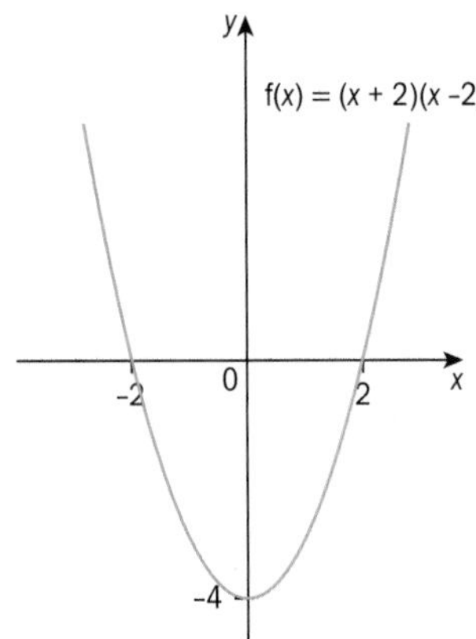

b) i)

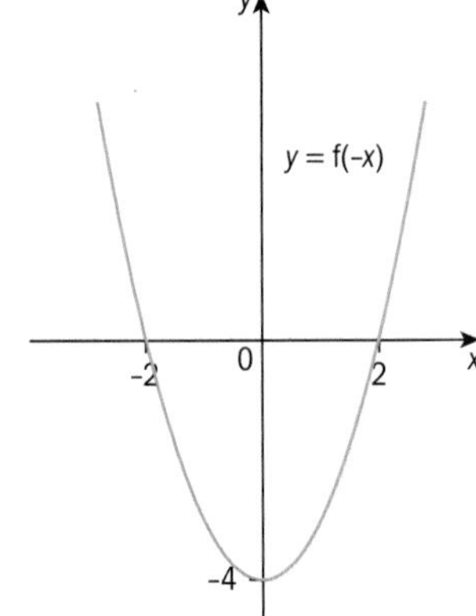

ii)

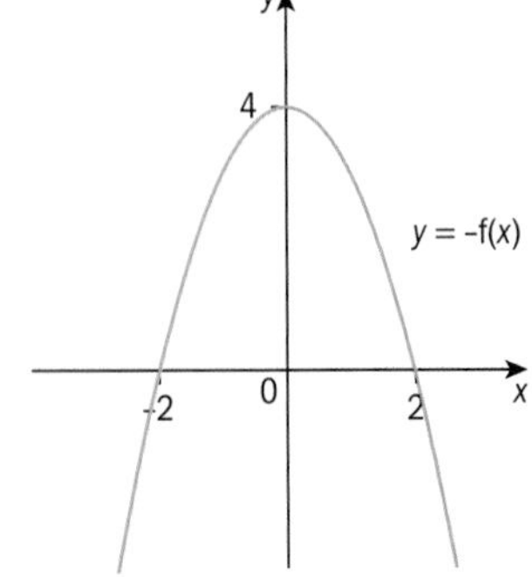

2. Reflection in the y-axis, then translation by the vector $\begin{pmatrix} 0 \\ 7 \end{pmatrix}$

3. a)

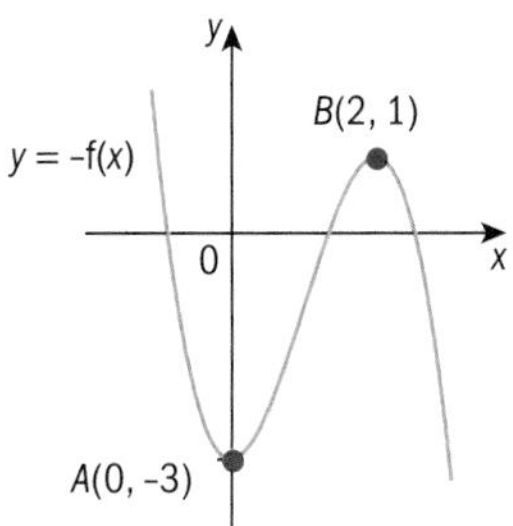

b)

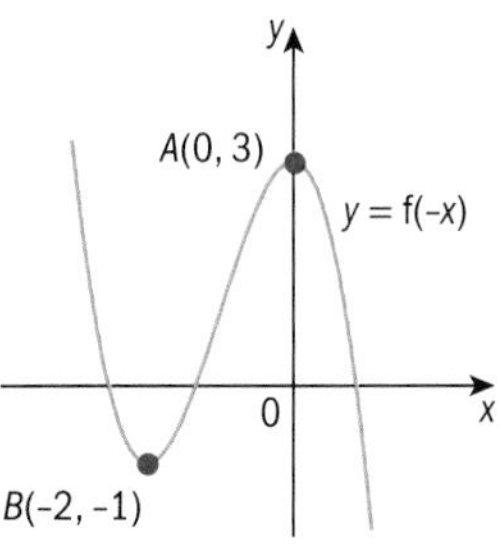

c)

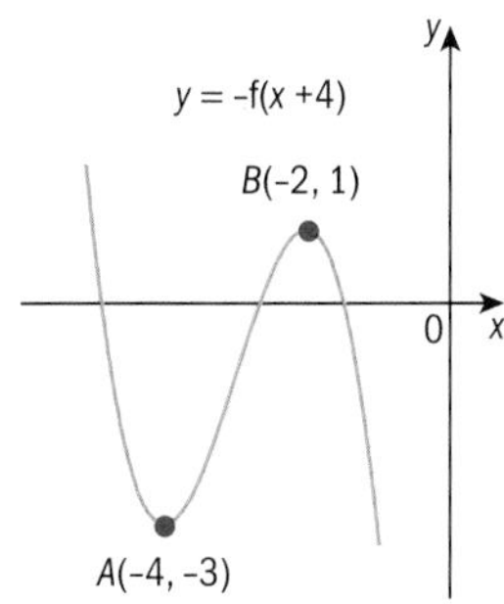

d)

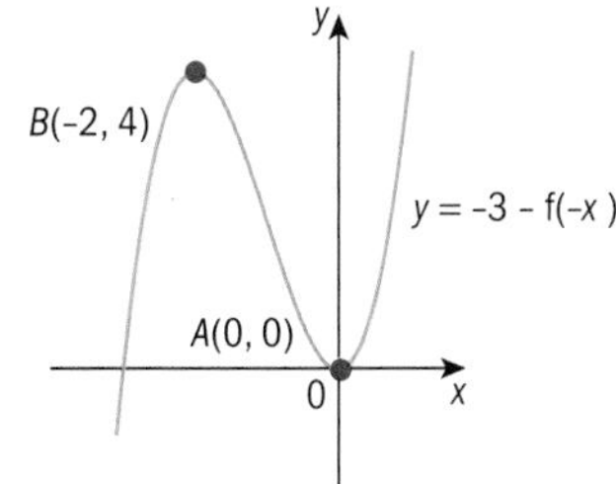

4. a) $P(3, 5)$, $Q(2, -8)$

b) $P(-3, -5)$, $Q(-2, 8)$

c) $P(2, 5)$, $Q(1, -8)$

d) $P(-3, -10)$, $Q(-2, 3)$

5. $y = -3x^2 - 2x + 8$

Exercise 2.6 page 42

1. a)

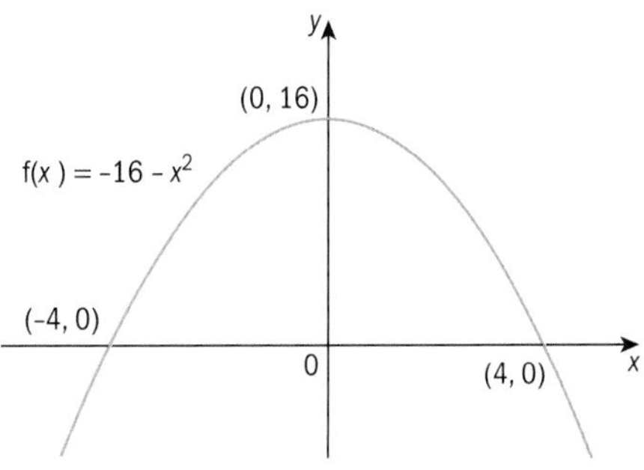

b) i)

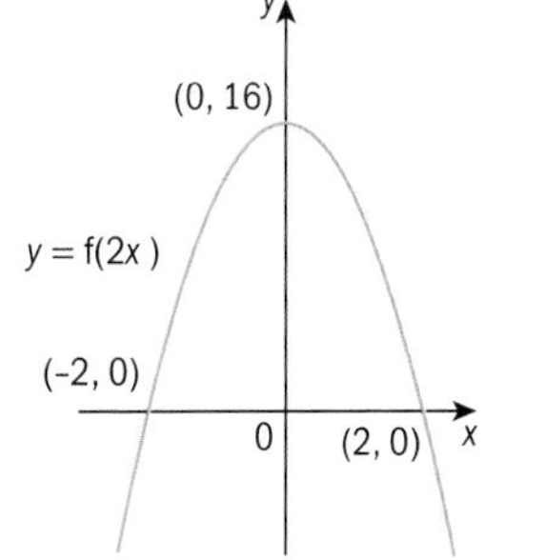

ii)

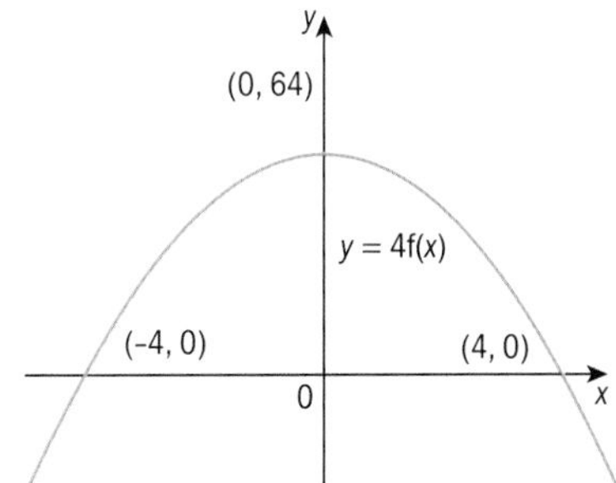

iii)

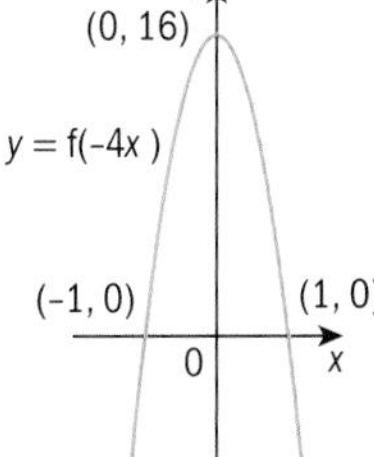

2. a)

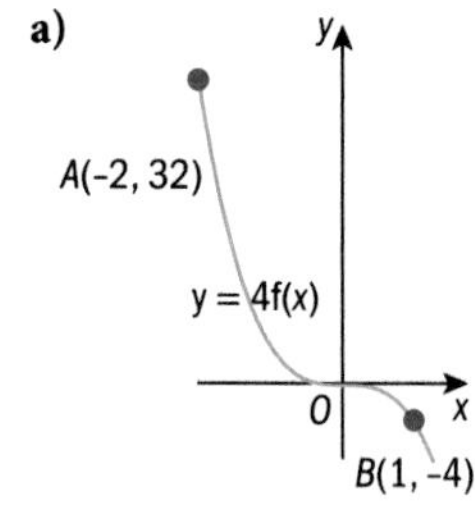
y
A(-2, 32)
y = 4f(x)
0
x
B(1, -4)

b)

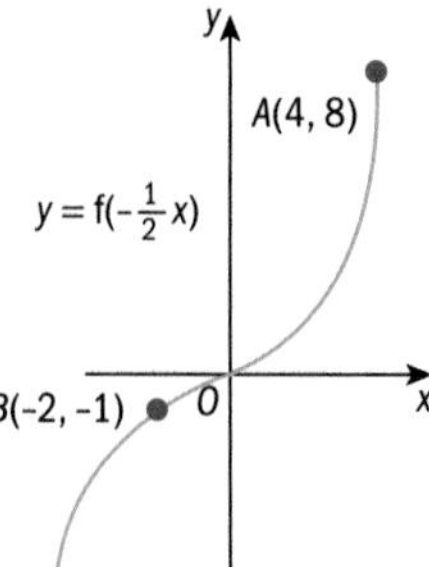
y
A(4, 8)
y = f(-½x)
B(-2, -1)
0
x

c)

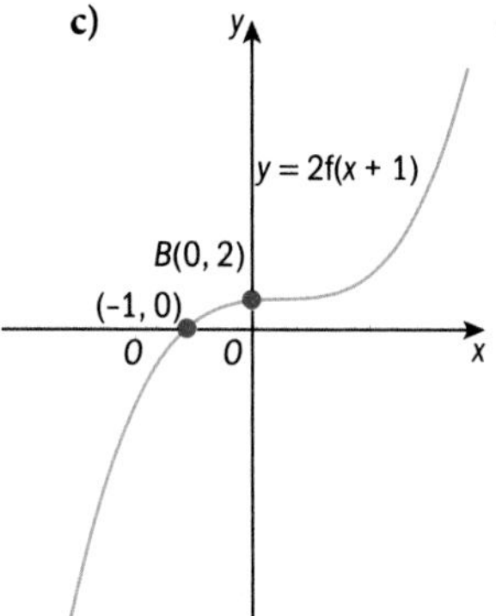
y
y = 2f(x + 1)
B(0, 2)
(-1, 0)
0
0
x
A(-3, -16)

d)

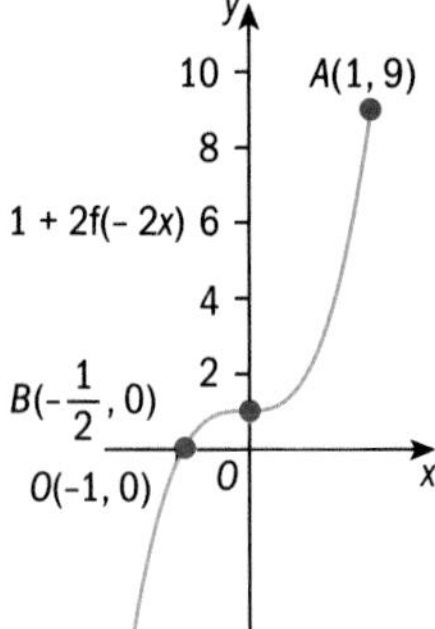
y
10
A(1, 9)
8
y = 1 + 2f(- 2x) 6
4
2
B(-½, 0)
0(-1, 0)
0
x

3. a)

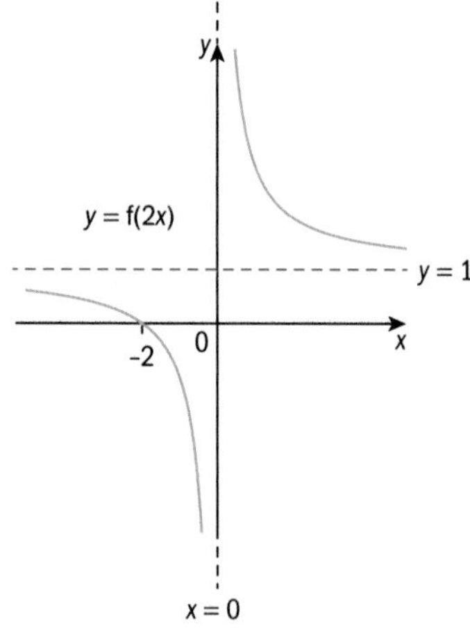
y
y = f(2x)
y = 1
-2
0
x
x = 0

b)

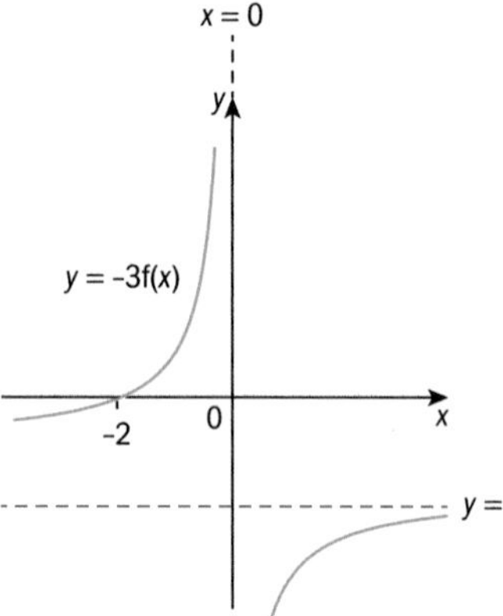
x = 0
y
y = -3f(x)
-2
0
x
y = -3

c)

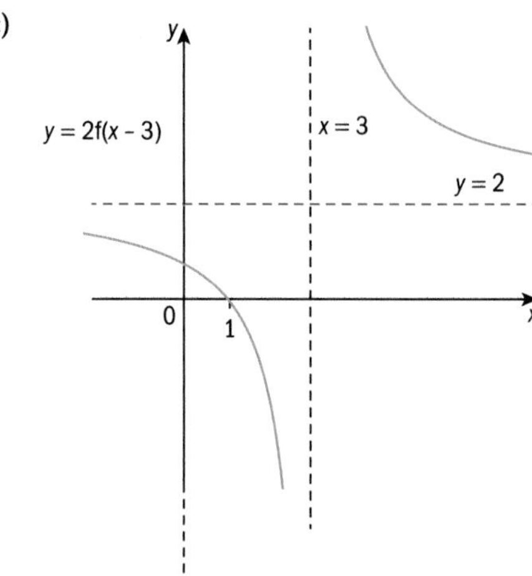
y
y = 2f(x - 3)
x = 3
y = 2
0
1
x

d)

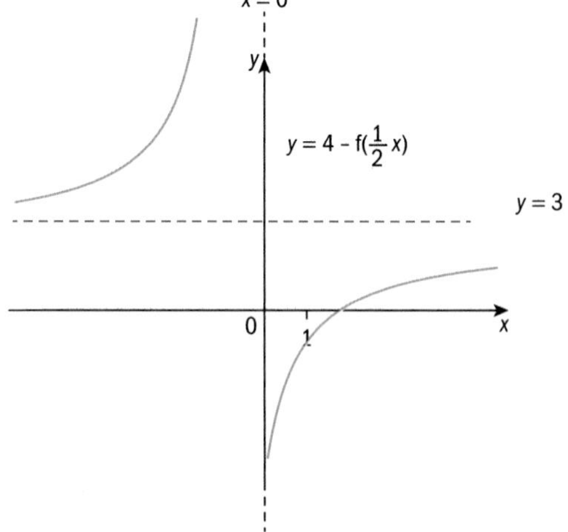
x = 0
y
y = 4 - f(½x)
y = 3
0
1
x

e)

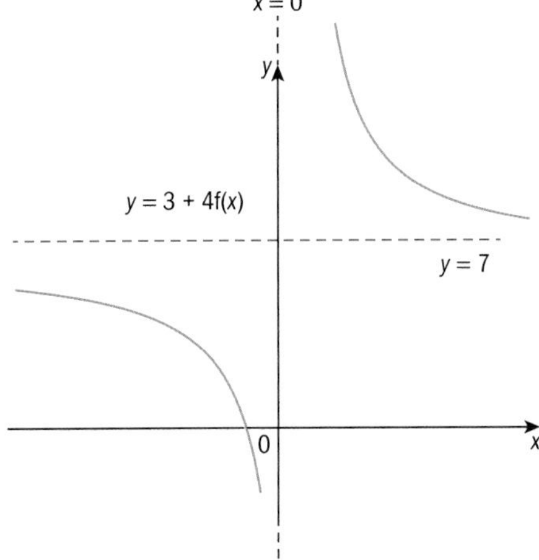

f)

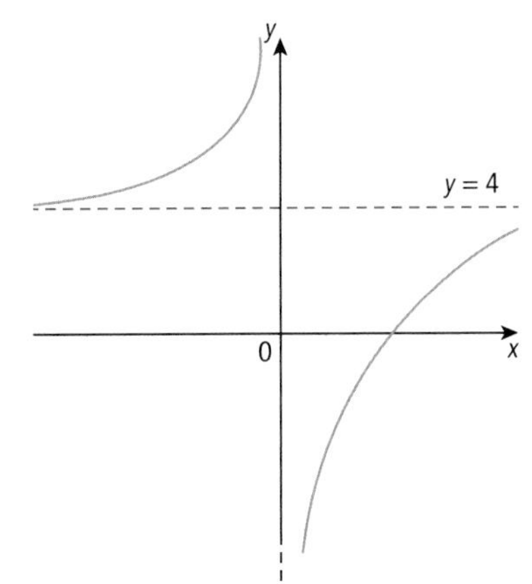

4. Stretch parallel to x-axis with stretch factor $\frac{1}{2}$, then reflection in the x-axis.

5. Translation by the vector $\begin{pmatrix}2\\0\end{pmatrix}$, then stretch parallel to the y-axis with stretch scale factor 3, then translation by the vector $\begin{pmatrix}0\\4\end{pmatrix}$.

Summary exercise 2 page 43

1. a) $1 < f(x) \leq 8$ b) $14 < g(x) \leq 21$
   c) $-\frac{1}{3} \leq h(x) \leq -\frac{4}{15}$ d) $\frac{1}{16} \leq f(x) < 1$
2. a) $f(x)$ is not a function because $f(-1) = 5$ and -6, i.e. f is one-to-many.
   b) One-to-one
3. Domain: $x \in \mathbb{R}, x \neq 1$ Range: $f(x) \in \mathbb{R}, f(x) \neq 0$
4. a) $(4x - 1)^2$ b) $x = 1, x = -3$
5. a) $x^2 - 5$ b) $\sqrt{3x-15}$ c) 49
6. $a = 1$ $b = -2$
7. $x = -\frac{1}{8}$
8. a) $-\frac{1}{3} \leq x \leq 3$
   b) Sketch graphs. Intersect at $\left(\frac{5}{4}, \frac{5}{4}\right)$
9. a) $fg(x) = 9x^2 + 24x + 15$ $gf(x) = 3x^2 + 1$
   b) $a = -\frac{7}{3}$ or $= -1$
10. $4(x + 2)^2 - 25$
11. f^{-1} = yes; $f^{-1}: x \mapsto \frac{x-5}{3}, 2 \leq x \leq 20$;

    g^{-1} = yes; $g^{-1}: x \mapsto \sqrt{\frac{4-x}{2}}, x \leq 4$

    h^{-1} = no as not one-to-one
12. a) $\sqrt{\frac{1-x}{2x-1}}$ b) $\frac{1}{2} < x \leq \frac{5}{9}$ c) Proof
13. a) $x = 0$ or $x = -2$
    b) $(fg)^{-1}(x) = \pm\sqrt{\frac{x-5}{3}}, x \in \mathbb{R}, x \geq 5$
    c) For all values of $x \geq 5$ we get two values for the inverse.
14. a) $f^{-1}(x) = \frac{x^2 - a}{2}, x \in \mathbb{R}, x \geq 0$ b) $a = -x$
15. a)

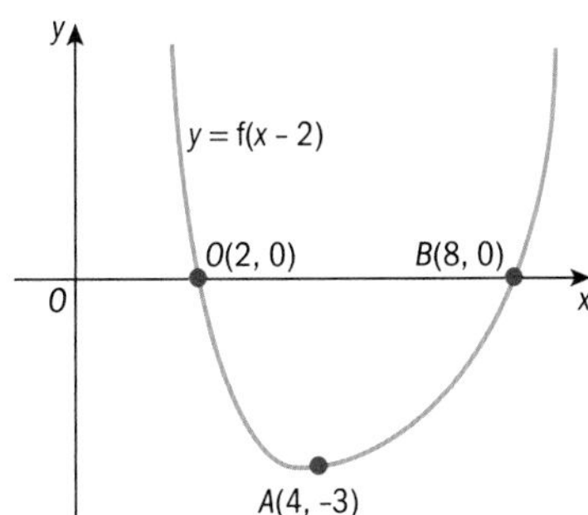

b)

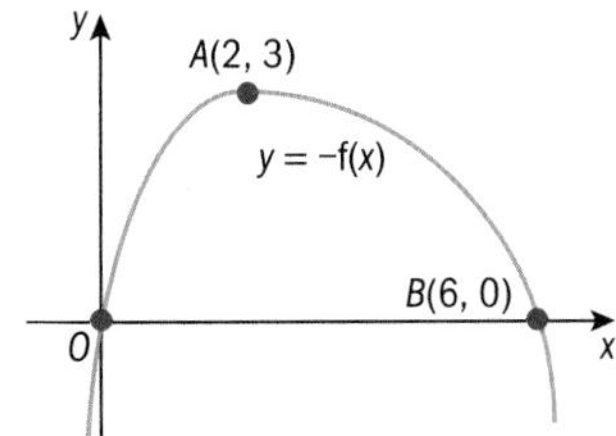

c)

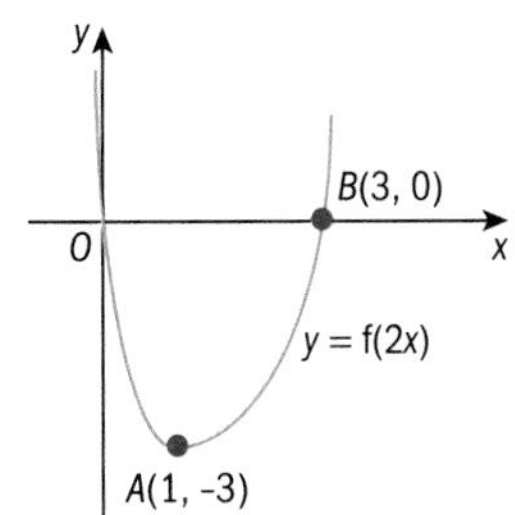

d)

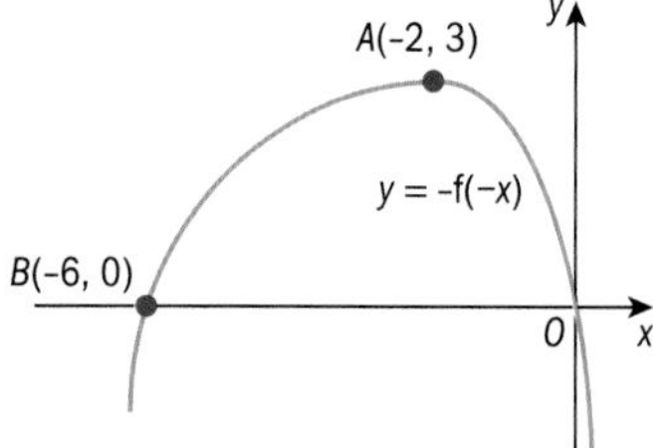

e)

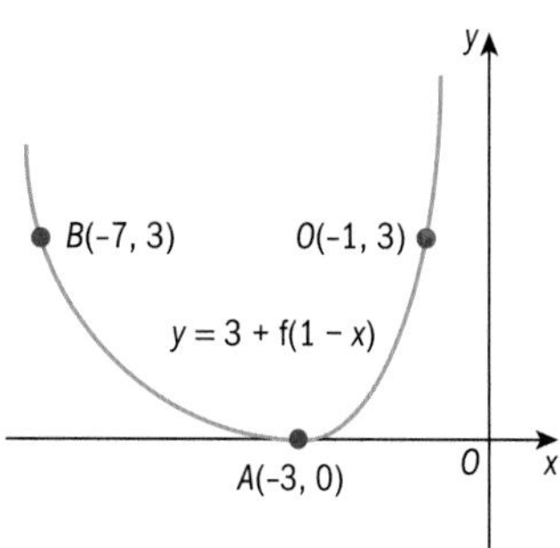

16. a)

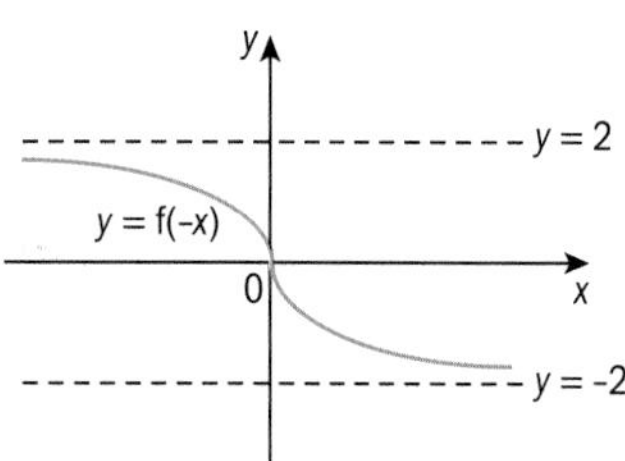

b)

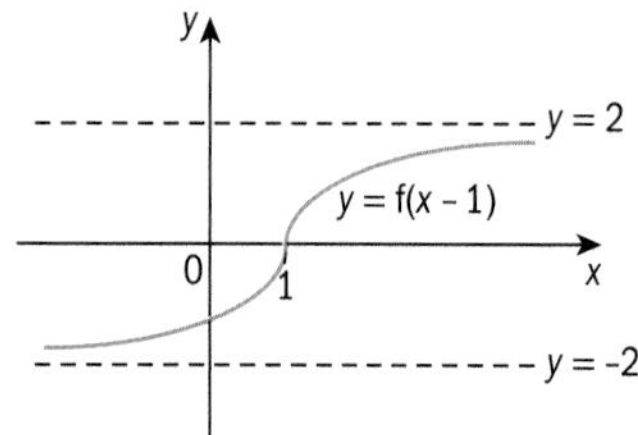

c)

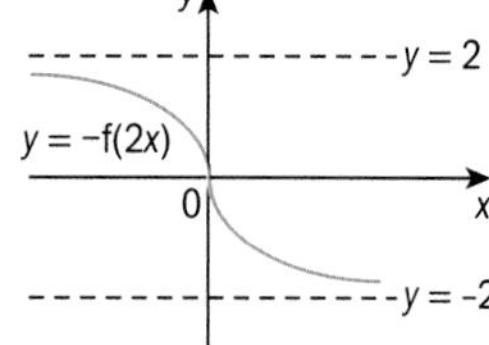

d)

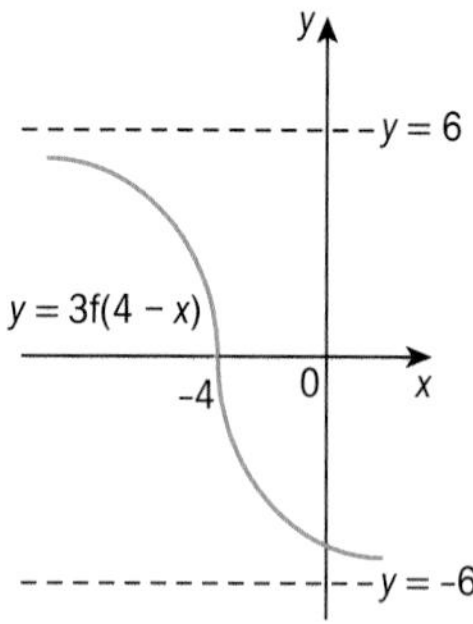

17. a) $y = f(4x)$

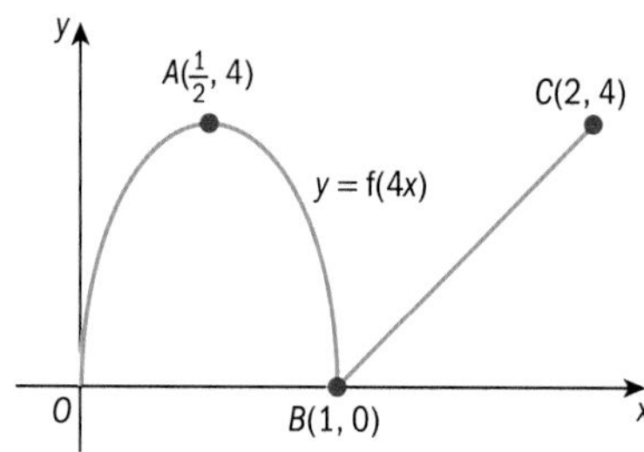

b) $y = f(3 - x)$

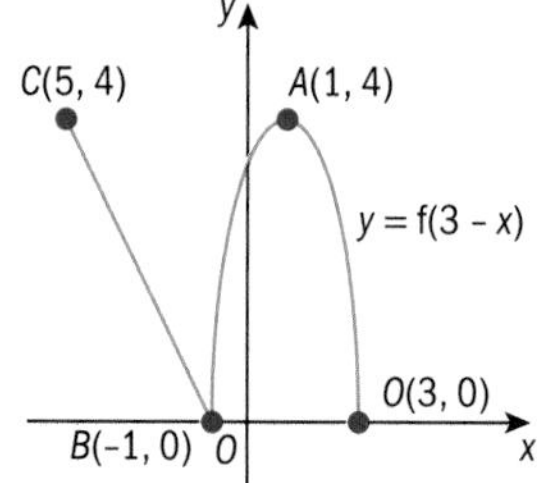

c) $y = -f(x + 1)$

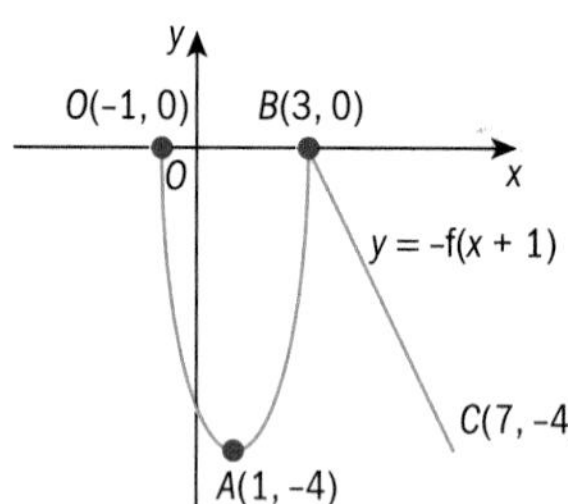

d) $y = 2f(-x)$

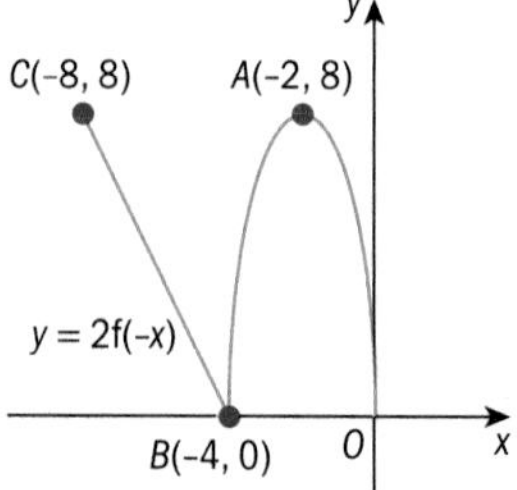

18. $y = -\frac{1}{x^2}$

19.

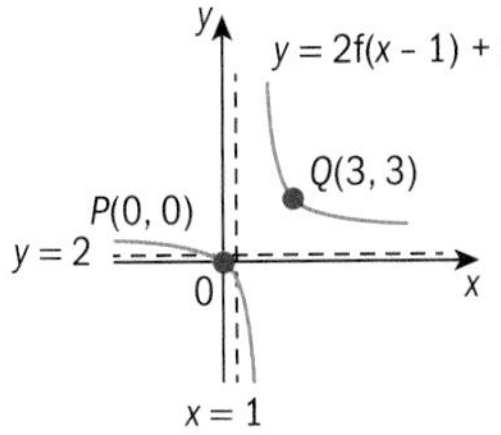

20. a) i)

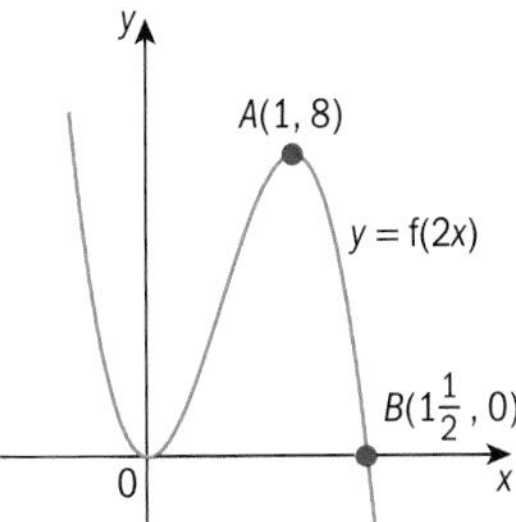

ii)

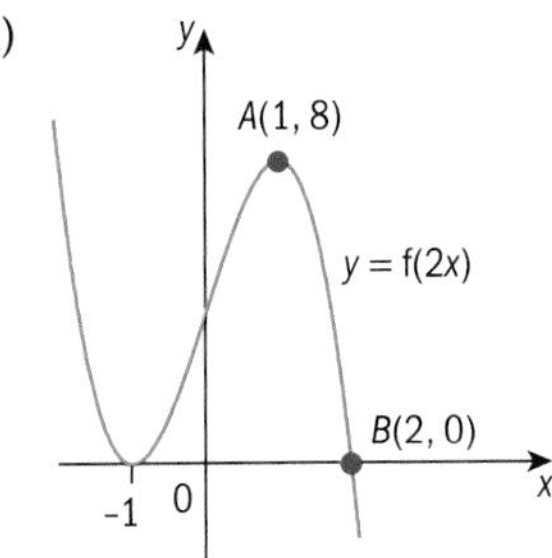

b) $c = -5$

c) $d = \frac{1}{2}$, $e = -2$

21. Reflection in the x-axis, then translation by the vector $\begin{pmatrix} 0 \\ 5 \end{pmatrix}$

22. a)

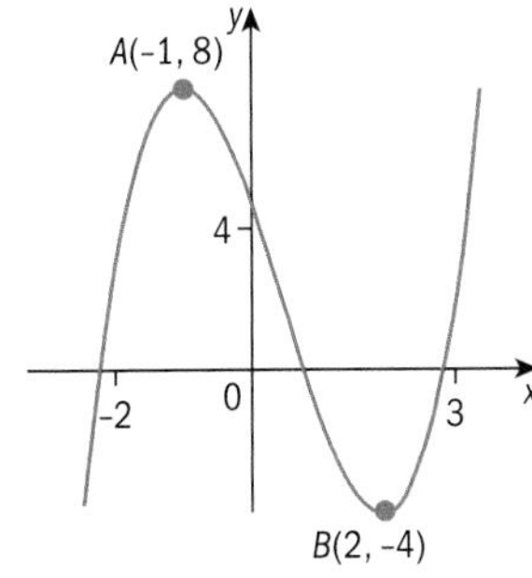

b)

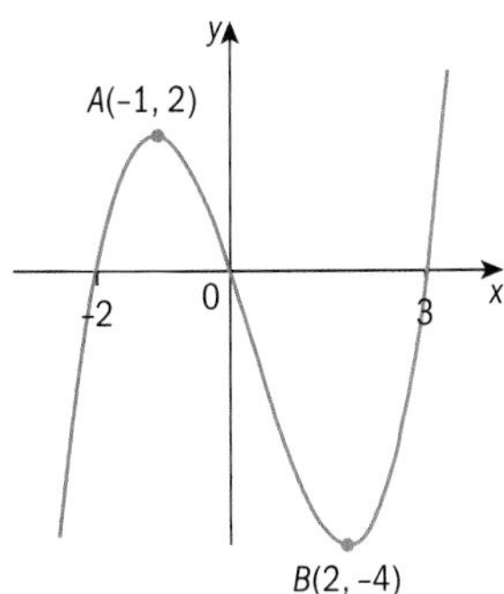

23. a)

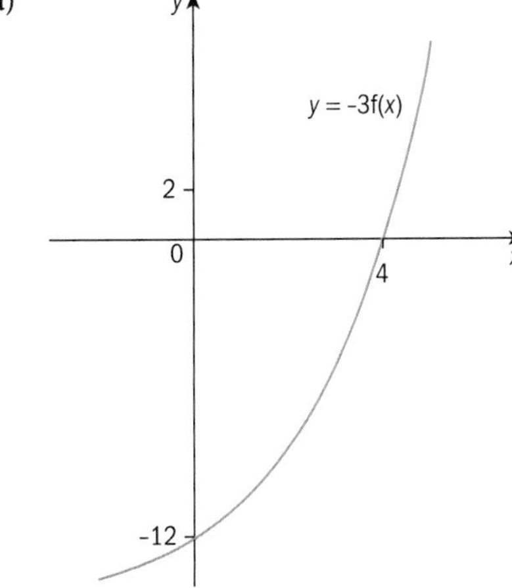

b) Reflection in the y-axis and then a stretch parallel to the x-axis, stretch factor 2

3 Coordinate geometry

Skills check page 48

1. a) 5.83 cm b) 3.87 cm c) 5.11 cm d) 6 cm
2. a) 2 b) −1 c) −3
   d) $\frac{1}{2}$ e) $\frac{1}{2}$ f) −2

Exercise 3.1 page 51

1. a) (4, 8) b) (−1, −3) c) (−4, 2)
   d) (−2.5, 1.5) e) $(-2k, -2)$ f) (p, p)
2. a) 1 b) −2 c) −2
   d) −0.4 e) 5 f) 4
3. a) 10 b) 5 c) 4.24
   d) 5.1 e) $5q$ f) $13p$
4. $p = -4$, $q = -5$

5. $k = 7$
6. $p = 4$ or $p = -2$
7. a) $\frac{9}{2}$ b) 9.22
8. $(3, -1)$
9. $a = 3, b = -10$
10. a) Proof b) 20
11. 5
12. $k = -3$
13. $p = 2$ or $p = 6$
14. Proof
15. 30

Exercise 3.2 page 54

1. a) i) -2 ii) $\frac{1}{2}$
   b) i) $-\frac{2}{7}$ ii) $\frac{7}{2}$
   c) i) 9 ii) $-\frac{1}{9}$
   d) i) $3\frac{1}{3}$ ii) $-\frac{3}{10}$
   e) i) $-3k$ ii) $\frac{1}{3k}$
   f) i) $-2\frac{1}{4}$ ii) $\frac{4}{9}$
2. a) i) 7 ii) $-\frac{1}{7}$
   b) i) -1 ii) 1
   c) i) -3 ii) $\frac{1}{3}$
   d) i) $\frac{4}{3}$ ii) $-\frac{3}{4}$
   e) i) 1 ii) -1
   f) i) -4 ii) $\frac{1}{4}$
3. a) Perpendicular b) Parallel
   c) Perpendicular d) Neither
4. $k = 8$
5. $x = -1$
6. $y = -4$
7. $x = 0$ or $x = -2$
8. Proof
9. Proof

10. $(3, 0)$

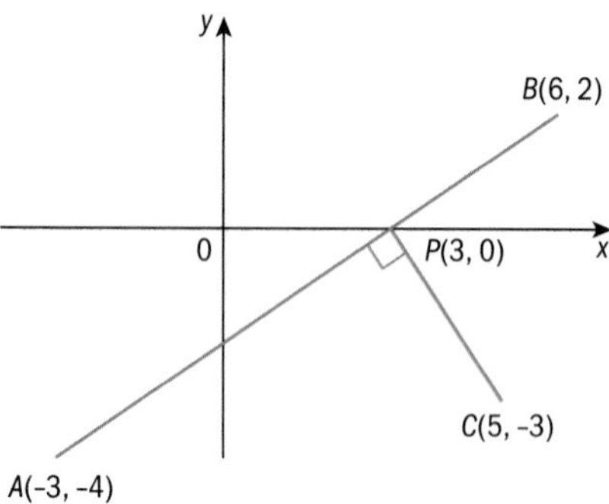

Exercise 3.3 page 58

1. a) $y = 2x - 3$ b) $y = -x - 3$
   c) $y = -3x - 11$ d) $y = 4x - 14$
   e) $3y = 4x + 21$ f) $5y = -x - 4$
2. a) $y = x + 4$ b) $y = -3x - 5$
   c) $y = -x + 2$ d) $y = x + 6$
   e) $2y = -x - 10$ f) $15y = -x - 36$
3. $y = -3x + 22$
4. $5y = -x - 3$
5. $5y = 2x + 17$
6. $2y = 3x + 9$
7. $4x + 3y = 28$
8. $4x + 3y + 25 = 0$
9. AB: $y = 18 - 2x$
   CD: $2y = x + 1$
   $P = (7, 4)$
10. AB: $2y = x - 5$
    CD: $y = 8 - 3x$
    $P = (3, -1)$
11. $(3, 4)$
12. a) $P(2, 0), Q(0, -3)$ b) $2x + 3y = 17$
13. AB: $y = x - 10$
    CD: $5y + 21 = 3x$
    $P = \left(\frac{29}{2}, \frac{9}{2}\right)$

Exercise 3.4 page 60

1. a) $(x - 9)^2 + (y - 1)^2 = 16$
   b) $(x + 5)^2 + (y - 3)^2 = 49$
   c) $(x + 4)^2 + (y + 7)^2 = 25$
   d) $(x - 6)^2 + (y + 2)^2 = 9$
2. a) Centre $(-2, 1)$, radius $= 3$
   b) Centre $(3, 8)$, radius $= 6$
   c) Centre $(5, -9)$, radius $= 4.47$ (3 s.f.)
   d) Centre $(-6, -7)$, radius $= 1$

e) Centre (3, −4), radius = 3.87 (3 s.f.)

f) Centre (−2, −1), radius = 2.45 (3 s.f.)

g) Centre $(\frac{1}{2}, 5)$, radius = 4.5

h) Centre $(-1, \frac{3}{2})$, radius = 3.20 (3 s.f.)

i) Centre $(2, -\frac{1}{2})$, radius = 1.80 (3 s.f.)

j) Centre (7, −1), radius = 7.16 (3 s.f.)

3. a) $(x + 2)^2 + (y + 1)^2 = 17$

b) $(x + 3)^2 + (y - 7)^2 = 10$

c) $(x - 5)^2 + (y + 4)^2 = 169$

d) $(x - 6)^2 + y^2 = 58$

4. $a = -4$

5. $(x + 3)^2 + (y + 1)^2 = 41$

6. a) Proof

b) Centre $(\frac{7}{2}, 4)$, radius = 3.35 (3 s.f.)

7. 12

Exercise 3.5 page 65

1. $k = -4$

2. a) $(x - 3)^2 + (y + 2)^2 = 5$

b) Centre (3, −2), radius $\sqrt{5}$

3. $k < -4, k > 4$

4. a) $k = 2$ or $k = -10$ b) (1, 3)

5. 1

6. a) $k = 3$ b) (−1, −2)

7. a) (−3, −4) and (1, 0) b) $k = -\frac{1}{3}$

8. a) C(4, 3), radius $3\sqrt{5}$ b) $y = 2x - 20$

9. $3x - y - 8 = 0$

10. a) $k = 2$ or $k = -2$

b) (3, 3) and (−3, 3); $y = 3$

11. $2x + 3y - 7 = 0$

12. Proof

13. a) $k = 4$ b) (2, 9)

14. Proof

15. $2x + y = 6$

16. a) $y = mx - m$ b) $m = 4$ c) $k = -12$

17. $k = 4, k = 6$

18. Proof

19. Proof

Summary exercise 3 page 67

1. $k = 7$ **2.** 7.21 **3.** $a = 28, b = 20$

4. Proof **5.** Proof **6.** Parallel

7. a) Proof b) (3, 3)

8. (1, 2) **9.** $2x - 5y = 9$

10. $y = x - 7$

11. a) 2 b) $2x - y + 9 = 0$

c) $y = 2x - 4$ d) $5\sqrt{2}$

12. $x + 2y - 5 = 0$

13. $2y = x - 5k$

14. $y = 2$

15. a) $k = -9$ b) $2x + 3y = 21$ c) $4\sqrt{5}$

16. $k = 13$

17. a) $Q = (-8, 41)$ b) $11\sqrt{10}$

18. $y = x - 5$

19. $(x - 1)^2 + (y - 2)^2 = 13$

20. a) $k = 5$ or -3 b) (2, 4) and (−2, 0)

21. $k < \frac{3}{4}$

22. $4\sqrt{2} - 2$

23. Proof

24. a) $k = -4$ or $k = 1$ b) $y = -2x + 3$

25. a) 20 b) 11.4 (3 s.f.)

26. a) C(3, −1), radius $\sqrt{5}$ b) $y = 2x - 12$

27. a) (−1, −2) b) $x + y - 5 = 0$

c) P(5, 0), Q(1, 4)

28. a) 5 b) 8.66 (3 s.f.)

c) 5

29. a) Proof

b) $(x - 2)^2 + (y - 9.5)^2 = 8.5^2$

30. a) (2, 3)

b) $(x - 2)^2 + (y - 3)^2 = 25$

c) $4y = 3x - 19$

31. a) C(−4, −2), radius 5

b) $y = 1$

c) Proof

32. a) Proof b) $(x + 2)^2 + (y - 3)^2 = 13^2$

33. a)

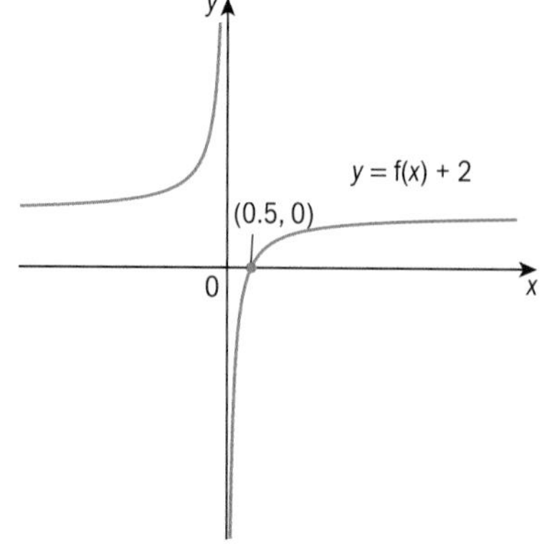

Domain $x \in \mathbb{R}, x \neq 0$; range $g(x) \in \mathbb{R}, g(x) \neq 2$

b) (0.5, 0)

c)

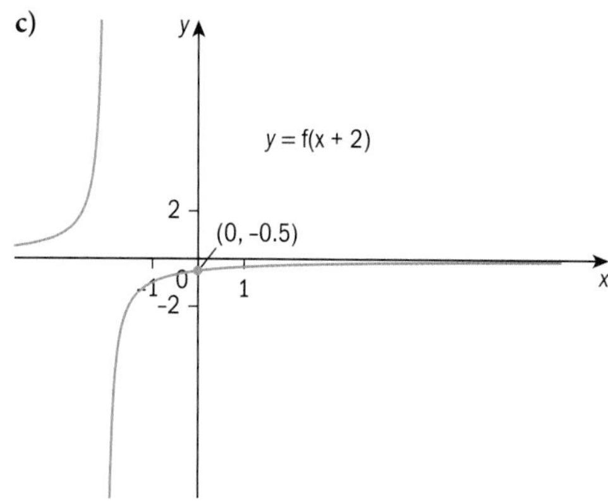

Domain $x \in \mathbb{R}, x \neq -2$; range $h(x) \in \mathbb{R}, h(x) \neq 0$

d) (0, −0.5)

34. a) $-10 < k < 2$
b) $k = -10, k = 2$
c) $k < -10, k > 2$

35. a) $x^2 + y^2 - 4x + 2y - 8 = 0$
b) $y = -\frac{2}{3}x + \frac{14}{3}$
c) $\left(0, \frac{14}{3}\right)$ and (7, 0)

36. $\sqrt{29}$

4 Circular measure

Skills check page 74

1. 21.4 cm, 7.73 cm^2 **2.** 5.24 cm, 26.2 cm^2

Exercise 4.1 page 76

1. a) $\frac{\pi}{12}$ **b)** $\frac{5\pi}{4}$ **c)** $\frac{3\pi}{4}$
d) $\frac{7\pi}{4}$ **e)** $\frac{7\pi}{20}$ **f)** $\frac{2\pi}{5}$

2. a) 0.436 **b)** 1.75 **c)** 4.36
d) 1.40 **e)** 2.39 **f)** 5.55

3. a) 18° **b)** 7° **c)** 67.5°
d) 150° **e)** 240° **f)** 40°

Exercise 4.2 page 77

1. a) 7.2 cm **b)** 110 mm
c) 7.33 m **d)** 29.3 mm

2. a) 0.5 m^2 **b)** 2645 mm^2
c) 1152 cm^2 **d)** 41.9 cm^2

3. a) 1.67 rad **b)** 3.2 rad
c) 25 m **d)** 6.51 cm

4. a) $\frac{28\pi}{3}$ cm^2 **b)** $12 + 4\pi$ cm

5. 18.6 cm **6.** 54.4 cm^2

Exercise 4.3 page 80

1. $r^2\left(\frac{\pi}{2} - 1\right)$ **2.** 74.8 cm **3.** 21.6 m^2

4. a) πr cm **b)** 158.6 cm^2

5. a) 1.17 rad, 1.77 rad **b)** 216.5 cm^2
c) 58.4 cm

6. a) 0.505 rad **b)** 142 cm^2 **c)** 142 cm

Summary exercise 4 page 82

1. a) 5.2 m **b)** 11.12 m^2

2. a) $\frac{3\pi}{4}$ rad **b)** $\frac{75\pi}{8}$ m^2

3. a) 1.05 rad **b)** 5.59 cm^2

4. a) 26.3 m^2 **b)** 8.38 m^2

5. a) 5 cm **b)** 14.4 cm **c)** 6.16 cm^2

6. a) 7.5 cm **b)** 37.5 cm^2 **c)** 66.8 cm^2

7. a) 1.98 rad **b)** 57.5 cm^2

8. $10\sqrt{\pi}$ m **9.** 172 cm

10. 40.6 m^2 (3 s.f.)

11. a) 6.93 cm **b)** 32.4 cm^2

12. Proof

13. i) $4\pi + 12\sqrt{3}$ **ii)** $36\sqrt{3} - 12\pi$

14. 8.36

15. i) 0.680 rad **ii)** 61.5 cm^2 **iii)** 222.5 cm^2

16. a) i) $\frac{1}{8}\pi r^2$ ii) r iii) $\frac{1}{2}r^2$
b) 30.5 cm

5 Trigonometry

Skills check page 86

1. a) 4.02 cm b) 5.29 mm c) 68.0° (= 1.19 rad)
2. a) i) $\frac{\pi}{2}$ (= 1.57) rad ii) $\frac{5\pi}{4}$ (= 3.93) rad
iii) $\frac{43\pi}{180}$ (= 0.750) rad
b) i) 135° ii) 252° iii) 143.2°
3. a) $x = -\frac{1}{2}, x = 4$ b) $x = 0.438, x = 4.56$

Exercise 5.1 page 89

1. a) $-\frac{1}{2}$ b) -1 c) $-\frac{\sqrt{3}}{2}$ d) $\frac{1}{\sqrt{2}}$
2. a) $\frac{1}{\sqrt{2}}$ b) $-\frac{\sqrt{3}}{2}$ c) -1 d) $-\sqrt{3}$
3. a) i) + ii) − iii) − iv) −
b) i) 0.766 ii) −0.643 iii) −0.966
iv) −0.259
4. a) sin 48° b) cos 50° c) tan 35°
d) −sin 40° e) −cos 27° f) −tan 32°
g) −cos 15° h) −sin 25°
5. a) 21° b) 159° 6. 73°, 287°
7. 40.0°, 320.0° 8. 201.3°, 338.7°
9. a) 64°, 116° b) 26°, 334° c) 217°, 323°
d) 71°, 289° e) 14°, 194° f) 156°, 336°
10. a) 0.412, 2.73 b) 1.16, 5.12 c) 4.07, 5.36
d) 1.78, 4.50 e) 0.644, 3.79 f) 2.80, 5.94

Exercise 5.2 page 92

1. a) 140° b) 400°, 500°
2. $\frac{\pi}{4}, \frac{5\pi}{4}, \frac{9\pi}{4}, \frac{13\pi}{4}, \frac{17\pi}{4}, \frac{21\pi}{4}$
3. 140.0°, 220.0° 4. 194.5°, 345.5°
5. a) 22.6°, 157.4° b) 63.3°, 243.3°
c) 67.7°, 292.3° d) 191.5°, 348.5°
6. a) $\frac{\pi}{6}, \frac{5\pi}{6}, \frac{13\pi}{6}, \frac{17\pi}{6}$ b) $\frac{3\pi}{4}, \frac{5\pi}{4}, \frac{11\pi}{4}, \frac{13\pi}{4}$
c) $\frac{3\pi}{2}, \frac{7\pi}{2}$ d) $\frac{3\pi}{4}, \frac{7\pi}{4}, \frac{11\pi}{4}, \frac{15\pi}{4}$
7. a) $-\sin x°$ b) $-\cos x°$ c) $\tan x°$ d) $-\tan x°$
8. a) $-\cos x$ b) $\sin x$ c) $-\sin x$ d) $\tan x$
9. 33.7°, 213.7°

Exercise 5.3 page 95

1. a) 0° b) −90° c) −60° d) 45°
e) 0° f) 30° g) Impossible
h) 180° i) 45° j) 0° k) 30°
l) −30°
2. a) 53.1° b) 14.5° c) 43.5°
d) 84.3° e) 146.1° f) −22.0°
3. a) i) 0.381 ii) −0.381 iii) 0.730
iv) −0.730 v) 0.902 vi) 2.24
b) $\tan^{-1}(-0.4) = -\tan^{-1}(0.4)$
c) $\sin^{-1}\left(-\frac{2}{3}\right) = -\sin^{-1}\left(\frac{2}{3}\right)$
d) $\cos^{-1}(0.62) + \cos^{-1}(-0.62) = \pi$
4. a) $\frac{1}{2}$ b) $\frac{\sqrt{3}}{2}$ c) $-\frac{1}{2}$
d) Not possible
5. a) i) $\frac{\pi}{3}$ ii) $\frac{\pi}{3}$ iii) $\frac{\pi}{4}$ iv) $\frac{\pi}{4}$
b) No, as shown by **(a)(ii)**
6. a) $0 \le x \le \frac{\pi}{2}$ b) $-1 \le f(x) \le 1$
c) $f^{-1}(x) = \frac{1}{2}\cos^{-1}x$
d) Domain: $-1 \le x \le 1$; Range: $0 \le f^{-1}(x) \le \frac{\pi}{2}$

Exercise 5.4 page 97

1. a) 180° b) 72° c) 360° d) 360°
e) 360° f) 60° g) 180° h) 180°
2. a)

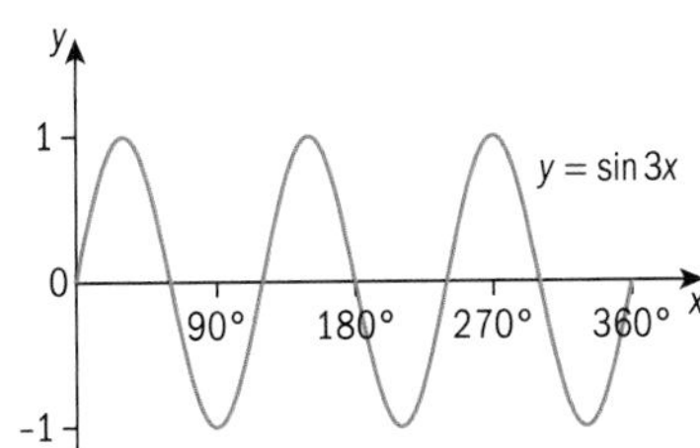

b)

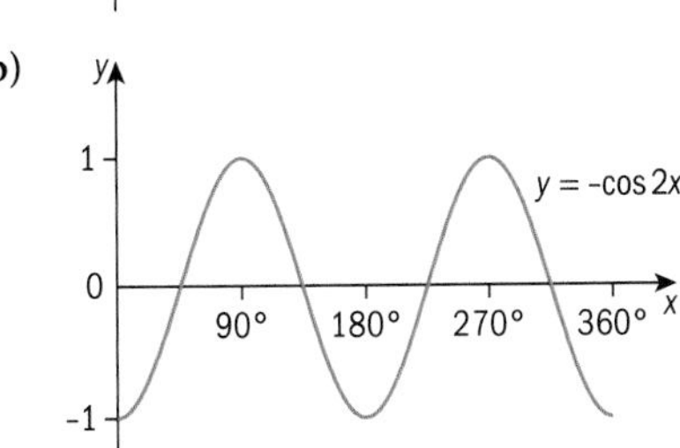

c)

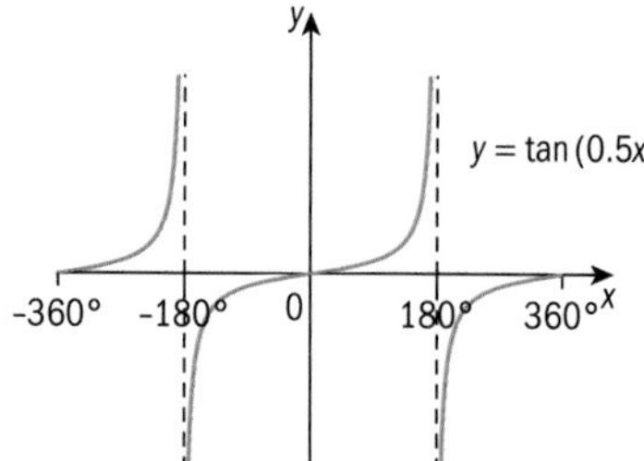

d)

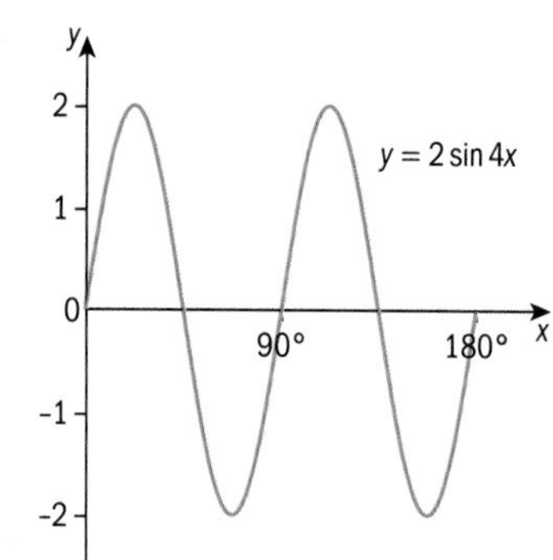

e)

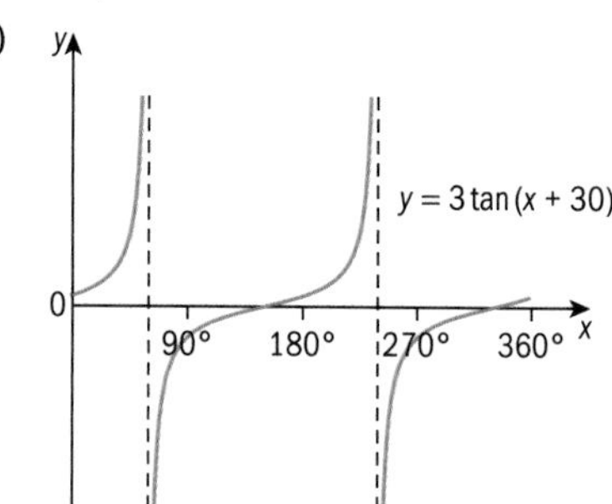

f)

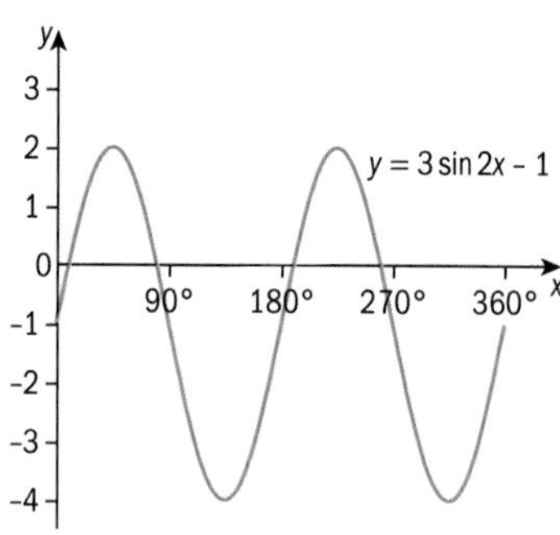

3. a)

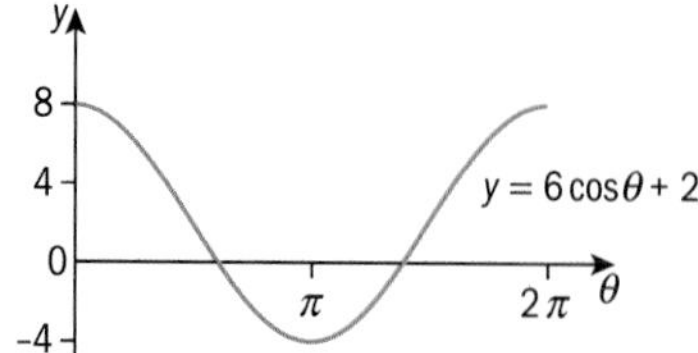

b)

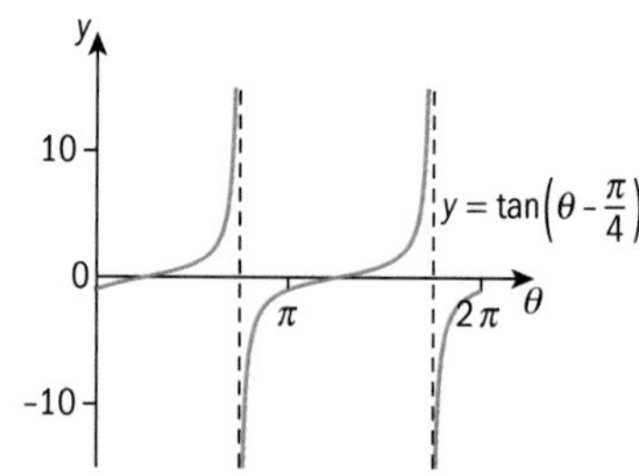

c)

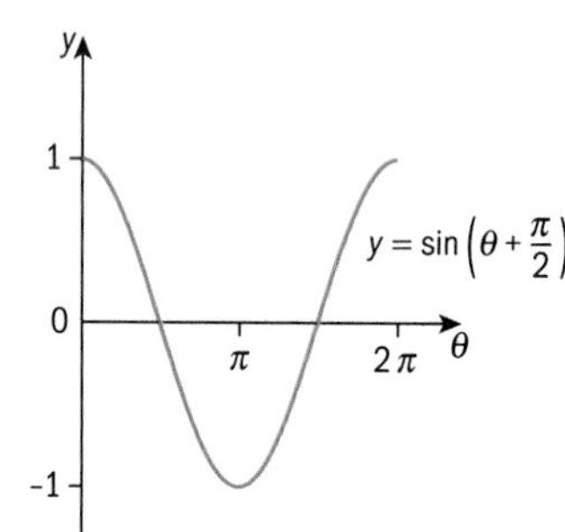

d)

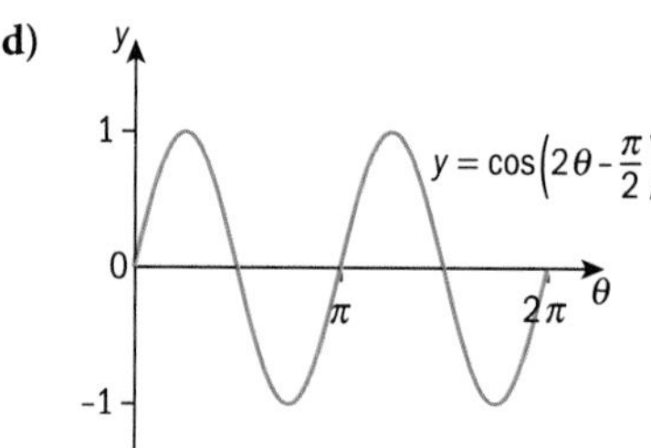

4. **a)** e.g. $y = \cos(x - 45)^\circ$ **b)** e.g. $y = \cos 3x^\circ$

c) e.g. $y = 4\sin 2x^\circ$ **d)** e.g. $y = \sin x^\circ - 1$

5. **a)**

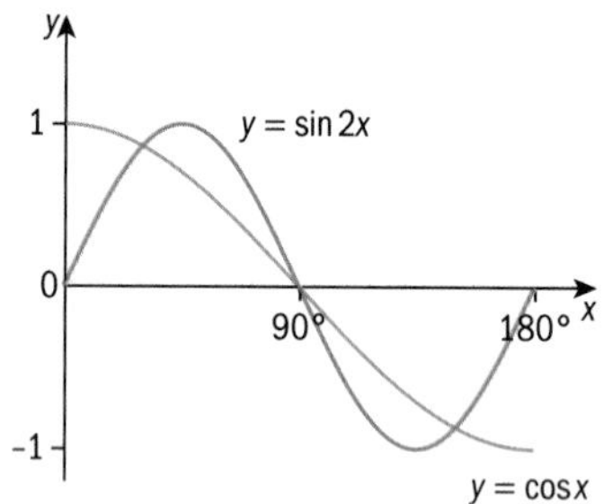

b) 3 **c)** 30°, 90°, 150°, 270°

Exercise 5.5 page 100

1. **a)** 48.6°, 131.4° **b)** 63.4°, 243.4°

c) 60°, 300° **d)** 18.7°, 121.3°

e) 70.5°, 289.5° **f)** 221.4°, 348.6°

2. a) 39.1°, 320.9° b) 100.7°, 169.3°, 280.7°, 349.3°
c) 21.1°, 81.1°, 141.1°, 201.1°, 261.1°, 321.1°
d) 224.5° e) 268.3°
f) 27.7°, 92.3°, 267.7°, 332.3°

3. a) $\frac{\pi}{4}, \frac{5\pi}{4}$ b) $\frac{\pi}{2}, \frac{5\pi}{6}$ c) $\frac{\pi}{12}, \frac{5\pi}{12}, \frac{13\pi}{12}, \frac{17\pi}{12}$
d) $\frac{\pi}{9}, \frac{5\pi}{9}, \frac{7\pi}{9}, \frac{11\pi}{9}, \frac{13\pi}{9}, \frac{17\pi}{9}$
e) 0.473, 1.26, 2.04, 2.83, 3.61, 4.40, 5.19, 5.97
f) 4.41

4. 2°, 10°, 14°, 22° 5. −120°, 60°

6. $\frac{-25\pi}{36}, \frac{-17\pi}{36}, \frac{-\pi}{36}, \frac{7\pi}{36}, \frac{23\pi}{36}, \frac{31\pi}{36}$

Exercise 5.6 page 101

1. $5\cos^2 x - 1$
2. a) $2\sin^2\theta - 1$ b) $2 - 4\sin\theta - 2\sin^2\theta$
3. a) 1 b) 2
4. Proof 5. a) – c) Proof
6. Proof
7. a) $\sin\theta + \cos\theta$ b) 1 c) $\sin\theta + \cos\theta$
8. a) – d) Proof
9. a) $(\cos x + 1)^2 - 2$ $p = 1, q = -2$
b) Maximum value = 2, minimum value = −2
10. a) Proof b) $\frac{3}{4}$

Exercise 5.7 page 104

1. 45°, 135°, 225°, 315° 2. 90°, 210°, 330°
3. 30°, 150°, 210°, 330° 4. 0°, 180°, 360°
5. 90° 6. 149.8°, 329.8°
7. 18.4°, 108.4°, 198.4°, 288.4°
8. 0°, 180°, 360° 9. $\theta = 1.19, 4.33$
10. $\theta = 2.90, 6.04$
11. a) $3\sin\theta$ b) $0, \frac{\pi}{6}, \frac{5\pi}{6}, \pi, 2\pi$
12. 0°, 63.4°, 135°, 180°

Summary exercise 5 page 105

1. a) $-\frac{3}{5}$ b) $\frac{4}{5}$ c) $-\frac{3}{4}$
2. 20.9°, 69.1° 3. 0.464
4. 72°, 108°, 252°, 288° 5. 30°, 150°, 228.6°, 311.4°
6. 74.4°, 165.6° 7. Proof 8. Proof

9. a)

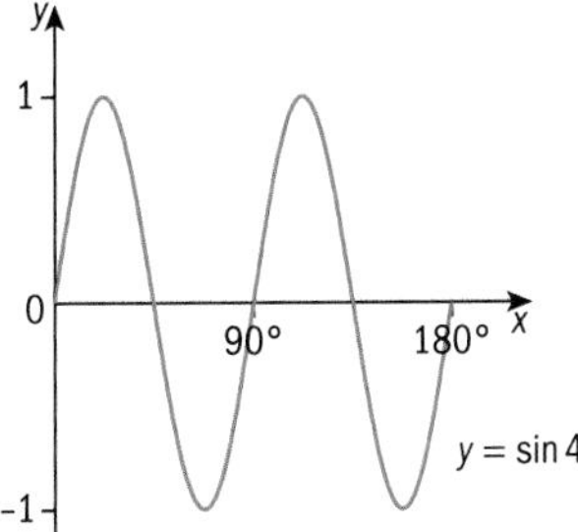

b) i) 15°, 30°, 105°, 120°
ii) 52.5°, 82.5°, 142.5°, 172.5°

10.

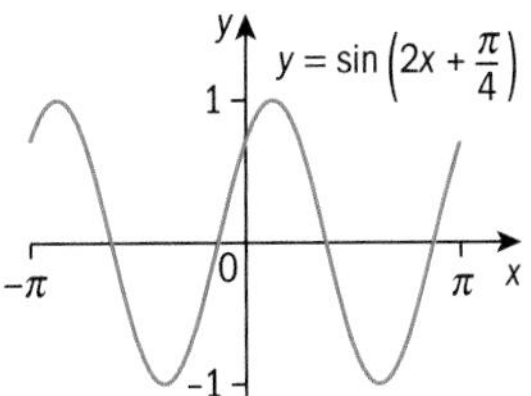

11. a) Proof b) 80°
12. a) 2.50, 8.78 b) 0, 1.25, 3.14, 4.39, 6.28
13. a) $3 - 7\cos^2 x$ b) 3, −4 c) 40.9°, 139.1°
14. a)

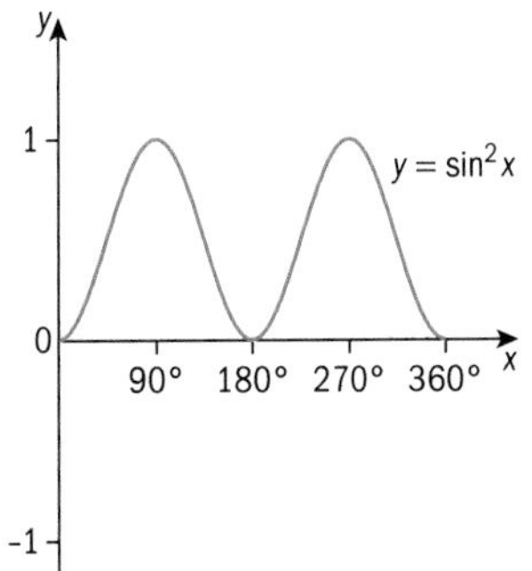

b) Proof c) $60° < x < 120°, 240° < x < 300°$
15. a) 109.5°, 250.5° b) 60°, 300°
c) 60°, 70.5°, 289.5°, 300°
d) 0°, 53.1, 180°, 306.9°, 360°
e) 45°, 135°, 225°, 315° f) 90°
16. a)

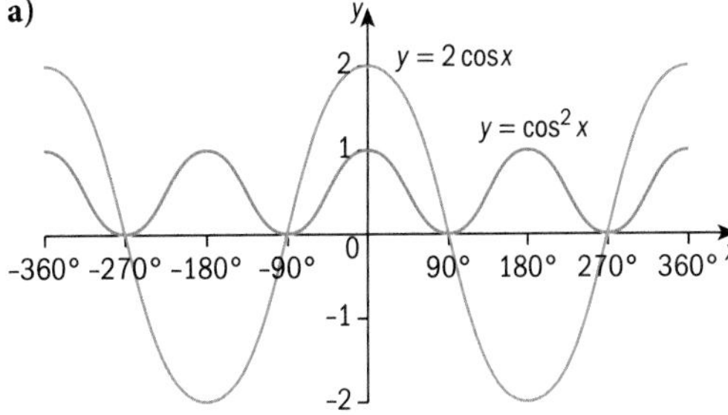

b) 4

17. a)

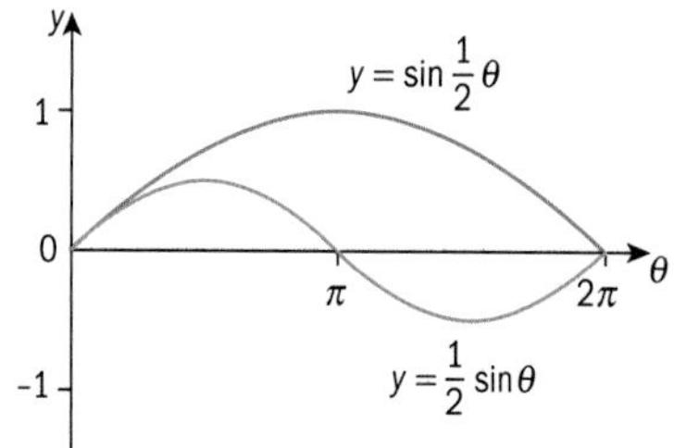

b) 2

18. a)

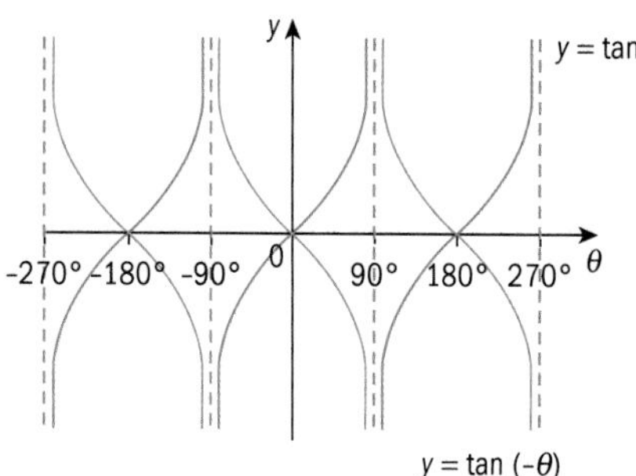

b) $-180°, 0°, 180°$

19. a)

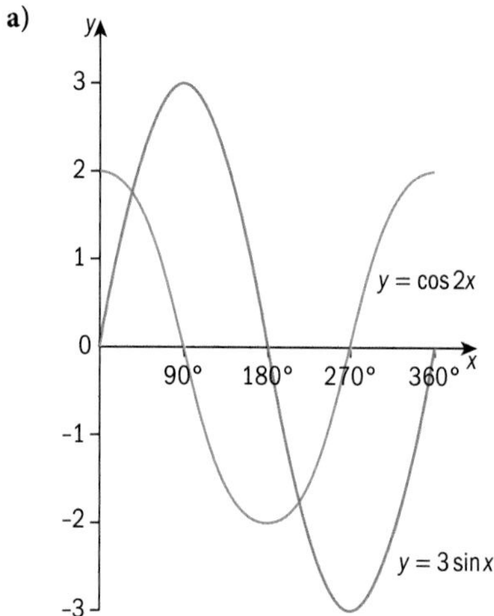

b) $33.7° \le x \le 213.7°$

20. a) i) Proof **ii)** 71.6°, 251.6°

iii) 91.6°, 271.6°

b) i) Proof

ii) 30°, 90°, 150°, 210°, 270°, 330°, 390°, 450°, 510°, 570°, 630°, 690°

21. i) Proof **ii)** 189.5° and 340.5°

22. i) Proof **ii)** 24.5° and 155.5°

iii) 54.5°

23. i) Proof **ii)** 10.9°, 100.9°

24. i) Proof **ii)** 60°, 180° **iii)** 120°

25. i) 30°

ii)

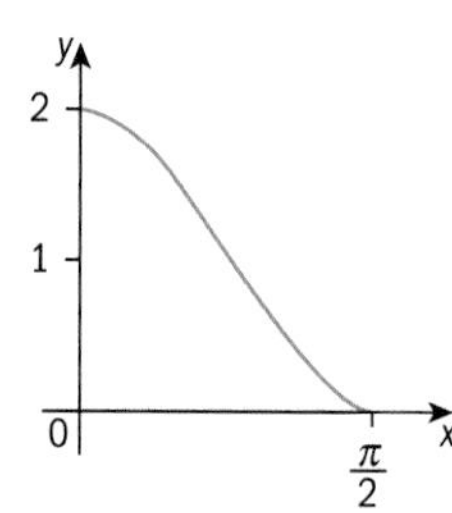

iii) $0 \le f(x) \le 2$

iv) $0 \le f^{-1}(x) \le \frac{\pi}{2}$

v)

6 Binomial expansion

Skills check page 109

1. a) $x^2 + 14x + 49$

b) $16x^2 - 8x + 1$

c) $8x^3 + 36x^2 + 54x + 27$

d) $25 - 10y + y^2$

e) $27x^3 - 270x^2 + 900x - 1000$

f) $1 - 6x + 12x^2 - 8x^3$

2. a) $16x^{40}$ **b)** $-27x^{12}$ **c)** $625x^{12}$

d) $-1024x^{35}$ **e)** $-3x^{45}$ **f)** $-135x^{18}$

Exercise 6.1 page 111

1. a) $16 + 96x + 216x^2 + 216x^3 + 81x^4$

b) $64x^6 + 192x^5 + 240x^4 + 160x^3 + 60x^2 + 12x + 1$

c) $1 - 20x + 160x^2 - 640x^3 + 1280x^4 - 1024x^5$

d) $x^7 - 21x^6 + 189x^5 - 945x^4 + 2835x^3 - 5103x^2 + 5103x - 2187$

2. $32 - 800x + 8000x^2$ **3.** $240x^4 + 24x^5 + x^6$

4. $n = 5$ **5.** 0.125 508 81

Exercise 6.2 page 112

1. a) 120 **b)** 3 628 800 **c)** 6840 **d)** 28

2. a) 84 **b)** 66 **c)** 7 **d)** 1

Exercise 6.3 page 114

1. $1 - 10x + 45x^2$

2. $6561 + 17\,496x + 20\,412x^2 + 13\,608x^3$

3. -7560

4. **a)** $81x^4 + 432x^3y + 864x^2y^2 + 768xy^3 + 256y^4$

b) $x^3 - 6x^2y + 12xy^2 - 8y^3$

c) $1 - \frac{5}{2}x + \frac{5}{2}x^2 - \frac{5}{4}x^3 + \frac{5}{16}x^4 - \frac{1}{32}x^5$

d) $16 + 160x + 600x^2 + 1000x^3 + 625x^4$

e) $\frac{x^3}{27} + \frac{x^2}{3} + x + 1$

f) $243 - \frac{1215}{x} + \frac{2430}{x^2} - \frac{2430}{x^3} + \frac{1215}{x^4} - \frac{243}{x^5}$

5. -8064 **6.** $90\,720$

7. $1 + 7ax + 21a^2x^2$ **8.** 20

9. $653\,184x^4$

10. **a)** $15\,625 - 18\,750x + 9375x^2$

b) $19\,683 + 118\,098x + 314\,928x^2$

c) $128 + \frac{448}{5}x + \frac{672}{25}x^2$

d) $1 - 8x + \frac{88}{3}x^2$

11. $96\,096$

12. **a)** $1 + 5x + \frac{45}{4}x^2 + 15x^3$ **b)** 1.63

13. **a)** $1 - 24x + 252x^2$

b) $2187 + 10\,206x + 20\,412x^2$

c) $32 + 16x + \frac{16}{5}x^2$

d) $1 - 4x + \frac{22}{3}x^2$

14. $1 - \frac{6}{a}x + \frac{15}{a^2}x^2 - \frac{20}{a^3}x^3$

15. $1.004\,006\,004\,001$ **16.** $-160y^3$

17. $48\,384$ **18.** 240

19. **a)** $1 + 5x + 10x^2 + 10x^3 + 5x^4 + x^5$ **b)** 15

20. **a)** $1 - 4x + 7x^2 - 7x^3$ **b)** 0.9227

Exercise 6.4 page 116

1. **a)** $2 + 17x + 64x^2$

b) $19\,683 + 78\,732x + 78\,732x^2$

2. $1 + 3x + \frac{5}{4}x^2$ **3.** $-160x^4$

4. 480 **5.** $\frac{1}{2}$

6. **a)** $2187 + 5103x + 5103x^2$

b) $2187 + 5103y + 10\,206y^2$

7. 240 **8.** 644

9. -48 **10.** $16 + 32x + 56x^2$

Summary exercise 6 page 117

1. $1 - 3x + \frac{15}{4}x^2$ **2.** $103\,680x^2$

3. 672 **4.** Proof

5. $p = -144 \quad a = \frac{3}{4}$ **6.** $n = 4$

7. $p = -4374 \quad q = \frac{5103}{4}$ **8.** $\frac{969}{2}x^5$

9. $8x + 8x^3$

10. **a)** $64 + 192x + 240x^2 + 160x^3 + 60x^4$

b) Proof

11. $1 : 3$ **12.** $-\frac{1760}{729x^7}$

13. **a)** $k = \frac{3}{2}, \quad a = 63, \quad b = 189$ **b)** 126

14. **a)** $1 + 6x + 12x^2 + 8x^3$

b) Proof **c)** $x = \pm\frac{1}{2}$

15. **a)** $1 + nax + \frac{n(n-1)}{2}a^2x^2$

b) Proof **c)** $\frac{3}{5}$

16. $a = 3, n = 8$

17. **a)** $1 + 5x + 10x^2 + 10x^3 + 5x^4 + x^5$

b) $p = 41, q = -29$

18. $38\frac{1}{2}$

19. $a = 5, \quad b = 8, \quad c = 44\,800$

20. $a = 10, \quad b = 2386$

7 Series

Skills check page 120

1. **a)** -1 **b)** 1 and -12

2. **a)** $5n - 3$ **b)** $-4n + 1$ **c)** $6n - 10$

d) $-9n + 8$

Exercise 7.1 page 122

1. **a)** 8, 11, 14 **b)** 3, 5, 13 **c)** $-4, -3, 0$

d) $-1, 3, \frac{3}{5}$ **e)** $7\frac{1}{2}, 7\frac{1}{5}, 7$ **f)** 1, 25, 5929

g) $-2, 4, -8$ **h)** $2, \frac{1}{2}, \frac{1}{5}$ **i)** 2, 25, 140

j) $-7, 28, -63$

2. 210

3. **a)** $6n + 1$ **b)** $10 - 3n$

c) $\frac{1}{n+1}$ **d)** $11n - 26$

4. 10

Exercise 7.2 page 123

1. a) 1, 4, 7 25 $3n - 2$
   b) 2, 4, 8 128 2^n
   c) 2, 6, 12 $n(n+1)$ $n(n+1)$
   d) 25, 36, 49 121 $(n+3)^2 + 2(n+3) + 1$
2. 31
3. a) $\sum_{r=1}^{\infty}(3r-1)$ b) $\sum_{r=1}^{11} r^2$ c) $\sum_{r=1}^{\infty}(11-4r)$
   d) $\sum_{r=1}^{6}(-1)^{r+1}x^{2r}$ e) $\sum_{r=1}^{8} r$ f) $\sum_{r=1}^{\infty}(3r+9)$
4. $n = 3$

Exercise 7.3 page 126

1. $a = 24 \quad d = -2$
2. 4
3. $a = 4 \quad S_{12} = 378$
4. 11
5. $3, 2\frac{1}{2}, 2 \quad u_{20} = -6\frac{1}{2}$
6. Proof
7. $d = 2 \quad S_{10} = 110$
8. 234
9. −2985
10. 16
11. Proof
12. 2400
13. −120
14. 150 cm
15. 20
16. a) 0 b) 150
17. a) 52, 50, 48 b) 53
18. a) $5x + 4$ b) $(5x+4)(8x+5)$
19. 32
20. 20 916

Exercise 7.4 page 130

1. a) $r = 3$ b) No c) No
   d) $r = \frac{1}{4}$ e) $r = -1$ f) $r = -0.2$
2. $S_{10} = 118\,096$
3. $a = \frac{5}{16}$; $r = -4$; $u_9 = 20\,480$
4. $r = 2$ or -2 and $a = 25$ or -25
5. 3.452…
6. $\frac{683}{2048}$
7. $r = \pm 0.2 \quad S_8 = 1.25$ (3 s.f.) or 0.833 (3 s.f.)
8. $\frac{45}{29}$
9. 364
10. $a = -\frac{1}{3}, r = -\frac{1}{3}$
11. $a = 4, u_8 = -512$
12. $(-2x)^{n-1}$
13. 20
14. 2, 6, 18, 54 $\quad r = 3$
15. 12 seconds
16. $10 737 418
17. 12
18. $r = 3, a = \frac{1}{4}$
19. $32 000
20. $\pm\sqrt{2}$

Exercise 7.5 page 132

1. $2\frac{1}{4}$
2. 23
3. $\frac{7}{10}$
4. $312\frac{1}{2}$
5. $\frac{4}{11}$
6. $x + 1$
7. a) −2, 4, −8 b) $|r| \geq 1$
8. $\frac{1}{2}$
9. $r = \frac{1}{3}$ or $r = -\frac{1}{3}$ $\quad S_\infty = 27$ or $13\frac{1}{2}$
10. $r = \frac{2}{3}$ or $r = -\frac{2}{3}$
11. $a = 60, r = \frac{2}{3}$

Summary exercise 7 page 133

1. −45, 120
2. $r = -\frac{1}{2}, \quad S_\infty = 53\frac{1}{3}$
3. 276
4. 16
5. $r = \frac{3}{4}$
6. $a = -4 \quad d = 5$
7. 49 205
8. $a = 12$ or $a = -2$
9. a) $a = \frac{2}{9}, r = 3$ b) Proof
10. $x = 12\frac{3}{4} \quad y = 81\frac{1}{4}$
11. $1, 4, \frac{10}{7}$
12. $624.32
13. $r = \frac{2}{5}$ and $S_\infty = -\frac{5}{7}$
14. $n = 34$
15. $r = \frac{3}{x}$
16. $n = 16$
17. 30 000
18. 795
19. a) 2, 6, 10
    b) i) $x = 1$ or $x = 7$
       ii) $|r| \geq 1$
20. $r = 4$
21. a) Proof b) $\frac{-1+\sqrt{5}}{2}$ c) $\frac{2a}{3-\sqrt{5}}$
22. a) $\frac{4}{5}$ b) 15 c) 10.7%
23. 3420
24. 4275
25. a) Proof b) Proof
    c) 15 345

8 Differentiation

Skills check page 138

1. a) $2x^{-3}$ b) $4x^{\frac{1}{2}}$ c) $2x^{-\frac{1}{2}}$
   d) $3x^{-\frac{1}{3}}$ e) $x^{\frac{5}{2}}$ f) $x^{-\frac{3}{2}}$

2. a) x^7 b) x^{-4} c) $x^{\frac{1}{2}}$

d) $\frac{1}{x^2}$ e) x^{-5} f) $x^{\frac{1}{2}}$

g) $x^{-\frac{2}{3}}$ h) $\frac{1}{x^3}$ i) $\frac{1}{\sqrt{x^9}}$

Exercise 8.1 page 139

1. a)

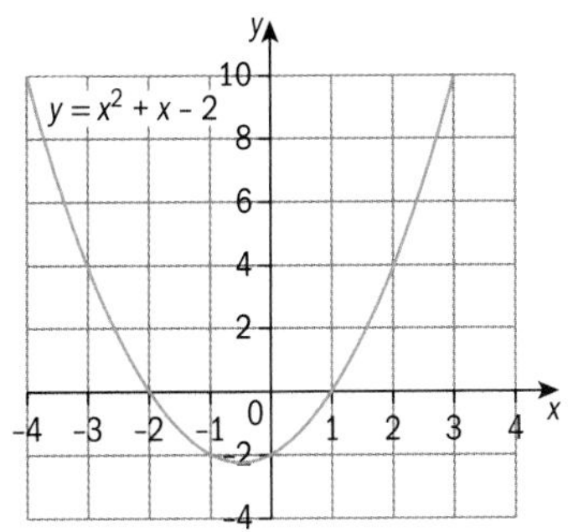

b) 5 c) -1

2) a) 3 b) $(1, -1)$, $(-1, 3)$

Exercise 8.2 page 141

1. $4x$ **2.** 8

3. $4x^3$ **4.** $2x + 3$

5. $-\frac{1}{x^2}$ **6.** nx^{n-1}

Exercise 8.3 page 142

1. a) $3x^2$ b) $9x^8$ c) $8x$ d) -15

e) $-28x^{-5}$ f) $-3x^{-2}$ g) $-\frac{1}{2}$ h) $8x^9$

i) $\frac{1}{2}x^{-\frac{1}{2}}$ j) $-2x^{-\frac{1}{2}}$ k) $-\frac{2}{x^2}$ l) $-\frac{14}{x^3}$

2. a) $x^{-\frac{1}{2}}$ b) $\frac{5}{x^2}$ c) $\frac{1}{6\sqrt{x}}$ d) $84x^6$

e) $\frac{-24}{x^4}$ f) $-\frac{3}{2}x^{-\frac{3}{2}}$ g) $-\frac{1}{3}x^{-\frac{4}{3}}$ h) $\frac{2}{9}$

i) $6x^5$ j) $28x^6$ k) $60x^2$ l) -4

3. a) $-5x^{-6}$ b) $\frac{1}{3}x^{-\frac{2}{3}}$ c) $-x^{-\frac{5}{4}}$ d) -2

e) $-\frac{2}{x^3}$ f) $\frac{5}{2}x^{-\frac{1}{2}}$ g) $\frac{3}{2x^4}$ h) $\frac{4}{3}x^{-\frac{1}{3}}$

i) $-5x^{-6}$ j) $6x^5$ k) $11x^{10}$ l) $32x^7$

4. a) $8x$ b) $-\frac{2}{3}$ c) $15x^{20}$ d) $8x^{-\frac{1}{2}}$

e) $-3x^{-\frac{4}{3}}$ f) $2x^{-\frac{3}{2}}$ g) -5 h) $5x^{-\frac{4}{5}}$

i) $100x^9$ j) 1 k) $15x^4$ l) -1

Exercise 8.4 page 144

1. a) $7x^6 + 3x^2$ b) $48x^7 - 4$

c) $6x + 5$ d) $12x - 13$

e) $-2x^{-3} + 2x$ f) $60x^3 - 105x^2 - 20x$

g) $3x^2 + 6x^5$ h) $36x^3 - 4x$

i) $2x - 8x^{-5}$ j) $-10 + 2x$

2. a) $4x^3 - \frac{1}{x^2} + \frac{3}{2x^{\frac{3}{2}}}$ b) $20x^3 - \frac{9}{2}x^2$

c) $-\frac{3}{x^2} - \frac{1}{2x^{\frac{3}{2}}}$ d) $1 - \frac{1}{x^2}$

e) $6x^2 - 32x^3 - 80x^4$ f) $2 + \frac{11}{2x^{\frac{1}{2}}}$

g) $-\frac{3}{2}x^{-\frac{1}{2}}$ h) $-\frac{1}{2}x^{-\frac{3}{2}} - 6x^{-\frac{5}{2}} - \frac{1}{2}x^{-\frac{1}{2}}$

i) $\frac{12}{x^2} - \frac{18}{x^3}$ j) $-\frac{4}{x^2} - \frac{6}{x^3}$

3. a) $12x^3 - 6x^2 + 10x - 1$ b) $\frac{2}{3}x^{-\frac{1}{3}} + \frac{1}{2}x^{-\frac{1}{2}}$

c) $12 + 6x^{\frac{1}{2}}$ d) $-\frac{14}{x^3} + \frac{12}{x^4} - 4x$

e) $4x^3 + 28x$ f) $2ax + b$

g) $\frac{5}{2}x^{\frac{3}{2}} + 4x^{-\frac{1}{2}}$ h) $-12x^{-4} + \frac{2}{3}x^{-3}$

4. a) $-2x^{-\frac{4}{3}} + 2x^{-\frac{1}{2}}$ b) $\frac{1}{2}x^{\frac{1}{2}} - \frac{1}{2}x^{\frac{3}{2}}$

c) $3x^2 - 8x + 4$ d) $\frac{1}{2}x^{-\frac{1}{2}} - \frac{3}{2}x^{-\frac{3}{2}}$

e) $-4x^{-2} + 5x^{-4}$ f) $1 - 4x^{-\frac{1}{2}}$

g) $5 + \frac{1}{4}x^{-\frac{3}{2}}$ h) $\frac{1}{2}x^{-\frac{1}{2}} - 2 + \frac{3}{2}x^{\frac{1}{2}}$

Exercise 8.5 page 146

1. a) $\frac{2}{\sqrt{4x-1}}$ b) $10x(x^2+4)^4$ c) $\frac{3}{(1-x)^4}$

d) $-\frac{10}{(2x+7)^2}$ e) $-3(5+2x)^{-\frac{3}{2}}$ f) $\frac{8}{(3-4x)^2}$

g) $-\frac{4}{3}(4x-3)^{\frac{-4}{3}}$ h) $\frac{9x^2}{2\sqrt{5+3x^3}}$ i) $-\frac{6(4x+3)}{(2x^2+3x)^2}$

j) $\frac{12x}{(1-3x^2)^5}$

2. a) $-\frac{9}{(9x+8)^2}$ b) $-\frac{7}{2\sqrt{8-7x}}$ c) $\frac{4x^3}{\sqrt{2x^4+7}}$

d) $\frac{3x}{\sqrt{(3-x^2)^3}}$ e) $-\frac{16}{(4x-5)^5}$ f) $-\frac{3}{2}(5+3x)^{-\frac{5}{4}}$

g) $24x(3x^2-2)^3$ h) $\frac{5-4x}{2\sqrt{5x-2x^2}}$ i) $\frac{9x^2-5}{(5x-3x^3)^2}$

j $-\frac{1}{2\sqrt{x}(1+\sqrt{x})^2}$

Exercise 8.6 page 148

1. 20 **2.** 44

3. (2, 4) and (−3, 9) **4.** 9 and −1

5. 12 **6.** 5 and −5

7. (2, −7) and (−2, 21) **8.** 4 and $-\frac{10}{3}$

9. (0, 0) and (−1.5, −1.6875) **10.** −1

11. 25; −10 **12.** 10

13. (0, 6) and (−2, −16) **14.** 5

15. (100, 2.05)

Exercise 8.7 page 150

1. a) $84x^5 + 24x$ **b)** $216x^7$ **c)** 4

d) $6x^{-4} + 60x^2$ **e)** $240x^3 - 72x^2 + 36x$

f) $20(9 - x)^3$ **g)** $6x + 2$ **h)** $96x^2$

i) $12x + \frac{6}{x^3}$

2. a) $8x^{-3} - 3x^{-4}$ **b)** $-10x^{-3} + \frac{3}{4}x^{-\frac{5}{2}}$

c) $2x^{-\frac{3}{2}}$ **d)** $\frac{1}{4}x^{-\frac{3}{2}}$ **e)** $\frac{5}{2}x^{-\frac{9}{4}}$

f) $\frac{-4}{\sqrt{(4x-3)^3}}$ **g)** $60x^{-6} + 36x^{-4}$

h) $56x^6$ **i)** $\frac{16}{3}x^{-\frac{7}{3}}$

3. −97 **4.** $\frac{27}{32}$ **5.** $-14x$

6. Proof **7.** $\frac{80}{(5-2x)^9}$

8. a) 129 **b)** 130 **c)** $-\frac{1}{2}$

9. 1 **10.** −11

Exercise 8.8 page 153

1. a) $y = 6x - 13$ **b)** $y = 45x + 48$

c) $x + y + 29 = 0$ **d)** $3x + y = 1$

e) $4y - 79x + 196 = 0$ **f)** $x - 12y + 16 = 0$

2. a) $x + 20y = 422$ **b)** $x + 13y + 79 = 0$

c) $x + 6y = 3$ **d)** $30y - x = 1495$

e) $8x + 7y + 9 = 0$ **f)** $y = 9x - 35$

3. $y - 6x = -5$ **4.** $y = 23x - 32$; $x + 23y = 324$

5. $7x + y + 21 = 0$; $y = 7x - 28$

6. a) (2, 4) and (−1, 4)

b) $y = 3x - 2$ and $y = -3x + 1$ **c)** Proof

7. a) $y = -4x + 6$ **b)** 4.5 square units

8. (1.5, 24.5)

9. $p = \frac{8}{3}$ and $q = \frac{16}{5}$ **10.** (−3, 3)

Summary exercise 8 page 154

1. a) $3x^2 + 12x + 9$ **b)** $3x^{-\frac{1}{2}} - x^{-\frac{4}{3}}$

c) $4x^3 - 4$ **d)** $-2x(3 - 2x^2)^{-\frac{1}{2}}$

2. a) $-3x^{-\frac{4}{3}} - \frac{1}{2}x^{-\frac{3}{2}}$ **b)** $2x^3 - 10x^4 + 6x$

c) $\frac{-30}{(5x-2)^3}$ **d)** $1 - \frac{1}{x^2}$

3. 0 **4.** $2y + 3x = 33$

5. $a = -9$ **6.** Proof

7. $2x + 3y - 9 = 0$ **8.** (−3, −4) and (1, 0)

9. $6ty + x = 18t^3 + t$

10. a) $y = 2x$ **b)** $\left(1\frac{1}{4}, \frac{15}{16}\right)$

11. $7 - 4x^{-\frac{1}{2}}$ **12.** Proof

13. $y = 3x + 6$; $y = 3x + 2$

14. a) $3y = 2x + 18$
$2y = -3x + 51$

b) 156 square units

15. a) −12 **b)** 12 **c)** −3

16. $-13\frac{3}{4}$

9 Further differentiation

Skills check page 156

1. a) −1 **b)** −2 **c)** $\frac{1}{2}$

d) $-\frac{2}{3}$ **e)** 9 **f)** $-\frac{2}{3}$

Exercise 9.1 page 157

1. a) $x > 3$ **b)** $x < \frac{3}{4}$

c) $x < -4$ and $x > 4$ **d)** $x < 1$ and $x > 2$

e) $0 < x < 1$ **f)** $x < -1$ and $x > 1$

2. a) $x < \frac{1}{2}$ **b)** $x > -1$ **c)** $x < -1$

d) $\frac{1}{3} < x < 3$ **e)** $-3 < x < 3$ **f)** $-1 < x < 5$

3. a) $x > -\frac{1}{2}$ **b)** $x < 2$ **c)** $x < -1$ and $x > 1$

d) $x > \frac{1}{2}$ **e)** $1 < x < 2$

f) $x < 2 - \sqrt{2}$ and $x > 2 + \sqrt{2}$

4. a) $x < 0$ and $x > 3$ **b)** $x > -\frac{1}{4}$

c) No value of x **d)** $-1 < x < 1$

e) $x < 0$ and $x > \frac{2}{9}$ **f)** $2 < x < \frac{20}{3}$

Exercise 9.2 page 161

1. a) $(1, 4)$ = minimum T/P
   b) $\left(\frac{1}{2}, 3\frac{1}{4}\right)$ = maximum T/P
   c) $(-1, -5)$ = minimum T/P
   d) $\left(\frac{1}{3}, \frac{40}{27}\right)$ = maximum T/P
   $(3, -8)$ = minimum T/P
   e) $(0, -2)$ = maximum T/P; $(1, -3)$ = minimum T/P; $(-1, -3)$ = minimum T/P
   f) $(2, 41)$ = maximum T/P
   $(-2, -23)$ = minimum T/P
2. a) $(4, -4)$ = minimum T/P
   b) $\left(\frac{2}{3}, \frac{4}{27}\right)$ = maximum T/P; $(0, 0)$ = minimum T/P
   c) $(0, 0)$ = inflexion; $(3, 27)$ = maximum T/P
   d) $(1, 6)$ = minimum T/P
   e) $(0, 0)$ = inflexion; $\left(\frac{3}{2}, -\frac{27}{16}\right)$ = minimum T/P
   f) $(3, 47)$ = maximum T/P;
   $\left(-5, -38\frac{1}{3}\right)$ = minimum T/P
3. a) $(0, 0)$ = maximum T/P $(1, -1)$ = minimum T/P
   b) $(1, 9)$ = maximum T/P $(2, 8)$ = minimum T/P
   c) $(-1, -2)$ = maximum T/P $(1, 2)$ = minimum T/P
   d) $(3, 3)$ = minimum T/P $(-3, -3)$ = maximum T/P
4. $\left(\frac{1}{4}, \frac{1}{4}\right)$ = maximum T/P
5. $\left(\frac{1}{3}, -4\right)$ maximum T/P; $(1, -8)$ minimum T/P
6. $\left(\frac{1}{3}, 27\right)$ = minimum T/P
7. $(0, -2)$ minimum T/P; $\left(\frac{1}{2}, -\frac{31}{16}\right)$ maximum T/P; $(1, -2)$ minimum T/P
8. Proof
9. a) $x = -\frac{1}{3}$ b) Sketch
10. $a = \frac{3}{2}$ $b = -\frac{3}{2}$ $c = -3$

Exercise 9.3 page 164

1. $800\,\text{m}^2$ 2. $500\,\text{cm}^3$ 3. $6\frac{2}{3}$ 4. -7
5. $p = 0$ $r = 222.8$ 6. $\frac{1024}{27}$
7. a) $\pi r^2 + \frac{1024\pi}{r}$ b) 192π
   c) $r = 8$ cm, height = 8 cm
8. $25\,\text{cm}^2$ 9. $r = h = 1$ m
10. $19.45\,\text{cm}^2$ 11. 250
12. $3 \times 3 \times 1.5$ m

Exercise 9.4 page 167

1. $64\pi\,\text{cm s}^{-1}$ 2. $10\pi\,\text{cm s}^{-1}$ 3. $9.6\,\text{m}^3\,\text{s}^{-1}$
4. $0.028\,\text{cm s}^{-1}$ 5. $14.4\,\text{cm}^2\,\text{s}^{-1}$ 6. $586.4\,\text{cm}^3\,\text{s}^{-1}$
7. $6.86\,\text{cm}^2\,\text{s}^{-1}$ 8. $0.375\,\text{m}^3\,\text{s}^{-1}$ 9. $18\,\text{cm}^3\,\text{s}^{-1}$
10. $0.08897\,\text{cm}^2\,\text{s}^{-1}$

Summary exercise 9 page 169

1. $x > \frac{1}{3}$ 2. $x < 0$ and $x > 1$
3. $-1 < x < \frac{1}{3}$ 4. $p = 6$
5. $(1, 3)$ = minimum T/P
6. a) $(1, -1)$ = minimum T/P; $(-2, 26)$ = maximum T/P
   b) $x < -2$ and $x > 1$
   c)

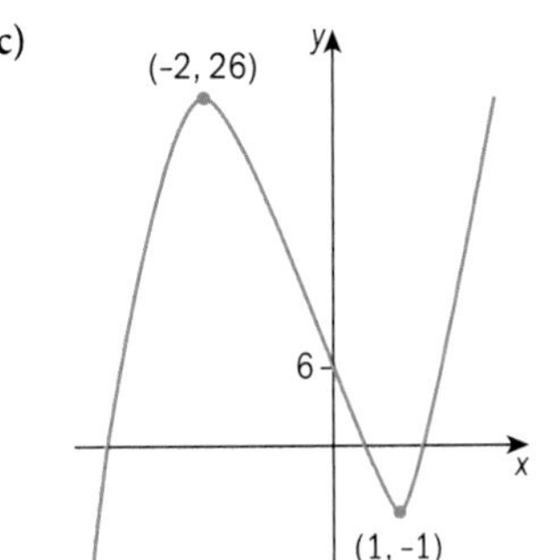

7. $\left(\frac{1}{2}, 0\right)$ = minimum T/P; $\left(-\frac{1}{2}, 2\right)$ = maximum T/P
8. a) Proof
   b) Maximum $A = \left(\frac{5000}{4+\pi}\right)\text{m}^2$ = approximately $700\,\text{m}^2$
9. Proof 10. $256\,\text{cm}^3$
11. a) Proof b) $553.6\,\text{cm}^2$
12. a) $k = 4$
   b) $(1, -9)$ is a minimum point; $\left(-\frac{4}{3}, \frac{100}{27}\right)$ is a maximum point
13. $128\pi\,\text{m}^3\,\text{s}^{-1}$ 14. $5\,\text{cm}^2\,\text{s}^{-1}$
15. $0.0212\,\text{cm s}^{-1}$
16. 40 cm
17. i) $(1, -11)$ and $\left(\frac{7}{3}, -\frac{329}{27}\right)$ ii) $y = 7x - 14$
   iii) $x < 1$ and $x > \frac{7}{3}$
18. i) Proof ii) $-\frac{3}{4} < x < \frac{2}{3}$
19. i) $y = 10x - \frac{3}{x^2}$ ii) 0.77 units per second
20. i) $f'(x) = -(4-3x)^{-\frac{2}{3}}$; $f''(x) = -2(4-3x)^{-\frac{5}{3}}$
   ii) $k = -\frac{1}{2}$

10 Integration

Skills check page 173

1. a) 33 b) 21 c) −1
   d) 54 e) 50
2. a) 0 and 2 b) 5 and −2
   c) 4 and −3 d) 0, −6 and 1
   e) 1 and $-\frac{4}{3}$

Exercise 10.1 page 175

1. a) $x^2 + c$ b) $\frac{x^8}{8} + c$
   c) $x^3 + c$ d) $-15x + c$
   e) $\frac{x^2}{2} - \frac{x^4}{4} + c$ f) $5x^2 + x^8 + c$
   g) $5x - \frac{x^2}{4} + c$ h) $\frac{x^6}{6} - 3x^2 + c$
   i) $\frac{2x^{\frac{3}{2}}}{3} + c$ j) $\frac{4}{3}x^3 - 6x^2 + 9x + c$
   k) $\frac{2x^5}{5} + 7x + c$ l) $-\frac{12}{x} + \frac{3}{x^2} + c$
2. a) $\frac{3x^{\frac{3}{2}}}{4} + c$ b) $-x^2 + c$
   c) $\frac{2x^{\frac{3}{2}}}{9} + c$ d) $2x^6 + c$
   e) $3x - \frac{x^2}{2} - \frac{x^6}{3} + c$ f) $\frac{-1}{3x^3} - \frac{10}{x} + 25x + c$
   g) $6x^{\frac{2}{3}} + c$ h) $8x^3 - 8x + c$
   i) $\frac{x^3}{3} + \frac{3x^2}{2} - 28x + c$ j) $-\frac{5}{6}x + c$
   k) $\frac{x^3}{3} - \frac{1}{x} + c$ l) $2x^7 + \frac{3}{x} + c$
3. a) $3x^2 + c$ b) $2x^2 + x + c$
   c) $8x^{\frac{1}{2}} + c$ d) $-x^{-7} + c$
   e) $\frac{x^3}{3} + 4x^2 + 16x + c$ f) $-4x^{-\frac{1}{3}} + c$
   g) $9x - 3x^2 + c$ h) $2x + \frac{5x^3}{3} + c$
   i) $x^2 - \frac{4x^3}{3} + \frac{x^4}{2} + c$ j) $\frac{8x^{\frac{5}{2}}}{5} + \frac{8x^{\frac{3}{2}}}{3} + 2x^{\frac{1}{2}} + c$
   k) $6x^{\frac{1}{2}} - \frac{2x^{\frac{5}{2}}}{5} + c$ l) $\frac{2x^{\frac{5}{2}}}{5} + 5x^2 + \frac{50x^{\frac{3}{2}}}{3} + c$

Exercise 10.2 page 178

1. a) $y = x^3 - 3x^2 + 2x + 14$ b) $y = x - 2x^2 + \frac{4x^3}{3} + \frac{23}{3}$
   c) $y = \frac{2x^3}{3} + \frac{5x^2}{2} - \frac{881}{6}$ d) $y = \frac{x^2}{2} - 2x^{\frac{3}{2}} + \frac{51}{2}$
   e) $y = 3x^3 - 3x + 364$ f) $y = 5x^3 + \frac{x^2}{2} - 2x + 298$
2. $y = 10x^{\frac{1}{2}} - 4x^{\frac{5}{2}} - 12$ 3. $y = -\frac{3}{x^2} + 2x - \frac{193}{49}$
4. $y = \frac{3x^2}{2} - \frac{x^3}{3} - \frac{29}{3}$ 5. $y = -\frac{4x^3}{3} + 4x + \frac{5}{3}$
6. $f(x) = 2x^4 - 4x + 6\sqrt{x} - 505$
7. $y = -\frac{x^3}{2} + x^2 + \frac{17x}{2} + 7$
8. a) $y = 2x^3 - 2x^2 + 3x - 35$ b) Proof
9. $y = x^3 - 3x - 20$ 10. $y = 3x^2 - 2x$

Exercise 10.3 page 180

1. a) $\frac{(2x-1)^7}{14} + c$ b) $\frac{(4-3x)^9}{-27} + c$
   c) $\frac{(5x+2)^6}{30} + c$ d) $\frac{1}{-12(3x+5)^4} + c$
   e) $\frac{1}{(1-3x)^5} + c$ f) $-\frac{1}{8(5+2x)^8} + c$
   g) $\frac{6}{7}\sqrt{7x+1} + c$ h) $-\frac{2}{\sqrt{6x-5}} + c$
   i) $-\frac{3(7-x)^{\frac{2}{3}}}{2} + c$ j) $-2(1-x)^{\frac{3}{2}} + c$
   k) $\frac{1}{3(1-2x)^6} + c$ l) $\frac{2(2+3x)^{\frac{7}{2}}}{21} + c$
2. $y = \frac{(3x-4)^6}{18} + \frac{89}{18}$ 3. $y = \frac{-(7-x)^5}{5} + \frac{17}{5}$
4. $f(x) = \frac{-1}{15(5x-3)^3} - 89\frac{119}{120}$
5. $y = \frac{2}{9}\left(\frac{1}{4}x + 1\right)^9 - 122x + 374\frac{2}{9}$

Exercise 10.4 page 181

1. a) 3 b) $\frac{117}{3}$ c) $12\frac{2}{3}$
   d) 0 e) $\frac{-9}{2}$ f) 2
2. a) 15 b) 17 c) $\frac{4}{5}$
   d) 0 e) 70 f) $\frac{182}{3}$
3. a) $4\frac{2}{3}$ b) $\frac{53}{72}$ c) $3\frac{1}{3}$
   d) $141\frac{1}{3}$ e) $\frac{2}{3}$ f) $\frac{33\sqrt{3} - 28\sqrt{2}}{15}$

Exercise 10.5 page 186

1. a) 8 b) 36 c) $\frac{3}{4}$
   d) $\frac{3}{2}$ e) $2\frac{2}{3}$ f) 27
2. a) 4 b) $\frac{37}{12}$ c) $145\frac{1}{12}$ d) $\frac{2}{3}$
3. $\frac{26}{3}$ 4. $\frac{17}{4}$ 5. $\frac{32}{3}$ 6. $\frac{104}{3}$
7. a) Draw graph, area = 12.5
   b) 12.5

8. $36\frac{2}{3}$

9 2

10. a) (3, 0), (5, 0) b) Proof c) $59\frac{3}{4}$

Exercise 10.6 page 190

1. a) $\frac{4}{3}$ b) $49\frac{1}{2}$

2. $\frac{1}{3}$ 3. $10\frac{2}{3}$ 4. $\frac{2}{15}$ 5. $10\frac{2}{3}$

6. a) Proof b) $9\frac{13}{27}$

7. $\frac{4}{3}$ 8. $\frac{1}{8}$ 9. $121\frac{1}{2}$ 10. 72

Exercise 10.7 page 194

1. a) Finite value of 8 b) No value

2. a) $\frac{p^4}{4} - 4$ b) $\frac{1}{8} - \frac{1}{2p^2}$

 c) No value d) Finite value of $\frac{1}{8}$

3. a) $-\frac{4}{b^{\frac{3}{2}}} + \frac{4}{a^{\frac{3}{2}}}$

 b) i) No value ii) Finite value of $\frac{4}{27}$

4. a) No value

 b) Finite value of 1

5. Integrand is not defined at $x = 0$

6. a) Finite value of 60 b) No value

 c) No value

Exercise 10.8 page 198

1. $\frac{512}{9}\pi$ 2. 8π 3. $\frac{25}{2}\pi$ 4. $\frac{1}{2}\pi$

5. 12π 6. 8.1π 7. $\frac{48}{5}\pi$ 8. $\frac{3}{2}\pi$

9. a) (0, 0), (1, 1) b) Sketch c) $\frac{\pi}{6}$

10. 104 11. $\frac{15}{4}\pi$ 12. $\frac{64}{3}\pi$ 13. $45\frac{5}{7}\pi$

Summary exercise 10 page 201

1. a) $4x^3 + c$ b) $\frac{-3}{x} - x^2 + c$

 c) $36x - 6x^2 + \frac{x^3}{3} + c$ or $-\frac{1}{3}(6 - x)^3 + c$

 d) $\frac{4}{3}x^{\frac{3}{2}} + 2x^2 + \frac{4}{5}x^{\frac{5}{2}} + c$ e) $\frac{-1}{x} + x^5 + c$

 f) $3x^{\frac{2}{3}} - 4x^{\frac{7}{2}} + c$

2. a) $y = 3x^3 - x^2 + x + 98$ b) $y = x^4 - 5x - 8$

 c) $y = \frac{x^2}{2} + 4x^{\frac{3}{2}} + 9x - 74$

 d) $y = 4x^2 + \frac{7}{x} - 15$

3. $\frac{2x^3}{3} - \frac{x^2}{2} + 2x^{\frac{3}{2}} - \frac{152}{3}$

4. $y = x^3 - 4x^2 + 5x - 50$

5. a) $\frac{40}{3}$ b) $\frac{23}{8}$ c) $-\frac{32}{3}$ d) $\frac{14}{3}$

6. $-\frac{85}{1024}$

7. a) Finite value of 8 b) No value

8. $\frac{1}{p^3} - \frac{1}{q^3}$

9. a) No value b) Finite value of 1

10. a) – c) Integrals not defined at $x = 0$

 d) i) No value

 ii) No value iii) Finite value of 54

11. 1 12. 5 13. $\frac{4}{3}$

14. $\frac{4}{3}$ 15. $\frac{125}{6}$ 16. 0

17. a) Sketch

 b) $\frac{1}{2}$ c) $\frac{8}{105}\pi$

18. $\frac{128\pi}{105}$

19. a) $\frac{9\pi}{2}$ b) $\frac{53\pi}{15}$

20. 6π 21. 21

22. a) Proof b) $\frac{32}{3}$ c) $81\frac{1}{15}\pi$

23. $\frac{52\pi}{5}$ 24. 2

Exam-style paper A – Pure 1 page 208

1. Centre (3, −2), radius 4

2. $\left(\frac{1}{2}, -3\right), \left(\frac{3}{2}, -1\right)$

3. a)

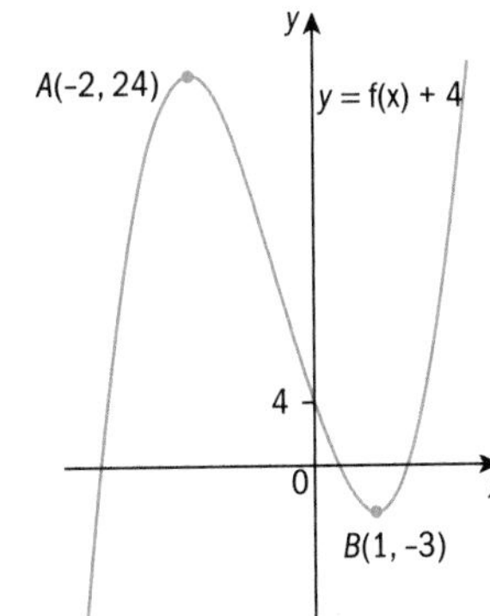

b)

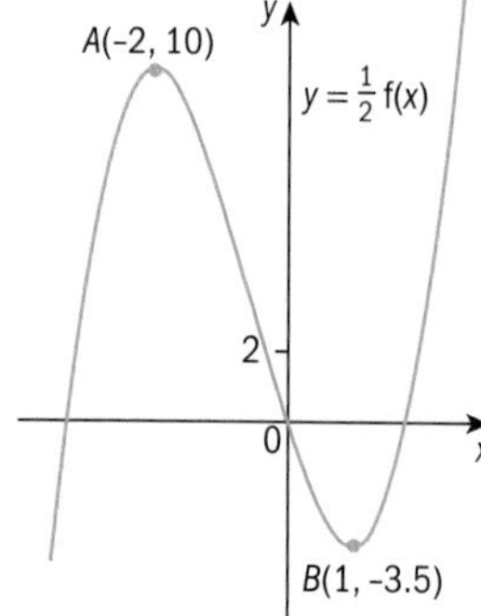

4. $\frac{3}{2}$

5. a) $729x^6 - 2916x^4 + 4860x^2$

b) -3888

6. a) $2 \le f(x) \le 12$

b)

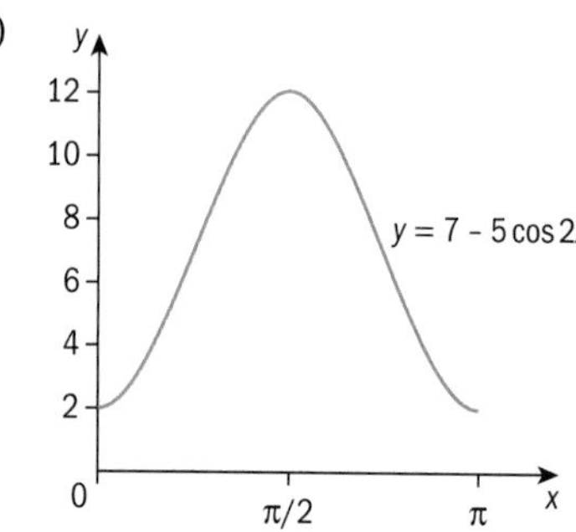

c) $fg(x) = 7 - 5\cos(2x + \pi)$

7. a) $\sqrt{29}$ b) $k = -1$

8. $\frac{r^2}{24}(6\sqrt{3} - \pi)$

9. a) 0°, 41.4°, 318.6°, 360° b) 21.4°, 298.6°, 340°

10. a) Proof

b) i) $a = -1, \quad b = -\frac{1}{2}, \quad c = 6\frac{1}{4}$

ii) $6\frac{1}{4}$ when $x = \frac{1}{2}$

11. a) $20x^3 + \frac{3}{\sqrt{x}} + \frac{3}{x^3}$ b) -2

12. a) $P = (6, 0)$ b) $(2, 32)$ c) 108

Exam-style paper B – Pure 1 page 211

1. $x < -\frac{1}{2}$ and $x > \frac{5}{3}$

2. $\frac{1}{7}(13 + 2\sqrt{2})$

3. a) Translation by vector $\begin{pmatrix} 0 \\ 3 \end{pmatrix}$

b) Stretch of factor $\frac{1}{4}$ parallel to the x-axis

4. $2n - 1$

5. a) $(x - 4)^2 + (y + 2)^2 = 25$

b) $y = -\frac{3}{4}x + \frac{29}{4}$

6. $k = 5$ or $k = -3$

7. $k = 2$ or -2

8. a) $\frac{1}{b^{\frac{3}{2}}} - \frac{1}{a^{\frac{3}{2}}}$

b) i) No finite value

ii) $-\frac{1}{8}$

9. a) Proof

b) 45°, 135°, 225°, 315°

10. a) Proof b) 200 m^2

11. a) Proof b) $15r - r^2$ c) 56.25 cm^2

12. a) $P = (5, 1)$; $Q = (10, 4)$

b) $5y = 4x - 20$

c) 187.5π

13. a) i) $f^{-1}(x) = \frac{2x-1}{x}, x \in \mathbb{R}, x \neq 0$

ii) $g^1(x) = \frac{x-4}{3}$

b) $(-2, -2)$

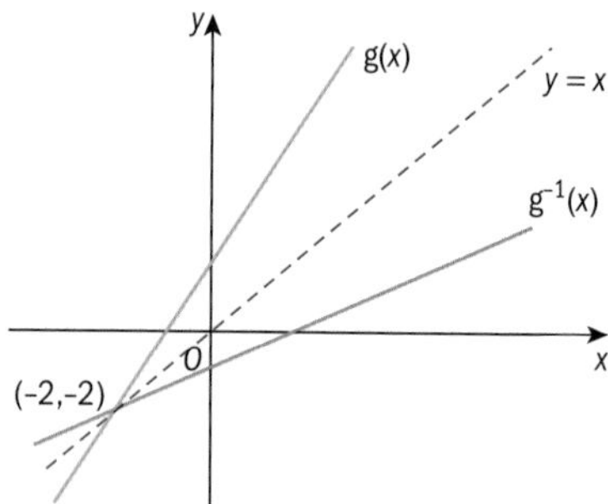

c) $k = \frac{7}{13}$

Command words

calculate Work out from given facts, figures and information.

describe Give the characteristics and main features.

determine Establish with certainty.

evaluate Judge or calculate the quality/importance/amount/value of something.

explain Set out purposes/reasons/mechanisms, or make the relationship between things clear, with supporting evidence.

identify Name/select/recognise.

justify Support a case with evidence/argument.

show that Provide structured evidence that leads to a given result.

sketch Make a simple freehand drawing showing the key features.

state Express in clear terms.

verify Confirm that a given statement/result is true.

Mathematical terms

arc Part of a circle.

arithmetic progression A sequence of numbers that has a common difference between each term.

asymptote A line that a curve approaches but never touches.

binomial expansion The result of multiplying out a two-term expression raised to power.

binomial theorem Quick method of expanding a two-term expression raised to power, e.g. $(x + y)^{10}$.

chain rule A rule for differentiating composite functions.

chord A line segment joining two points on a circle (or any curve).

coefficient A constant used to multiply a variable (e.g. 4 is the coefficient of $4x^3$).

collinear Three or more points lying on the same straight line.

common difference The difference between successive terms in an arithmetic progression.

common ratio The multiplier between one term and the next in a geometric progression.

completing the square Expressing a quadratic expression $ax^2 + bx + c$ in the form $p(x - q)^2 + r$.

composite function The result of applying one function and then a second function to the result of the first function.

constant of integration A constant that is added to the function obtained by integrating the terms of a given function.

cosine (cos) In a right-angled triangle, the value obtained when the length of the adjacent side is divided by the length of the hypotenuse.

cosine rule In a triangle ABC, $a^2 = b^2 + c^2 - 2bc \cos A$.

decreasing function A function with a negative gradient: $f(x)$ decreasing for $a < x < b$ if $f'(x) < 0$ for $a < x < b$.

definite integral The difference between the values of the integral at the upper and lower limits of the variable.

derivative An expression representing the rate of change of a function.

differentiation Determining the instantaneous rate of change of a function with respect to one of its variables.

discriminant The value of $b^2 - 4ac$ in the quadratic formula.

domain The input set of numbers to a mapping.

expansion Multiplying out the brackets in an expression.

factorise Write an expression as a product of factors.

finite series A series that stops after a finite number of terms.

function A mapping where every element of the domain is mapped onto exactly one element of the range.

general term The term of a sequence expressed in terms of its position in the sequence.

geometric progression A sequence of numbers that has a common multiplier between successive terms.

gradient The slope of a line, or for a curve, the slope of the tangent at a point on the curve.

identity A result that is true for any value of the variable(s).

improper integral An integral where either at least one of the limits is infinite, or the function to be integrated is not defined at a point in the interval of integration.

increasing function A function with a positive gradient: f(x) increasing for $a < x < b$ if $f'(x) > 0$ for $a < x < b$.

indefinite integral An integral with no upper or lower values of the variable.

inequality The relationship between two quantities that are not equal.

infinite series A series that continues indefinitely.

integration The process of finding a function given its derivative – the reverse of differentiation.

inverse function The function that maps the range back onto the domain.

line of symmetry The imaginary line where you could fold the image and have both halves match exactly.

line segment A line that connect two points.

linear An equation or expression where the highest power of the variable is one, (e.g. $5x + 3$).

many-to-one function A function where every element in the domain is mapped onto exactly one element in the range, but some elements in the range arise from more than one element in the domain.

mapping The relationship between two sets of numbers.

maximum A turning point on a curve where the gradient changes from positive to negative.

member (of a set) A number that forms part of a set of numbers.

mid-point The point halfway along a line segment.

minimum A turning point on a curve where the gradient changes from negative to positive.

normal A line that is at right angles to a curve at a point.

one-to-one function A function where every element in the domain corresponds to exactly one element in the range and where every element in the range corresponds to exactly one element in the domain.

parallel (lines) Lines that are the same distance apart along their entire length.

Pascal's triangle The triangle of numbers where each number is the sum of the two numbers above it.

period For a repeating function, the length of one cycle of the function.

perpendicular At right angles to.

perpendicular bisector The line that divides a line segment into two equal parts and makes a right angle with the line segment.

point of inflexion A point on a curve where the gradient is zero, but is neither a minimum nor a maximum.

polynomial An expression with constants and variables where the power(s) of the variables are all positive integers.

Pythagoras' theorem In a right-angled triangle, the square of the longest side is equal to the sum of the squares of the two shorter sides.

quadratic An equation or expression where the highest power of the variable is two, (e.g. $5x^2 + 3$).

quadratic formula The formula used to solve quadratic equations.

radian The angle subtended at the centre of a circle, radius r, by an arc of length r.

range (of a function) The output set of numbers from a mapping.

real root A root that is a real number.

reflection An image or shape as it would be seen in a mirror.

root A solution of an equation.

second derivative An expression obtained by differentiating the first derivative of a function.

sector A shape enclosed by two radii and an arc of a circle.

self-inverse function A function where the inverse function is identical to the function itself, i.e. $f(x) = f^{-1}(x)$.

sequence A set of numbers following a pattern.

series The sum of the terms of a sequence.

simultaneous equations A set of equations in more than one variable.

sine (sin) In a right-angled triangle, the value obtained when the length of the opposite side is divided by the length of the hypotenuse.

sine rule In a triangle ABC, $\frac{a}{\sin A} = \frac{b}{\sin B} = \frac{c}{\sin C}$.

solid of revolution The solid formed by rotating a curve through 360° about one of the coordinate axes.

stationary point A point on a curve where the gradient is zero.

stretch Make a shape bigger in one direction.

surd A number that is left in root form (√) so that its value is exact.

tangent (line) A straight line that just touches a curve at a single point.

tangent (tan) In a right-angled triangle, the value obtained when the length of the opposite side is divided by the length of the adjacent side.

term (of a sequence or series) A number that forms part of a sequence or series.

transformation Changing a shape or a graph by translating, reflecting, rotating, stretching or enlarging.

translation Movement of a shape which does not change its size or orientation.

trigonometry The study of triangles and angle functions.

vertex (of a curve) A minimum or maximum turning point.

y-intercept The value where a graph cuts the y-axis.

Index

V

Z